Practice of Discrete Element Method in Soil-Structure Interface Modelling

Wan-Huan Zhou · Zhen-Yu Yin

Practice of Discrete Element Method in Soil-Structure Interface Modelling

同济大学出版社
TONGJI UNIVERSITY PRESS

Springer

Wan-Huan Zhou
Department of Civil and Environmental Engineering
University of Macau
Macau, China

Zhen-Yu Yin
Department of Civil and Environmental Engineering
The Hong Kong Polytechnic University
Hong Kong, China

ISBN 978-981-19-0046-4 ISBN 978-981-19-0047-1 (eBook)
https://doi.org/10.1007/978-981-19-0047-1

Jointly published with Tongji University Press
The print edition is not for sale in China (Mainland). Customers from China (Mainland) please order the print book from: Tongji University Press.

This Springer imprint is published by the registered company Springer Nature Singapore Pte Ltd.
The registered company address is: 152 Beach Road, #21-01/04 Gateway East, Singapore 189721, Singapore

Dedicated to our families, teachers, and students.

Foreword

Soil-structure interface (SSI) is usually confronted in geotechnical engineering, including geosynthetic reinforcement, foundation engineering, infrastructure construction, and retaining walls. The SSI shearing behavior, localized within a thin zone, plays a critical role in the soil-structure interaction system when relative displacements occur at the interface. A range of experimental interface shear tests have pointed out that the SSI mechanism is influenced by the interface roughness, initial soil density, particle geometry, particle grading, and loading conditions. The experimental interface shear test is capable of providing the macroscopic observations of SSI shearing behavior. Besides, the emergence of particle image velocimetry (PIV) technique, being incorporated into the interface shear test apparatus, makes it possible that the soil particle movement is recorded and that the soil deformation is characterized from a microscopic perspective. However, this technique is tedious, and it is still challenging to characterize the microscopic contact behavior of inter-particles using the PIV.

Alternatively, the discrete element method (DEM), which is able to simulate the mechanical behavior of a soil-structure interaction system using rigid entities (i.e., particles and walls) and soft contact, is a suitable technique to investigate the constitutive properties from particle-scale to macro-scale and to correlate the micro behavior with the macro behavior. Obtaining the kinematic state and the microscopic contact behavior of inter-particles enriches the understanding of microscopic interface shearing behavior and enables the improvement of SSI design as well as the development of advanced constitutive models.

I came to know Dr. Yin in 2010 when he was working with my friend, Prof. Pierre-Yves Hicher in Ecole Centrale de Nantes, on micromechanical modeling of soils. Later on, I was invited to be a reviewer of his HDR (Habilitation to supervise Ph.D. students) in France. Dr. Yin's rigorous approach, deep understanding of soil behaviors, and solid numerical and analytical skills have left deep impressions on me. Since he joined Ecole Centrale de Nantes as Associate Professor in 2013, we began the academic discussion and cooperation on micromechanics of granular materials by co-supervising a Ph.D. student (Dr. Hao Xiang financially supported by Chinese Scholarship Council) and co-working with another Ph.D. student (Miss Na Deng).

Because of family reasons, Dr. Yin moved to Hong Kong in 2018, but our collaboration has never been suspended. Based on his good connection with Asian researchers, he has been helping our association "GDRI Multi-Physics and Multiscale Couplings in Geo-environmental Mechanics" organizing Franco-Chinese workshops, doctoral schools, expanding, therefore, the GDRI network. I am also very happy to see that within three years, Dr. Yin has successfully developed a research team in PolyU-HK with more than 20 Ph.D. students, postdoc fellows, and research assistant professors, and has published many well-cited peer-reviewed international journal articles. Benefited from his five years of experience as an engineer before returning to academia, he has developed many application tools or methods from theory to practice. This book is a testament to his outstanding ability.

I have known Dr. Zhou since my former Ph.D. student Dr. Huaxiang Zhu applied to work with her as a postdoc in 2016. At that time, she started to work on modeling soil-structure interface in collaboration with Dr. Yin. In 2017, on behalf of the GDRI conference committee, I invited her to give a talk at the GDRI annual workshop in Arpino, Italy. Since then, we have been developing more connections in research. I should note that Dr. Zhou has a strong background in geomechanics. Her Ph.D. work was focused on soil nail reinforcement issue with physical and numerical modeling for landslide prevention under the supervision of Prof. Jianhua Yin. The modeling of SSI with DEM presented along this book is a fine continuity of her previous research.

This book consists of a summary and a refinement of all the results obtained by the authors on the topic of micromechanical analysis of SSI using DEM. It gathers the authors' original results in the effect of soil properties and structure properties on soil-structure interaction micromechanics, including deformation mechanism of shear banding, evolutions of void ratio and coordination number, soil fabric, and force chain characteristics. It allows the reader to quickly and accurately grasp the basic SSI fundamentals from a micromechanical perspective.

In summary, this book provides an informative exposition of the current advances in the micromechanical analysis of soil-structure interaction using DEM and will benefit both scientists and practitioners. The students will also find in this volume a stimulating introduction to the mechanics of geomaterials.

François Nicot
Directeur de Recherche/Research
Professor—Université Savoie Mont-Blanc
Laboratoire EDYTEM—USMB/CNRS
UMR 5204, Editor-in-chief of the European
Journal of Environmental and Civil Engineering

Preface

Many problems in geotechnical engineering, such as pile foundation, geosynthetics-soil reinforcement, soil nailing, retaining wall, etc., involve local interaction between granular materials and a stiff or soft boundary of structures. The whole system consists of granular soil and structure. Shear failure normally occurs within a localized zone near the structural surface. In other words, the soil-structure interface (SSI) is generally considered to be a thin layer of soil that is adjacent to the construction material. Even though the SSI includes the properties of the thin layer of soil, its shearing behavior still differs from that of pure soil due to the complex surface morphology and material properties of the structure. The simulation results are very likely to be unrealistic if the SSI is directly solved using the constitutive model of granular material. Thus, the accurate description and evaluation of the shearing behavior of SSI is crucial for understanding the performance of SSI as well as the development of a constitutive model. In this regard, it is crucial to investigate how the microscopic interaction mechanism governs the overall behavior of an SSI system during quasi-static shearing. It is of considerable significance to civil engineering, material science, geophysics, and granular physics.

Whether the structure is made of geosynthetics, steel, concrete, or a new material yet to be developed, the mechanical behavior of structure—soil particles interactions must be governed by underlying physical principles that are related to the surface and material characteristics of the structure, loading, and boundary conditions. The roles of external and internal properties on the interface shearing behavior have been extensively and experimentally investigated. Commonly, the initial relative density and stress level of the granular assembly controls the soil dilatancy, while the shear strength of the soil governs the shear resistance ability at the interface. Other than the macroscopic behavior, the deformation pattern and soil fabric change significantly under interface shearing. From a microscopic viewpoint, heterogeneity arises in soil subjected to interface shearing. Strain localization is significant in "a thin interface shearing zone," which exhibits a distinct deformational pattern compared to the bulk soil. Moreover, fabric anisotropy changes significantly in the interface zone under shearing, resulting in the changes in macroscopic mechanical behavior of the granular material. Therefore, it is worthwhile to evaluate the microscopic interaction mechanism for SSI shearing behavior. DEM is a suitable technique to evaluate the mechanism from microscale and macroscale and to correlate the micro-behavior with the macro-behavior.

Hence, research works in this book have been conducted to investigate the micromechanism of SSI using DEM. In Chapter 1, recent research achievements on the experimental interface shearing behavior and experimental devices are summarized. The localization effect and deformational patterns of SSI are investigated in Chapter 2, while Chapter 3 is dedicated to evaluating the fabric evolution to a common micromechanical state at the SSI. Chapter 4 focuses on the micro-mechanical analysis of SSI behavior under constant normal stiffness conditions. Chapters 5, 6, and 7 are dedicated to investigate the effects of roughness, particle shape, initial fabric, and shear direction on the interface shearing behavior. In Chapter 8, we present a metaheuristic model for the interface shear strength between granular soil and structure considering complex surface morphology. In Chapter 9, a new model is presented to evaluate the effects of normal stress and surface hardness on sand-smooth geomembrane interface shearing behavior.

This book begins with the recent development of laboratory interface shear test and focuses on the micromechanical analysis of SSI using DEM, which can help readers quickly and accurately understand how the localized interface zone is formed, how soil heterogeneity arises subjected to the interface shearing, and how to correlate the micro behavior with the macro behavior. We hope to improve our understanding of SSI from a micro-scale by sharing with readers the microscopic observations under different soil properties, various structure properties, and different loading conditions.

Macao, China — Wan-Huan Zhou

Hong Kong, China — Zhen-Yu Yin
November 2021

Acknowledgements

In the process of writing this book, we received much sincere guidance and help from experts and colleagues. Here, I would like to express my sincere gratitude to Professor Pierre-Yves Hicher from Ecole Centrale de Nantes in France, Prof. Jian-Hua Yin from the Hong Kong Polytechnic University, and Prof. Francois Nicot from the University of Savoie Mont-Blanc in France, for their careful guidance and help during the book-writing process. And also, we thank Dr. Huaxiang Zhu, Dr. Xueying Jing, Dr. Linshuang Zhao, and Ms. Xixi Jie currently or formerly from the University of Macau, Dr. Xiang Wang, Dr. Yinfu Jin, and Dr. Hanlin Wang from the Hong Kong Polytechnic University, in particular, Dr. Chen Weibin from the University of Macau and Dr. Pei Wang from the Hong Kong Polytechnic University, for their participating in the research projects and hard work in the process of organizing and reviewing the book. We are also grateful to Springer Nature and Tongji University Press for performing detailed work on the publication of this book.

Some of the results and publications in this book have been funded by the Science and Technology Development Fund, Macau SAR (File no. FDCT/125/2014/A3, FDCT/193/2017/A3, SKL-IOTSC(UM)-2021-2023), the University of Macau Research Fund (MYRG2017-00198-FST) and the Research Grants Council (RGC) of Hong Kong Special Administrative Region Government (HKSARG) of China (Grant No.: 15217220).

About This Book

This book systematically introduces the practice of the discrete element method (DEM) on micromechanical analysis of soil-structure interaction. Based on DEM observation, four aspects are mainly discussed: (1) the localized band formation and how to determine the localized band thickness; (2) the evolution of soil and force heterogeneities subjected to the interface shearing; (3) how to correlate the micro behavior with the macro behavior according to the macro- and micro observations; (4) how to take the random morphology parameters into consideration determine the interface shear strength.

This book can be used as a postgraduate textbook for civil engineering, hydraulic engineering, transportation, railway, engineering geology, and other majors in colleges and universities, and as an elective course for senior undergraduates. It can also be used as a reference for relevant professional scientific researchers and engineers.

Contents

About the Authors

Wan-Huan Zhou is currently Head and Professor of the Department of Civil and Environmental Engineering at the University of Macau. Prof. Zhou received her B.Eng. degree in 2002 and M.Sc. degree in 2005 from Zhejiang University in China and Ph.D. in Geotechnical Engineering from The Hong Kong Polytechnic University in 2008. After graduation, she worked as a lecturer at The Hong Kong Polytechnic University and later joined the University of Macau in September 2009. She has acted as the Principal Investigator (PI) on 15 research projects, including the 2020 Excellent Young Scientists Fund (HK and Macau) from the National Natural Science Foundation of China. She has published more than 70 SCI indexed articles in top-ranked geotechnical journals. She also serves as an editorial board member in several top geotechnical engineering journals, such as the Canadian Geotechnical Journal and the International Journal of Geomechanics.

Zhen-Yu Yin is currently Professor of Geotechnical Engineering at The Hong Kong Polytechnic University. Prof. Yin received his B.Eng. in Civil Engineering from Zhejiang University in 1997, followed by five years of engineering consultancy at the Zhejiang Jiahua Architecture Design Institute. Then, he obtained his MSc and Ph.D. in Geotechnical Engineering at Ecole Centrale de Nantes (France) in 2003 and 2006, respectively. Prof. Yin has published over 220 articles in peer-reviewed international journals with an H index of 48 according to Web of Science. He is Associate Committee Member of Granular Materials Committee of American Society of Civil Engineers; Associate Editor of "European Journal of Environmental and Civil Engineering" and "Geotechnique Letters"; and Editorial Board Members of some top journals in the field of soil mechanics and geotechnical engineering (Canadian Geotechnical Journal, International Journal of Geomechanics ASCE, Soils and Foundations, Acta Geotechnica, Transportation Geotechnics, Computers and Geotechnics, etc.).

Chapter 1
Overview of Experimental Granular Soil–Structure Interface Behavior

1.1 Introduction

Soil–structure interfaces, such as infrastructure construction, underground pipeline system, and reinforcement structure, exist extensively in geotechnical engineering. The investigation timeline of the SSI includes three stages, as summarized in Fig. 1.1, [1–7]. The topics of SSI have gained more and more attention over time. According to Wang et al. [8], the interface is reckoned as an amalgam of the surface and an adjacent thin–layer soil [9], as presented in Fig. 1.2. Even though the SSI includes several soil properties, it still differs from the shearing behavior of pure soil, which will bring about significant differences if the issue of SSI is directly solved based on the constitutive model of granular material. Thus, the accurate description and evaluation of the shearing behavior of SSI is crucial for understanding the performance of SSI as well as the development of a constitutive model.

This chapter attempts to conduct a critical review investigating SSI shearing behavior based on experimental and numerical observations. Firstly, we present the experimental devices and improvements. Secondly, the experimental studies regarding the effect of granular material properties, structure properties, boundary conditions, and testing environment on the interface shearing behavior are reviewed. Conclusions are drawn, and future study topics are discussed in the last section.

1.2 Experimental Devices and Improvements

A diverse range of laboratory test devices are widely used to deal with the monotonic and cyclic interface shearing behavior (i.e., stress–displacement relationship and volumetric strain). Some improvements have been made based on the conventional devices for the purpose of investigation.

W.-H. Zhou and Z.-Y. Yin, *Practice of Discrete Element Method in Soil-Structure Interface Modelling*, https://doi.org/10.1007/978-981-19-0047-1_1

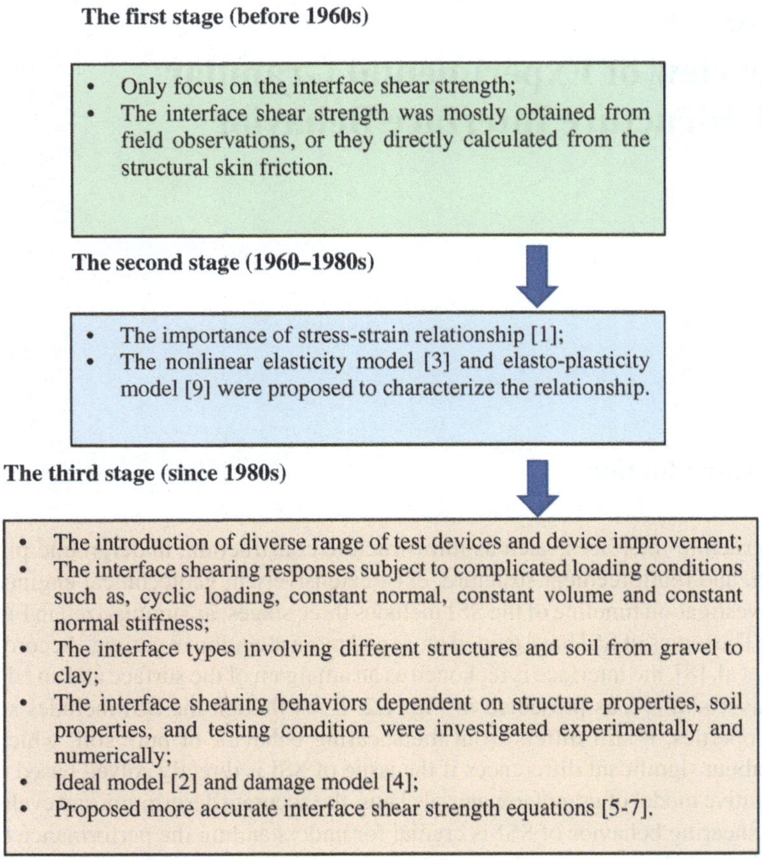

Fig. 1.1 The research history of soil-structure interface

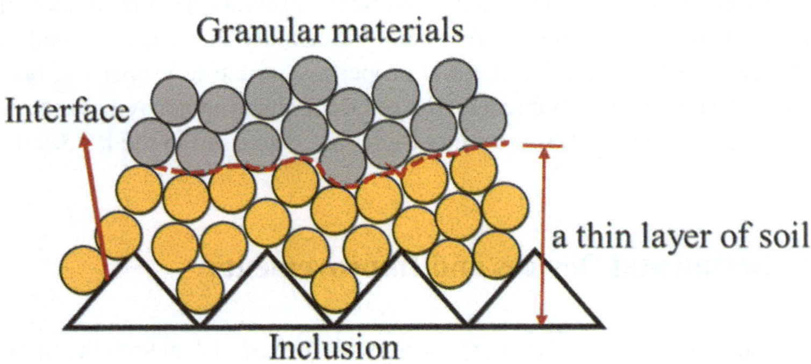

Fig. 1.2 Schematics of a composite system

1.2.1 Conventional Laboratory Test Devices

Schematic and main features of the conventional laboratory devices are illustrated in Table 1.1, which includes direct shear test, simple shear test, ring torsion test, inclined shear test, and annular shear test. Apparently, the direct shear test is most well-liked due to its simplicity and easy operation. Although the other devices have obvious advantages, the difficulties in sample preparation and operation preclude their wide applications.

Table 1.1 Summary of experimental devices for soil–structure interface

Device schematic	Advantages and disadvantages	References
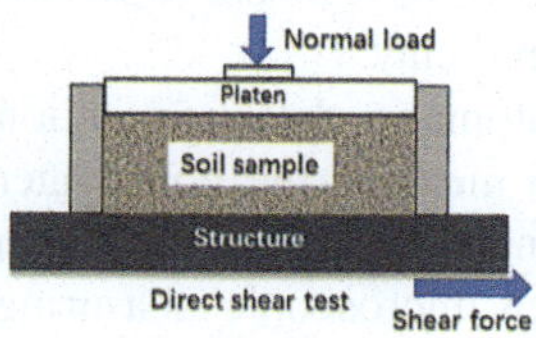 Direct shear test	• Easy sample preparation • Easy test procedure • Non-homogeneous distribution of the stress or strain along the interface • Distribution of deformation within the specimen cannot be detected	[12, 17, 18, 28, 34–42]
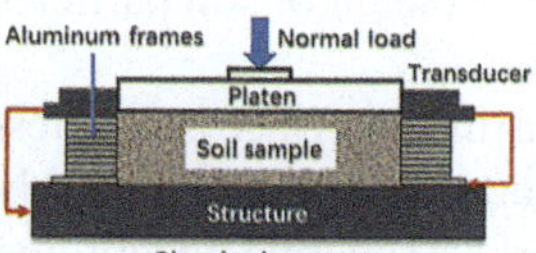 Simple shear test	• Both sliding and shear deformation within the specimen can be measured • Stress concentration at both ends of the specimen	[27, 30, 43–47]
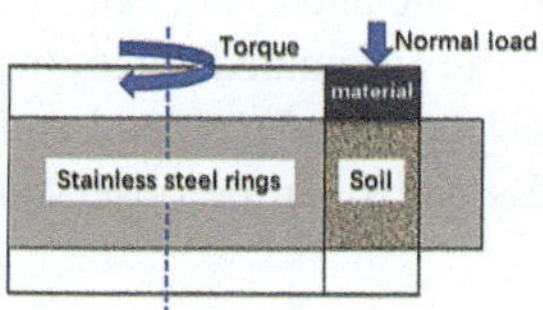 Ring torsion shear test	• The constant contact area • No stress concentration • Soil deformation can be measured • Tedious sample preparation • Complicated test procedure	[48–52]
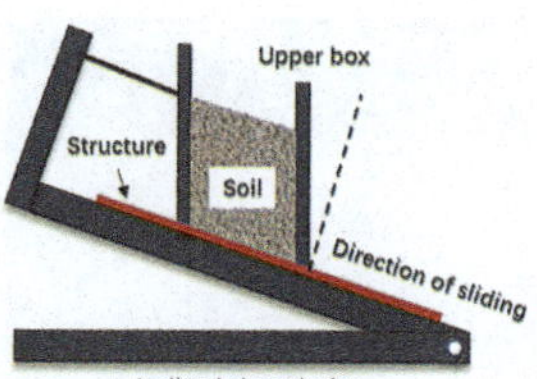 Inclined plane device	• It is common to be used in some situ, such as reservoir dams, lined canals, and scrapheaps cover • Only can be conducted under very low normal stress • Normal stress decreases in the process of shearing, which is different from the conventional shear test	[53–56]
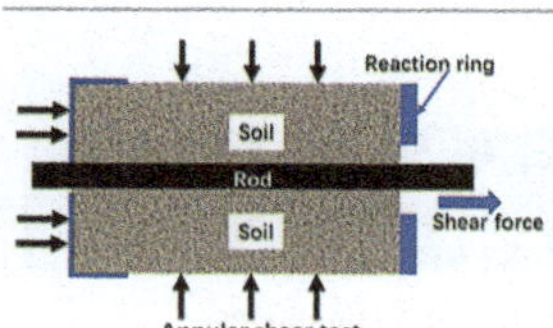 Annular shear test	• A useful device to investigate the interface between soil and pile foundation • Intricacy in specimen preparation and deformation measurement	[57–59]

1.2.2 Device Improvements

To more precisely understand both macro–and micro–shearing mechanisms, the conventional devices have been improved. The improvements can be categorized into three types:

(1) Enlarge sample size: the dimension of the shear box has an impact on the testing results due to its boundary confinements, as demonstrated by Ingold [10]. To mitigate the boundary effect, the size of the shear box was enlarged for that purpose. A large–scale shear test apparatus (50 cm in length and 36 cm in width) was developed to investigate the monotonic and cyclic behavior of SSI, with the focus on the interface between a structure and gravelly soil [11]. Chen et al. [12] developed a large–scale shear setup (50 cm × 50 cm × 50 cm) with a focus on the interface between structure and red clay.
(2) Add measurement units: other than the basic automatic data acquisition systems, there have been a number of attempts to measure the pore water pressure [13], the environmental temperature [14], and the deformation from a microscopic point of view [15–19]. For example, a new microscopic measuring system was established to observe and measure the movement of soil particles during an SSI test, as shown in Fig. 1.3.
(3) Perfect loading condition: the loading condition in the interface shear test becomes more assorted and more proximate to the practical situation. It mainly can be summarized as follows: (a) to simulate the interface shearing undergoing wind or wave loads, an automated apparatus for monotonic and cyclic tests of the soil–structure interface was developed [20–26]; (b) an automated apparatus for three–dimensional tests of the soil–structure interface was developed and the response due to the application of different stress paths can be observed

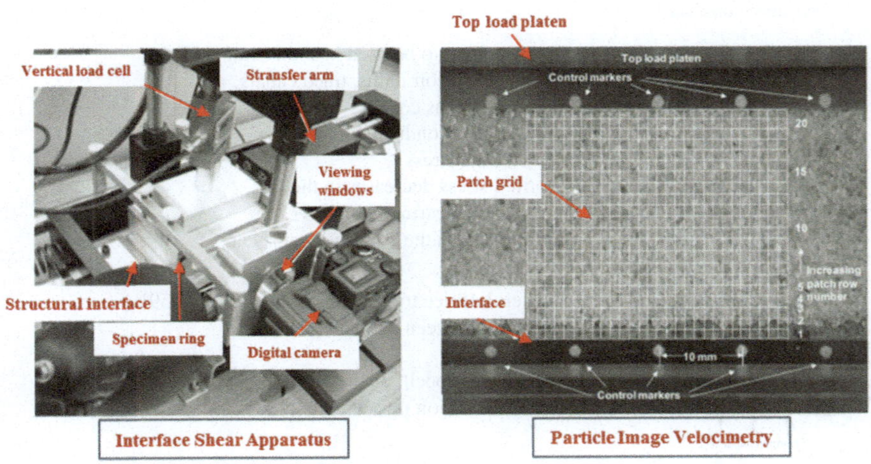

Fig. 1.3 Modified direct shear test setup with particle image velocimetry [16, 17]

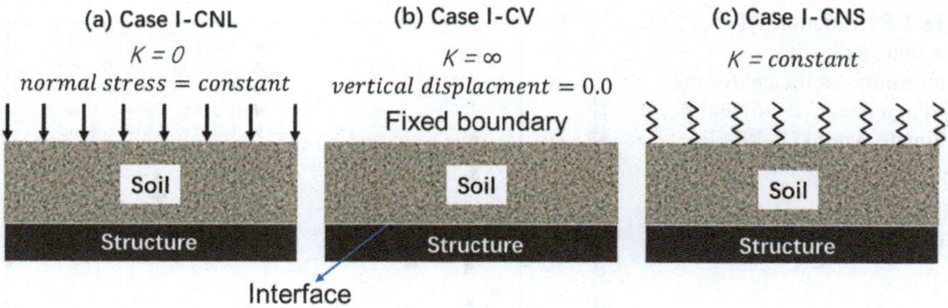

Fig. 1.4 Boundary conditions in the SSI: Case I: CNL; Case II: CV and Case III: CNS

[27–29]; (c) to replicate the field condition of the axially loaded pile in the laboratory tests, the apparatus was designed and fabricated with the capability of maintaining the constant normal stress (CS), constant normal stiffness (CNS) and constant volume (CV) during shearing [20, 22, 30–33], as shown in Fig. 1.4.

In summary, the direct shear test is most popular due to its simplicity and easy operation. In addition, installing additional micro and macro measurement units, enlarging the loading capacity, and enriching the loading path should be taken into account when a multifunctional and comprehensive apparatus is developed.

1.3 Effect of Soil Properties

The roles of soil properties on interface shearing behavior are of great concern. A wide variety of static and cyclic granular material–structure (steel, concrete, or geomembrane) interface shear tests were conducted with focuses on the roles of soil density [16, 20, 30, 60], particle geometry [16–18, 37, 61–63], soil particle inherent anisotropy [64–66], soil particle hardness and breakage [16, 67–69], soil types [15, 70], and particle size and gradation [61, 71–73]. The general uniform findings showed that (1) the specimens with more angular or elongated particle shapes possess higher peak interface friction angles; (2) the dense specimens are able to develop higher shear stresses and more distinct dilatancy; (3) similar to the pure soil shearing, interface shear test (IST) with smaller uniformity coefficient tends to present a higher peak shear stress (see Fig. 1.5a). No consistent result exists at the ultimate state between pure soil and IST (see Fig. 1.5b); (4) the shear stress for dense specimen increases with cyclic number, whereas it shows opposite trend for loose specimen; (5) particle breakage is significant in the granular assembly-structure interface under repeated loading; (6) Peak friction angle of pure soil grows with the bedding plane and peaks at a bedding plane of about 100°. However, as shown in Fig. 1.6, there is no consistent trend regarding the effect of bedding angle on peak friction angle of IST.

In addition to incohesive soil or coarse-grained soil interacting with geotechnical structure, some researchers also focus on marine clay [12, 80–82], fine-grained

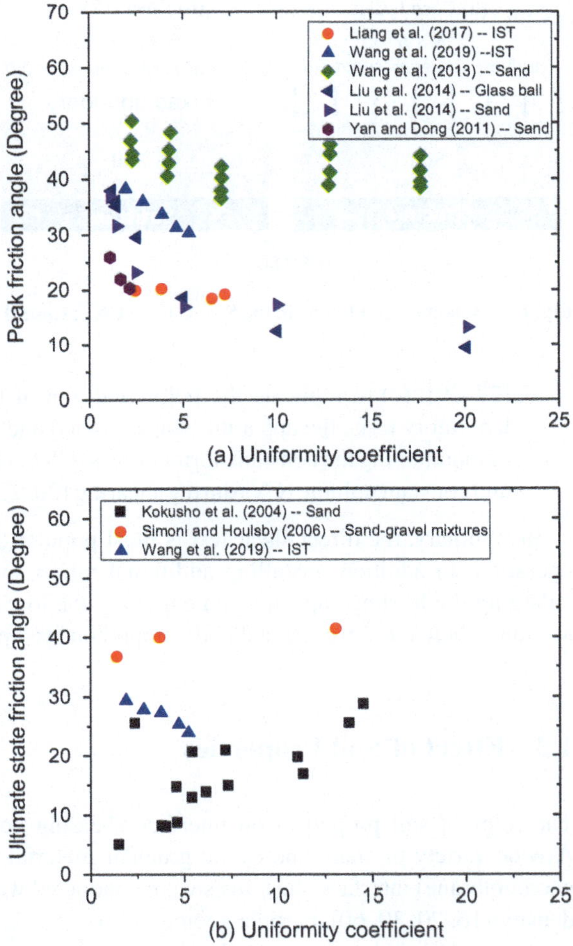

Fig. 1.5 Variations of friction angle with uniformity coefficient for the SSI at **a** peak [72–76] and **b** ultimate states [77–79]

soils [83], organic soil [84], and silty sand soil [85]. The experimental observations demonstrated that the dry density, water content, and void ratio play a significant role in SSI. More details regarding fine-grained soil-structure interface are out of scope because granular soils are focused on in the state-of-the-art.

Generally, the experimental study on the granular soil interacting with geotechnical structure has been extensively conducted and summarized, except that the effect of particle grading and inherent soil anisotropy.

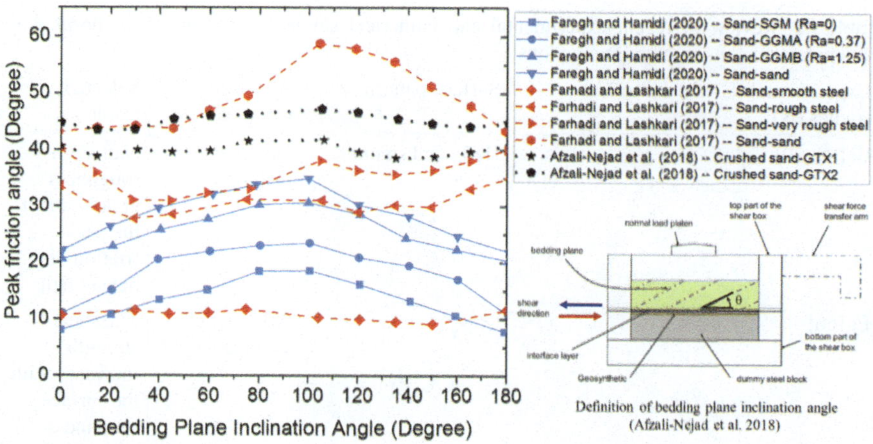

Fig. 1.6 Variation of peak friction angle with bedding plane inclination angle for the selected studies [64–66]

1.4 Effect of Structure Properties

Great efforts have been made on the influence of structure properties on the SSI shearing behavior. The surface texture, as one of structure surface characteristics, occupies an important role in the interface shearing behavior because many construction materials possess randomly textured surfaces [40]. A wide variety of studies [36, 86–88] have illustrated that the interface shearing behavior can be considerably ameliorated by producing ordered texturing patterns on the surface. Of all structure properties, surface morphology is regarded as the major affecting factor in the interface shearing behavior, and the relevant studies are listed in Table 1.2. Review of recent experimental and numerical studies on the effect of morphology parameters on soil-structure interactions. Through all the literature, it is concluded that the investigated surface used to be simplified as regular patterns in order to facilitate analysis. Besides, several studies characterized structure roughness using a qualitative technique [89, 90]. Specifically, the morphology of surface was roughly categorized from very smooth to very rough on inspection. The subjective of this approach is self-evident. Alternatively, the quantitative method with one or four roughness parameters was adopted to evaluate the surface. The commonly used parameter is the relative roughness ($R_n = R_{max}/D_{50}$, where R_{max} is the maximum altitude of surface and D_{50} is the mean particle diameter), which was suggested by Uesugi and Kishida [44]. However, R_n is a local parameter and incapable of reflecting the partial distribution of the surface profile as the same R_n can represent massive diverse profiles, as shown in Fig. 1.7. The studies regarding the effect of R_n are summarized in Table 1.3 Summary of the effect of R_n on interface shear strength. Some key points from Table 1.3 can be drawn as follows: (a) no consistent result was achieved excluding that the effect of R_n is segmented. The interface shear resistance increases

Table 1.2 Review of recent experimental and numerical studies on the effect of morphology parameters on soil-structure interactions

Experimental (E) and numerical studies (N)	Surface pattern	Quantified parameter	Selected results
E: [89, 90]	Helix and smooth	Qualitative method	Nail surface roughness greatly affects the peak friction of helical nail
E: [66]	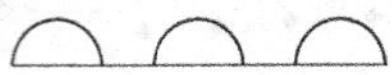	Arithmetic average height	The shear strength increases with the surface roughness
E: [40]	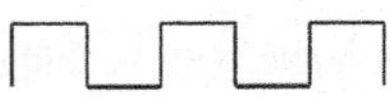	Relative roughness Surface texture angle	The maximum interface shear strength is obtained for the surface with ruled ribs inclined at right angle to the shear direction
E: [92, 93]	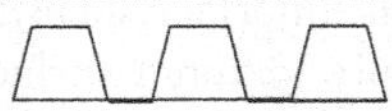	Roughness factor proposed by Horvath et al. [94] or relative roughness	Interface roughness on the shear behavior was discussed
E: [12]	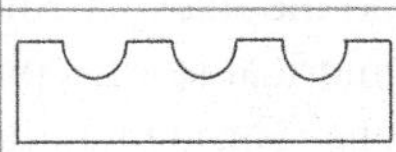	Average depth of the poured sand	The rougher surface creates higher interface shear strength; The confining pressure and surface roughness affect the shear failure plane
E: [95]	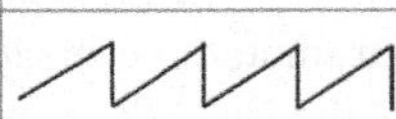	Roughness angle	The surface roughness is able to affect the interface shear strength

(continued)

Table 1.2 (continued)

Experimental (E) and numerical studies (N)	Surface pattern	Quantified parameter	Selected results
E: [86] DEM: [96]	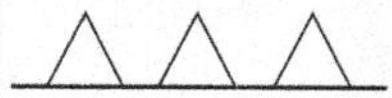	Asperity angle Root spacing Asperity spacing Asperity height	The geometric properties have great impacts on the interface behavior; The mechanical equation was proposed, which can be used for complex manufactured surfaces
N: [47, 97, 98]		Relative roughness	Rougher geomembrane shearing leads to stronger interaction between soil and geomembrane
E: [35, 72, 99–102] N: [8, 103–108]	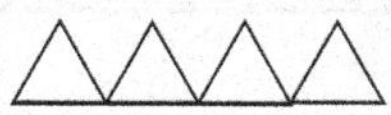	Relative roughness	A critical roughness was found to affect the interface shear strength
N: [106, 109]	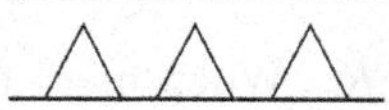	Relative roughness	For smoother interfaces, the contact normal force attenuates more quickly compared to rougher interfaces
E: [110, 111]		Flat width Intersection angle Groove depth	The interface shear strength increases with normal stress and the intersection angle

(continued)

Table 1.2 (continued)

Experimental (E) and numerical studies (N)	Surface pattern	Quantified parameter	Selected results
E: [14, 36, 61, 71, 84, 100, 102, 112–115] N: [6, 96, 105]	Random	Normalized average of maximum roughness Average roughness 13 morphological parameters	The rougher surface creates larger interface shear strength
E: [34, 70, 116–122]	Smooth	The parameter of interface element	The model implemented in a FEM or FDM can describe the SSI shearing behavior satisfactorily

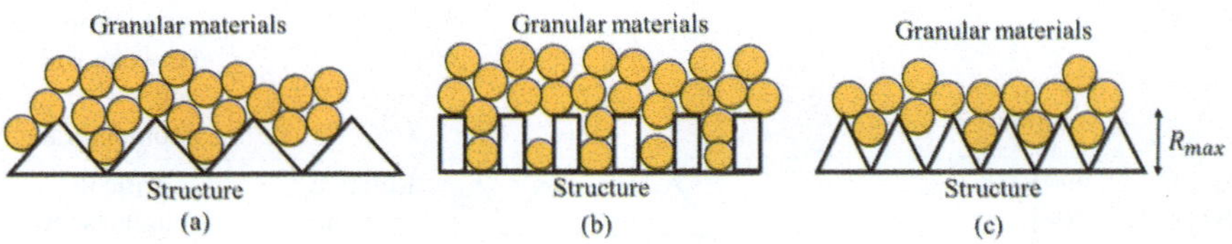

Fig. 1.7 Same $\boldsymbol{R_n}$ with different interfaces

R_n within a certain range of R_n. When more than this range, the effect of R_n is negligible; (b) the way how R_n affects the interface shear resistance depends on the soil density and surface pattern.

The localized band is visualized through recording the movement of soil particles using a microscopic measuring unit, as stated in Sect. 1.2.2. Some relevant studies are listed in Table 1.4. Summary of localized band thickness based on recent experimental and numerical studies. It is concluded that the thickness of the localized band $\boldsymbol{T}_{\rm s}$ under IST ranges from 0 to 15 times of D_{50}, which is influenced by the soil types and interface roughness. Spearman's rank correlation coefficient $Corr$, which ranges from -1.0 to 1.0, was used to quantify the correlation between the R_n and $\boldsymbol{T}_{\rm s}/\boldsymbol{D}_{50}$. Kowalczyk et al. [91] supposed that the data investigated are valid when the $Corr$ ranges from either -1.0 to -0.4 or from 0.4 to 1.0. It can be seen from Fig. 1.8 the $Corr$ for R_n is 0.763, which indicates that $\boldsymbol{T}_{\rm s}/\boldsymbol{D}_{50}$ has a positive affinity with R_n. The numerical results are also considered herein for convenience of comprehensive analysis and more detailed will not be shown in the following section.

Table 1.3 Summary of the effect of R_n on interface shear strength

Experimental (E) and numerical (N) studies	Surface type			Interface type	Peak interface shear resistance > Peak soil shear resistance
	Smooth	Intermediate	Rough		
E: [44]-Dr = 90%	–	$0.051 \leq R_n < 0.172$	$0.172 \leq R_n$	Sand-steel	Yes
E: [44]-Dr = 45%	–	$0.048 \leq R_n < 0.117$	$0.117 \leq R_n$	Sand-steel	Yes
E: [123]-Dr = 30%	$R_n < 0.02$	$0.02 \leq R_n < 0.5$	$0.5 \leq R_n$	Sand-solid	No information
E: [34]	$R_n < 0.1$	$0.1 < R_n$	–	Sand-steel	No information
E: [124]	$R_n < 0.03$		$0.26 \leq R_n$	Sand-pile	No information
E: [35]	–	$0 < R_n \leq 0.25$	$0.25 \leq R_n$	Sand-steel	No
E: [125]		$0.01 \leq R_n < 0.1$	$0.1 \leq R_n$	Sand-steel	No
N: [108]	$R_n < 0.125$	$0.1 < R_n < 0.375$	$0.375 \leq R_n$	Sand-solid	No
E: [42]	$R_n = 0.0$	$0.0 < R_n$	–	Clay-steel	Yes
N: [126]	$0.0 < R_n < 0.1$	$0.1 < R_n < 0.75$	$0.75 \leq R_n$	Sand-solid	No
E: [127]	$R_n < 0.006$	$R_n = 0.059$	$0.066 < R_n$	Sand-pile	No information

Apart from the rigid construction materials (i.e., steel or concrete), great devotions were also made on the soft or deformable construction materials, which are widely available in geosynthetics reinforcement. Compared with the rigid interfaces, the shearing mechanism for the deformable interfaces is more complex. The failure mechanism for the interface between smooth geomembrane and structure mainly depends on particle sliding along the surface with limited particle rolling or plowing into the surface [37, 129]. The influence of material hardness was evaluated by O'rourke et al. [38] and by Frost and Karademir [130] based on the IST on smooth HDPE and PVC geomembranes, and by Dove and Frost [131] through taking into particle–interface contacts account. The results suggested that the interface strength decreases with the increase of material hardness. Last several years, Abuel-Naga et al. [125] further pointed out that there is a coupling effect of surface roughness and hardness on the interface shearing behavior, and a critical surface hardness exists in affecting interface shear resistance. When less than the critical surface hardness, the material hardness is unable to resist the soil particle ploughing anymore, leading to increasing the interface shear resistance, as shown in Fig. 1.9.

Table 1.4 Summary of localized band thickness based on recent experimental and numerical studies

Studies	Interface type	R_n	τ_p/σ_n	T_s/D_{50} (Times)
[16]	Ottawa sand – steel	0.074	0.85	2.03
	Q-Rok – steel	0.073	0.76	5.33
	Ottawa sand-steel	0.95	1.25	6.76
		0.95	2.30	10.13
		0.95	0.84	6.76
Experimetal study: [128]	Sub-Angular 20–30-textured sleeve	0.01	0.39	0.0
		0.35	0.77	3.0
		0.69	0.96	5.0
		1.39	1.19	5.3
		2.78	1.39	6.2
	Sub-Angular 50–70-textured sleeve	0.03	0.45	4.2
		0.96	0.78	9.6
		1.92	1.02	11.8
		3.85	1.26	14.8
		7.69	1.44	15.0
	Sub-rounded 20–30-textured sleeve	0.01	0.36	0.0
		0.35	0.84	3.6
		0.69	1.19	5.3
		1.39	1.35	6.2
		2.78	1.50	6.9
Numercial study: [108]	Sand-rigid wall	0.0	0.25	0.0
		0.125	0.30	0.0
		0.25	0.45	1.7
		0.30	0.75	2.8
		0.375	1.00	4.0.33
		0.5	0.98	4.32
		0.75	1.18	4.3
		1.0	1.13	3.8
		0.0	0.30	0.0
		0.125	0.33	0
		0.25	0.63	1.7
		0.30	0.80	2.3
		0.375	1.07	3.7
		0.5	1.05	3.4
		0.75	1.13	3.75

(continued)

Table 1.4 (continued)

Studies	Interface type	R_n	τ_p/σ_n	T_s/D_{50} (Times)
		1.0	1.12	2.75
		0.0	0.25	0.0
		0.125	0.30	0.0
		0.25	0.68	1.3
		0.30	0.88	1.7
		0.375	1.06	3.15
		0.5	1.08	3.15
		0.75	1.15	3.1
		1.0	1.08	2.75
		0.0	0.28	0.0
		0.125	0.30	0.0
		0.25	0.65	1.3
		0.30	0.88	1.7
		0.375	1.00	3.15
		0.5	1.05	3.15
		0.75	1.18	2.75
		1.0	1.06	2.75
Numercial study: [98]	Sand-woven	0.19	0.54	3.0
		0.19	0.49	2.6
		0.19	0.44	2.4
		0.03	0.49	2.2
		0.30	0.53	2.6
		0.19	0.52	3.0
Numercial study: [106]	Sand-rigid wall	0.01	0.32	1
		0.1	0.42	2
		0.25	0.87	6
		0.5	0.97	10
		0.75	1.15	14
		1	0.93	14
		2.0	1.15	14

1.5 Other Influence Factors

As mentioned in Sect. 1.2.2, the boundary condition acting on the soil–pile interface could be more precisely duplicated by the CNS, CV, and CNS during shearing. Fakharian and Evgin [30] assumed these three boundary conditions to be a series of springs connecting between soil sample and the loading system in interface shear test apparatus, which can be realized when $k_n = 0$, $k_n = \infty$, and $k_n =$constant,

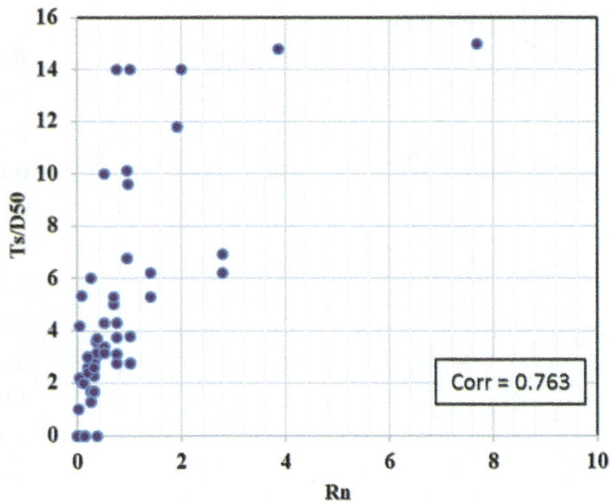

Fig. 1.8 Correlation between relative roughness and localized band thickness for all selected studies

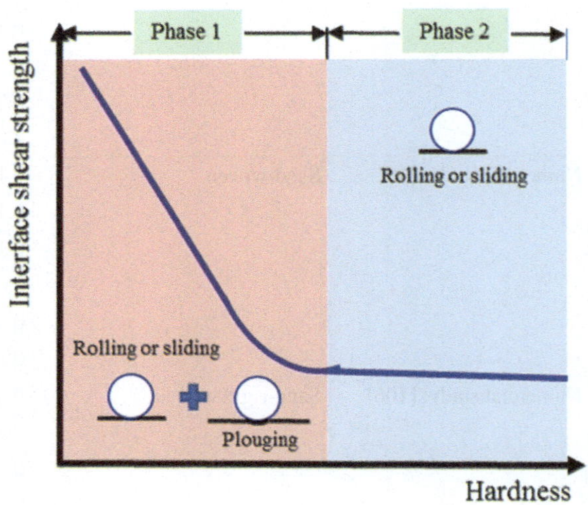

Fig. 1.9 Relationship between interface shear strength and hardness

respectively, where k_n is the normal stiffness, as shown in Fig. 1.4. A multifunctional interface shear test device was fabricated by Porcino et al. [132], Jiang et al. [33], DeJong et al. [17], and Di Donna et al. [114] to obtain more precise design parameters for soil–pile interaction. The results illustrated that the normal stress and shear stress grow with the dilation until the residual state is achieved. In addition, a CV boundary condition is normally the best choice to replicate the completely undrained condition.

Interface shearing behavior is also a great concern in thermoactive geostructures. Di Donna et al. [114] experimentally investigated the pile–soil interface shearing behavior at different temperatures. The result appeared that the sand–concrete interface under CNS and CNL is not affected directly by temperature, which is identical to the observation by Maghsoodi et al. [101].

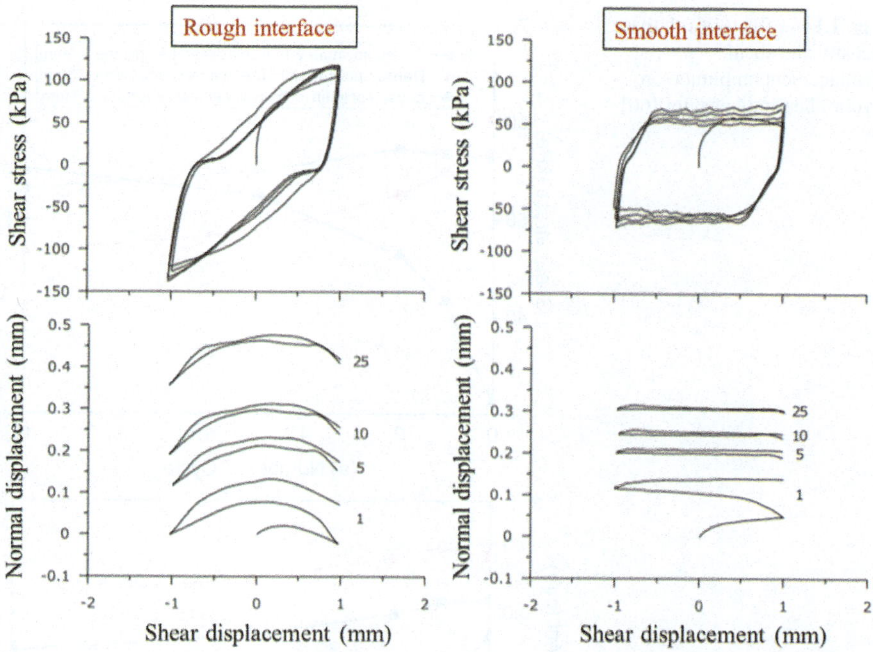

Fig. 1.10 Comparison between cyclic shear tests on rough interface and smooth interface [22]

All sorts of experimental tests have been performed to investigate the SSI shearing behavior subject to the cyclic loading using direct shear tests [20–22, 102, 133, 134] and simple shear tests [22, 30, 47, 60, 69, 135]. Experimental results indicated that (1) the cyclic interface shearing behavior for the rough interface is different from that for smooth interface (see Fig. 1.10). The densification leads to an increment in the maximum shear stress during the cycle loading for both tests; (2) the mobilized interface shear stress increases for loose specimen but decreases for dense specimen with an increase in the cycle number. Eventually, the mobilized shear stress becomes irrelevant to the initial density, as shown in Fig. 1.11; (3) It can be seen from Fig. 1.12 that the maximum shear stress under the same cycle number increases with the decreasing irregularity of particles, whereas it increases as normal stress increases. Other than the focus on the cyclic interface shearing behavior, Wang et al. [136] and Liu et al. [26] analyzed the post–cyclic interface shearing behavior and compared it to the interface shearing behavior under monotonic loading. However, no uniform result at the critical state was achieved, as demonstrated in Fig. 1.13.

Fig. 1.11 Effects of relative density and shear displacement amplitude on cyclic shearing behavior [60]

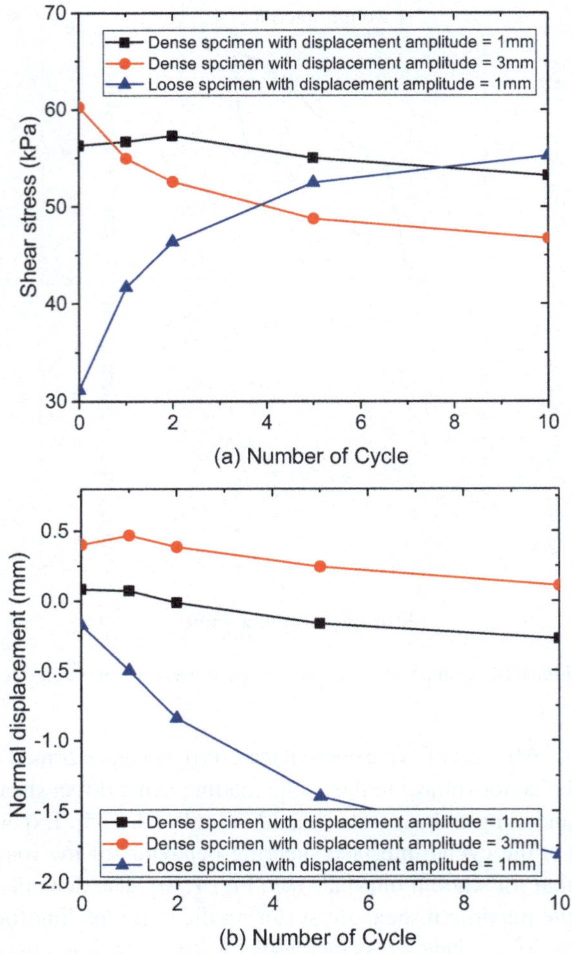

1.6 Conclusions

Since the stress-strain relationship of SSI has been concerned, various studies have been performed experimentally to investigate the shearing behavior of SSI. Through a systematic review, some conclusions and suggestions are as follows:

(1) The direct shear test is the most pervasive in virtue of its simplicity and easy operation. Although the other devices have obvious advantages, the difficulties in sample preparation and operation constrain their wide applications. Installing additional micro and macro measurement units, enlarging the loading capacity, and enriching the loading path should be taken into consideration when a multifunctional and comprehensive apparatus is developed;

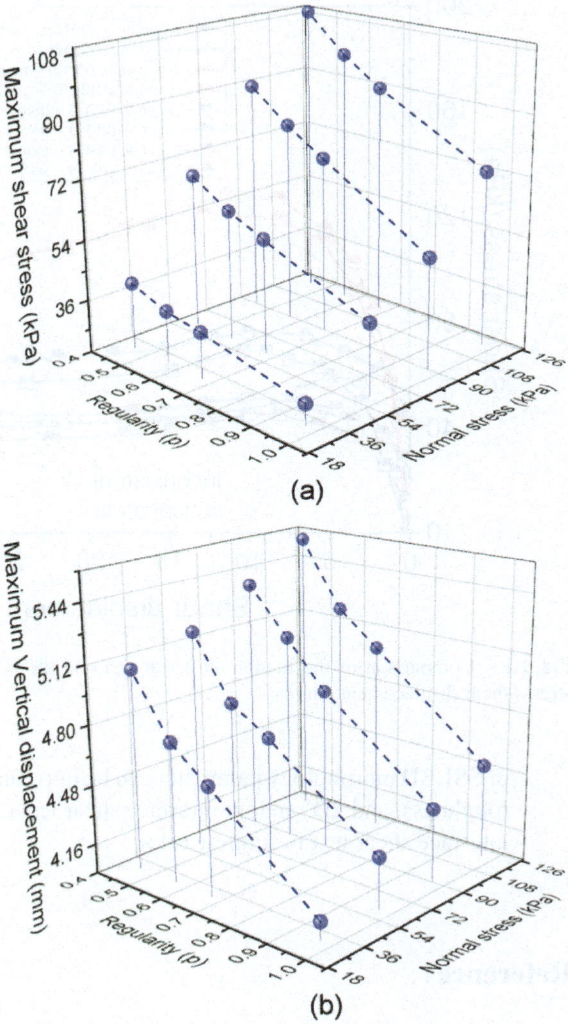

Fig. 1.12 **a** Variation of maximum peak stress with normal stress and regularity, and **b** Variation of maximum vertical displacement with normal stress and regularity [63]

(2) The experimental studies on the sand or coarse-grained soil interacting with geotechnical structure have been systematically conducted and summarized. However, the effect of particle grading and inherent soil anisotropy need further experimental and numerical explorations;

(3) It is widely accepted that the granular material–concrete interface is independent of the environmental temperature. Nevertheless, the microscopic influencing mechanism was rarely involved;

4) No consistent result was achieved except that the effect of R_n (commonly used) is segmented. Although irregular surfaces and a complete set of 2D morphology parameters were adopted, there is still a way to apply them to the 3D problem

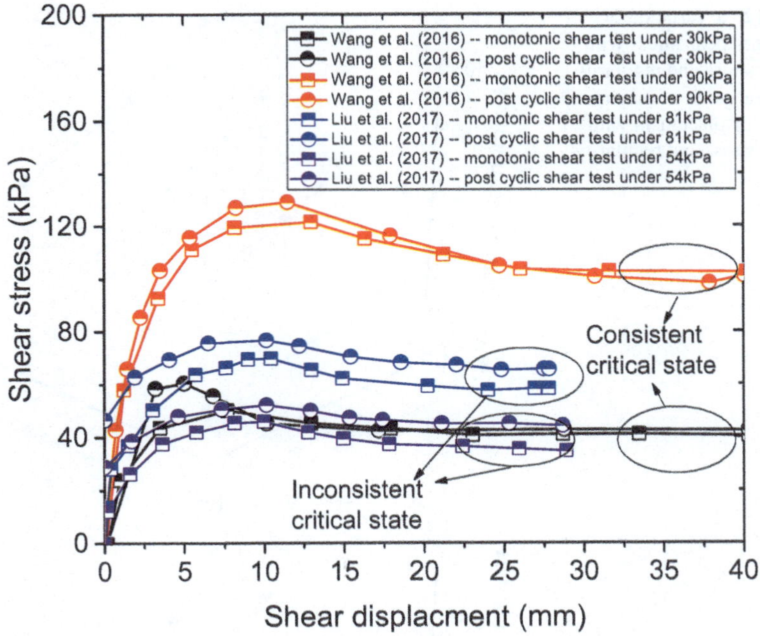

Fig. 1.13 Comparison of the monotonic shear test and the post cyclic shear tests for the shear stress–shear displacement curves

of SSI. 3D morphology parameter can be introduced to characterize the surface roughness, and 2D random interface shear test can be extended to 3D random interface shear test in future works.

References

1. Potyondy JG (1961) Skin friction between various soils and construction materials. Geotechnique 11:339–353
2. Brandt JRT (1985) Behaviour of soil-concrete interfaces
3. Clough GW, Duncan JM (1971) Finite element analyses of retaining wall behavior. J Soil Mech Found Div 97:1657–1673
4. Desai C, Drumm E, Zaman M (1985) Cyclic testing and modeling of interfaces. J Geotech Eng 111:793–815
5. Cabalar AF (2016) Cyclic behavior of various sands and structural materials interfaces. Geomech Eng 10:1–19
6. Chen WB, Zhou WH, Sadowski Ł, Yin ZY (2021) Metaheuristic model for the interface shear strength between granular soil and structure considering surface morphology. Comput Geotech 135:104141
7. Chou J-S, Yang K-H, Lin J-Y (2016) Peak shear strength of discrete fiber-reinforced soils computed by machine learning and metaensemble methods. J Comput Civ Eng 30:04016036

8. Wang J, Gutierrez MS, Dove JE (2007) Numerical studies of shear banding in interface shear tests using a new strain calculation method. Int J Numer Anal Meth Geomech 31:1349–1366
9. Ghaboussi J, Wilson EL, Isenberg J (1973) Finite element for rock joints and interfaces. J Soil Mech Found Div 99:833–848
10. Ingold TS (1982) Reinforced earth
11. Zhang G, Zhang J-M (2006) Monotonic and cyclic tests of interface between structure and gravelly soil. Soils Found 46:505–518
12. Chen X, Zhang J, Xiao Y, Li J (2015) Effect of roughness on shear behavior of red clay–concrete interface in large-scale direct shear tests. Can Geotech J 52:1122–1135
13. Desai CS, Rigby DB (1997) Cyclic interface and joint shear device including pore pressure effects. J Geotech Geoenvironmental Eng 123:568–579
14. Yavari N, Tang AM, Pereira J-M, Hassen G (2016) Effect of temperature on the shear strength of soils and the soil–structure interface. Can Geotech J 53:1186–1194
15. Zhang G, Liang D, Zhang J-M (2006) Image analysis measurement of soil particle movement during a soil–structure interface test. Comput Geotech 33:248–259
16. DeJong JT, Westgate ZJ (2009) Role of initial state, material properties, and confinement condition on local and global soil-structure interface behavior. J Geotech Geoenvironmental Eng 135:1646–1660
17. DeJong JT, White DJ, Randolph MF (2006) Microscale observation and modeling of soil-structure interface behavior using particle image velocimetry. Soils Found 46:15–28
18. Vangla P, Latha GM (2015) Influence of particle size on the friction and interfacial shear strength of sands of similar morphology. Int J Geosynthetics Ground Eng 1:6
19. Martinez A, Stutz HH (2019) Rate effects on the interface shear behaviour of normally and overconsolidated clay. Géotechnique 69:801–815
20. Pra-ai S, Boulon M (2017) Soil–structure cyclic direct shear tests: a new interpretation of the direct shear experiment and its application to a series of cyclic tests. Acta Geotech 12:107–127
21. Mortara G, Ferrara D, Fotia G (2010) Simple model for the cyclic behavior of smooth sand-steel interfaces. J Geotech Geoenvironmental Eng 136:1004–1009
22. Mortara G, Mangiola A, Ghionna VN (2007) Cyclic shear stress degradation and post-cyclic behaviour from sand–steel interface direct shear tests. Can Geotech J 44:739–752
23. Wang J, Ying M-J, Liu F-Y, Fu H-T, Ni J-F, Shi J (2020) Effect of particle-size gradation on coarse sand-geotextile interface response in cyclic and postcyclic direct shear test. Adv Civil Eng 2020
24. Cen W, Wang H, Yu L, Rahman M (2020) Response of high-density polyethylene geomembrane-sand interfaces under cyclic shear loading. Int J Geomech 20:04019166
25. Cen W, Wang H, Sun Y, Wen L (2018) Monotonic and cyclic shear behaviour of geomembrane-sand interface. Geosynth Int 25:369–377
26. Liu F-Y, Wang P, Geng X-y, Wang J, Lin X (2016) Cyclic and post-cyclic behaviour from sand–geogrid interface large-scale direct shear tests. Geosynth Int 23:129–139
27. Evgin E, Fakharian K (1997) Effect of stress paths on the behaviour of sand steel interfaces. Can Geotech J 33:853–865
28. Feng D, Zhang J, Hou W (2021) Role of normal stiffness in 3D cyclic behavior of gravel–steel interface from large-scale direct shear tests. Acta Geotech 16:151–165
29. Feng D, Zhang J, Hou W (2021) Three-dimensional behaviors of a gravel-steel interface considering initial shear stress. J Geotech Geoenvironmental Eng 147:04020158
30. Fakharian K, Evgin E (1997) Cyclic simple-shear behavior of sand-steel interfaces under constant normal stiffness condition. J Geotech Geoenvironmental Eng 123:1096–1105
31. Feng D, Zhang G, Zhang J, W-j HOU (2009) Experimental study on 3D cyclic behaviors of soil-structure interface under constant normal stiffness condition. Chin J Geotech Eng 31:1571–1577
32. Tabucanon JT, Airey DW, Poulos HG (1995) Pile skin friction in sands from constant normal stiffness tests. Geotech Test J 18:350–364
33. Jiang Y-j, Xiao J, Tanabashi Y, Mizokami T (2004) Development of an automated servo-controlled direct shear apparatus applying a constant normal stiffness condition. Int J Rock Mech Min Sci 41:275–286

34. Hu L, Pu J (2004) Testing and modeling of soil-structure interface. J Geotech Geoenvironmental Eng 130:851–860
35. Su L-J, Zhou W-H, Chen W-B, Jie X (2018) Effects of relative roughness and mean particle size on the shear strength of sand-steel interface. Measurement 122:339–346
36. Martinez A, Frost J (2017) The influence of surface roughness form on the strength of sand–structure interfaces. Géotechnique Letters 7:104–111
37. Vangla P, Gali ML (2016) Shear behavior of sand-smooth geomembrane interfaces through micro-topographical analysis. Geotext Geomembr 44:592–603
38. O'rourke T, Druschel S, Netravali A (1990) Shear strength characteristics of sand-polymer interfaces. J Geotechn Eng 116:451–469
39. Zettler T, Frost J, DeJong J (2000) Shear-induced changes in smooth HDPE geomembrane surface topography. Geosynth Int 7:243–267
40. Samanta M, Punetha P, Sharma M (2018) Influence of surface texture on sand–steel interface strength response. Géotechnique Letters 8:40–48
41. Xiong M, He P, Mu Y, Na X (2021) Modeling of concrete-frozen soil interface from direct shear test results. Adv Civil Eng 2021
42. Li YH, Lv MF, Guo YC, Huang M-S (2021) Effects of the soil water content and relative roughness on the shear strength of silt and steel plate interface. Measurement:109003
43. Uesugi M, Kishida H (1986) Influential factors of friction between steel and dry sands. Soils Found 26:33–46
44. Uesugi M, Kishida H (1986) Frictional resistance at yield between dry sand and mild steel. Soils Found 26:139–149
45. Uesugi M, Kishida H, Uchikawa Y (1990) Friction between dry sand and concrete under monotonic and repeated loading. Soils Found 30:115–128
46. Yang D, Niu S, Cai Y, Feng W, Zhang D (2021) Analysis of shear characteristics of cohesive soil and rigid base: a case study of concrete base. Geotech Geol Eng 39:135–143
47. Feng DK, Zhang JM, Deng LJ (2018) Three-dimensional monotonic and cyclic behavior of a gravel–steel interface from large-scale simple-shear tests. Can Geotech J 55:1657–1667
48. Yoshimi Y, Kishida T (1981) A ring torsion apparatus for evaluating friction between soil and metal surfaces. Geotech Test J 4:145–152
49. Desai CS, Pradhan SK, Cohen D (2005) Cyclic testing and constitutive modeling of saturated sand–concrete interfaces using the disturbed state concept. Int J Geomech 5:286–294
50. Tan S, Chew S, Wong W (1998) Sand–geotextile interface shear strength by torsional ring shear tests. Geotext Geomembr 16:161–174
51. Stark TD, Poeppel AR (1994) Landfill liner interface strengths from torsional-ring-shear tests. J Geotech Eng 120:597–615
52. Vaid Y, Rinne N (1995) Geomembrane coefficients of interface friction. Geosynth Int 2:309–325
53. Chung M, Seo MW, Kim KS, Park JB (2006) Protective effect of overlying geosynthetic on geomembrane liner observed from landfill field tests and inclined board laboratory experiments. Waste Manage Res 24:250–259
54. Ling H, Burke C, Mohri Y, Matsushima K (2002) Shear strength parameters of soil-geosynthetic interfaces under low confining pressure using a tilting table. Geosynth Int 9:373–380
55. Pitanga HN, Gourc J-P, Vilar OM (2009) Interface shear strength of geosynthetics: evaluation and analysis of inclined plane tests. Geotext Geomembr 27:435–446
56. Narejo DB (2003) A simple tilt table device to measure index friction angle of geosynthetics. Geotext Geomembr 21:49–57
57. Brumund W, Leonards G (1973) Experimental study of static and dynamic friction between sand and typical constuction materials. J Test Eval 1:162–165
58. Leonards G (1965) Experimental study of static and dynamic friction between soil and typical construction materials. Purdue Univ Lafayette in School of Civil Engineering
59. Alenzi A, Marinack M, Higgs C, McCarthy J (2013) DEM validation using an annular shear cell. Powder Technol 248:131–142

60. Oumarou TA, Evgin E (2005) Cyclic behaviour of a sand steel plate interface. Can Geotech J 42:1695–1704
61. Han F, Ganju E, Salgado R, Prezzi M (2018) Effects of interface roughness, particle geometry, and gradation on the sand–steel interface friction angle. J Geotech Geoenvironmental Eng 144:04018096
62. Afzali-Nejad A, Lashkari A, Shourijeh PT (2017) Influence of particle shape on the shear strength and dilation of sand-woven geotextile interfaces. Geotext Geomembr 45:54–66
63. Liu FY, Ying MJ, Yuan GH, Wang J, Gao ZY, Ni JF (2021) Particle shape effects on the cyclic shear behaviour of the soil–geogrid interface. Geotext Geomembr
64. Farhadi B, Lashkari A (2017) Influence of soil inherent anisotropy on behavior of crushed sand-steel interfaces. Soils Found 57:111–125
65. Afzali-Nejad A, Lashkari A, Farhadi B (2018) Role of soil inherent anisotropy in peak friction and maximum dilation angles of four sand-geosynthetic interfaces. Geotext Geomembr 46:869–881
66. Faregh AHA, Hamidi A (2020) Impact of inherent anisotropy on the shear behavior of sand-textured geomembrane interface. Int J Civil Eng 18:1113–1123
67. Sweta K, Hussaini SKK (2018) Effect of shearing rate on the behavior of geogrid-reinforced railroad ballast under direct shear conditions. Geotext Geomembr 46:251–256
68. Saberi M, Annan C-D, Konrad J-M (2017) Constitutive modeling of gravelly soil–structure interface considering particle breakage. J Eng Mech 143:04017044
69. Uesugi M, Kishida H, Tsubakihara Y (1989) Friction between sand and steel under repeated loading. Soils Found 29:127–137
70. Zhang G, Zhang J-M (2009) Numerical modeling of soil–structure interface of a concrete-faced rockfill dam. Comput Geotech 36:762–772
71. Rui S, Wang L, Guo Z, Zhou W, Li Y (2021) Cyclic behavior of interface shear between carbonate sand and steel. Acta Geotech 16:189–209
72. Wang H-L, Zhou W-H, Yin Z-Y, Jie X-X (2019) Effect of grain size distribution of sandy soil on shearing behaviors at soil–structure interface. J Mater Civ Eng 31:04019238
73. Liang Y, Yeh T-CJ, Wang J, Liu M, Hao Y (2017) Effect of particle size distribution on soil-steel interface shear behavior. Soil Mech Found Eng 54:310–317
74. Wang J-J, Zhang H-P, Tang S-C, Liang Y (2013) Effects of particle size distribution on shear strength of accumulation soil. J Geotech Geoenvironmental Eng 139:1994–1997
75. Liu C-N, Yang K-H, Nguyen MD (2014) Behavior of geogrid–reinforced sand and effect of reinforcement anchorage in large-scale plane strain compression. Geotext Geomembr 42:479–493
76. Yan W, Dong J (2011) Effect of particle grading on the response of an idealized granular assemblage. Int J Geomech 11:276–285
77. Kokusho T, Hara T, Hiraoka R (2004) Undrained shear strength of granular soils with different particle gradations. J Geotech Geoenvironmental Eng 130:621–629
78. Simoni A, Houlsby GT (2006) The direct shear strength and dilatancy of sand–gravel mixtures. Geotech Geol Eng 24:523
79. Wang H-L, Chen R-P, Liu Q-W, Kang X, Wang Y-W (2019) Soil–geogrid interaction at various influencing factors by pullout tests with applications of FBG sensors. J Mater Civ Eng 31:04018342
80. Tsubakihara Y, Kishida H, Nishiyama T (1993) Friction between cohesive soils and steel. Soils Found 33:145–156
81. Miller GA, Hamid TB (2007) Interface direct shear testing of unsaturated soil. Geotech Test J 30:182–191
82. Taha A, Fall M (2013) Shear behavior of sensitive marine clay-concrete interfaces. J Geotech Geoenvironmental Eng 139:644–650
83. Subba Rao K, Allam M, Robinson R (2000) Drained shear strength of fine-grained soil–solid surface interfaces. Proc Inst Civil Eng Geotech Eng 143:75–81
84. Canakci H, Hamed M, Celik F, Sidik W, Eviz F (2016) Friction characteristics of organic soil with construction materials. Soils Found 56:965–972

85. Al-Emami O (2018) Investigation of soil-steel interface behavior of Iraqi soil by direct shear apparatus. In: MATEC Web of Conferences. EDP sciences, p 01003
86. Dove JE, Jarrett JB (2002) Behavior of dilative sand interfaces in a geotribology framework. J Geotech Geoenvironmental Eng 128:25–37
87. Hryciw RD, Irsyam M (1993) Behavior of sand particles around rigid ribbed inclusions during shear. Soils Found 33:1–13
88. Irsyam M, Hryciw R (1991) Friction and passive resistance in soil reinforced by plane ribbed inclusions. Géotechnique 41:485–498
89. Junaideen S, Tham L, Law K, Lee C, Yue Z (2004) Laboratory study of soil nail interaction in loose, completely decomposed granite. Can Geotech J 41:274–286
90. Sharma M, Samanta M, Sarkar S (2017) Laboratory study on pullout capacity of helical soil nail in cohesionless soil. Can Geotech J 54:1482–1495
91. Kowalczyk T, Pleszczynska E, Ruland F (2004) Grade models and methods for data analysis: with applications for the analysis of data populations. Springer Science & Business Media
92. Fei Y, Yasufuku N, Zhao C, Zhao C, Xu K (2020) Shear behaviour of sand-concrete interface with side post-grouting considering the unloading effect. Soils Found
93. Zhao C, Zhang R, Zhao C, Wang W, Wang Y (2021) A three-dimensional evaluation of interface shear behavior between granular material and rough surface. J Test Eval:49
94. Horvath R, Kenney T, Kozicki P (1983) Methods of improving the performance of drilled piers in weak rock. Can Geotech J 20:758–772
95. Chu LM, Yin JH (2005) Comparison of interface shear strength of soil nails measured by both direct shear box tests and pullout tests. J Geotech Geoenvironmental Eng 131:1097–1107
96. Wang J, Dove J, Gutierrez M (2007) Determining particulate–solid interphase strength using shear-induced anisotropy. Granular Matter 9:231–240
97. Gu X, Chen Y, Huang M (2017) Critical state shear behavior of the soil-structure interface determined by discrete element modeling. Particuology 35:68–77
98. Feng SJ, Chen JN, Chen HX, Liu X, Zhao T, Zhou A (2020) Analysis of sand–woven geotextile interface shear behavior using discrete element method (DEM). Can Geotech J 57:433–447
99. Zhang ZL, Cui ZD, Zhao LZ (2020) Shear strength of sand–clay interfaces through large-scale direct shear tests. Arab J Sci Eng:1–15
100. Feng ZK, Xu WJ, Meng QS (2020) Mechanical behaviors of interaction between coral sand and structure surface. J Central South Univ 27:3436–3449
101. Maghsoodi S, Cuisinier O, Masrouri F (2020) Thermal effects on mechanical behaviour of soil–structure interface. Can Geotech J 57:32–47
102. Maghsoodi S, Cuisinier O, Masrouri F (2020) Effect of temperature on the cyclic behavior of clay-structure interface. J Geotech Geoenvironmental Eng 146:04020103
103. Wang J, Liu S, Cheng YP (2017) Role of normal boundary condition in interface shear test for the determination of skin friction along pile shaft. Can Geotech J 54:1245–1256
104. Zhou WH, Jing XY, Yin ZY, Geng X (2019) Effects of particle sphericity and initial fabric on the shearing behavior of soil–rough structural interface. Acta Geotechnica:1–18
105. Wang J, Dove JE, Gutierrez MS (2007) Anisotropy-based failure criterion for interphase systems. J Geotech Geoenvironmental Eng 133:599–608
106. Grabowski A, Nitka M, Tejchman J (2021) 3D DEM simulations of monotonic interface behaviour between cohesionless sand and rigid wall of different roughness. Acta Geotech 16:1001–1026
107. Zhu H, Nguyen HN, Nicot F, Darve F (2016) On a common critical state in localized and diffuse failure modes. J Mech Phys Solids 95:112–131
108. Jing XY, Zhou WH, Zhu HX, Yin ZY, Li Y (2018) Analysis of soil-structural interface behavior using three-dimensional DEM simulations. Int J Numer Anal Meth Geomech 42:339–357
109. Zhang N, Evans TM (2018) Three dimensional discrete element method simulations of interface shear. Soils Found 58:941–956
110. Wang X, Wang X-Z, Zhu C-Q, Meng Q-S (2019) Shear tests of interfaces between calcareous sand and steel. Mar Georesour Geotechnol 37:1095–1104

111. Guo J, Wang X, Lei S, Wang R, Kou H, Wei D (2020) Effects of groove feature on shear behavior of steel-sand interface. Adv Civil Eng
112. Rui S, Wang L, Guo Z, Cheng X, Wu B (2020) Monotonic behavior of interface shear between carbonate sands and steel. Acta Geotech
113. Samanta M, Punetha P, Sharma M (2018) Effect of roughness on interface shear behavior of sand with steel and concrete surface. Geomech Eng 14:387–398
114. Di Donna A, Ferrari A, Laloui L (2016) Experimental investigations of the soil–concrete interface: physical mechanisms, cyclic mobilization, and behaviour at different temperatures. Can Geotech J 53:659–672
115. Yazdani S, Helwany S, Olgun G (2019) Influence of temperature on soil–pile interface shear strength. Geomech Energy Environ 18:69–78
116. Tejchman J, Wu W (1995) Experimental and numerical study of sand–steel interfaces. Int J Numer Anal Meth Geomech 19:513–536
117. Kozicki J, Niedostatkiewicz M, Tejchman J, Muhlhaus H-B (2013) Discrete modelling results of a direct shear test for granular materials versus FE results. Granular Matter 15:607–627
118. Vlcek J, Valaskova V (2019) Investigation of the soil-geosynthetic interaction using direct shear testing and FEM method. In: IOP conference series: materials science and engineering. IOP Publishing, p 052052
119. Desai C, Zaman M, Lightner J, Siriwardane H (1984) Thin-layer element for interfaces and joints. Int J Numer Anal Meth Geomech 8:19–43
120. Yin Z-Z, Zhu H, Xu G-H (1995) A study of deformation in the interface between soil and concrete. Comput Geotech 17:75–92
121. Katona MG (1983) A simple contact–friction interface element with applications to buried culverts. Int J Numer Anal Meth Geomech 7:371–384
122. Hegde A, Roy R (2018) A comparative numerical study on soil–geosynthetic interactions using large scale direct shear test and pullout test. Int J Geosynthetics Ground Eng 4:2
123. Paikowsky SG, Player CM, Connors PJ (1995) A dual interface apparatus for testing unrestricted friction of soil along solid surfaces. Geotech Test J 18:168–193
124. Tehrani F, Han F, Salgado R, Prezzi M, Tovar R, Castro A (2016) Effect of surface roughness on the shaft resistance of non-displacement piles embedded in sand. Géotechnique 66:386–400
125. Abuel-Naga H, Shaia H, Bouazza A (2017) Effect of surface roughness and hardness of continuum materials on interface shear strength of granular materials. J Test Eval 46:826–831
126. Grabowski A, Nitka M, Tejchman J (2020) 3D DEM simulations of monotonic interface behaviour between cohesionless sand and rigid wall of different roughness. Acta Geotechnica:1–26
127. Tovar-Valencia RD, Galvis-Castro A, Salgado R, Prezzi M (2018) Effect of surface roughness on the shaft resistance of displacement model piles in sand. J Geotech Geoenvironmental Eng 144:04017120
128. Hebeler G, Martinez A, Frost J (2016) Shear zone evolution of granular soils in contact with conventional and textured CPT friction sleeves. KSCE J Civ Eng 20:1267–1282
129. Houlihan R, Williams K (1987) Recent enhancements in mined oil sands bitumen extraction technology. J Can Pet Technol 26
130. Frost J, Karademir T (2016) Shear-induced changes in smooth geomembrane surface topography at different ambient temperatures. Geosynth Int 23:113–128
131. Dove JE, Frost JD (1999) Peak friction behavior of smooth geomembrane-particle interfaces. J Geotech Geoenvironmental Eng 125:544–555
132. Porcino D, Fioravante V, Ghionna VN, Pedroni S (2003) Interface behavior of sands from constant normal stiffness direct shear tests. Geotech Test J 26:289–301
133. DeJong JT, Randolph MF, White DJ (2003) Interface load transfer degradation during cyclic loading: a microscale investigation. Soils Found 43:81–93
134. Shahrour I, Rezaie F (1997) An elastoplastic constitutive relation for the soil-structure interface under cyclic loading. Comput Geotech 21:21–39

135. Evgin E, Fakharian K (1998) Cyclic rotational simple-shear behaviour of sand-steel interfaces. Soils Found 38:191–199
136. Wang J, Liu F, Wang P, Cai Y (2016) Particle size effects on coarse soil-geogrid interface response in cyclic and post-cyclic direct shear tests. Geotext Geomembr 44:854–861

Chapter 2
Shear Banding in 2D Numerical Interface Tests

2.1 Introduction

As a typical boundary value problem, heterogeneity arises in soil subjected to interface shearing and somehow develops along loading history. In this case, the strain will gradually localize into a band-shaped shear zone covering the rough surface, while the soil outside this zone, as a whole, undergoes quasi-elastic unloading. Out of the escalating degree of strain localization, a shear band will finally form at the large strain and show its characteristic highly concentrated soil deformation. Once this shear band stabilizes, the soil appears to be structurized into two domains with distinct deformational patterns: (1) the inside shear-band area, featured by a large and non-affine deformation and the remainder, with affine or even vanished shear deformation (the term "affine" refers to a constant displacement or velocity gradient). Meanwhile, macro- and micro-mechanical properties also structurize in association with this deformational heterogeneity [1–3]. Following these facts, on the one hand, SSI is essentially the interaction between the structure surface and the soil layer nearby. SSI's non-elastic nature and shear resistance derive from this mechanical system. On the other hand, the shear-banding process, therefore, plays a significant role in SSI's mechanical performance. Clarifying how strain localization accumulates and develops to form the shear band is crucial in explaining some of SSI's significant constitutive behaviors during the shearing, for example, stress hardening, softening, and shear-dilatancy.

In fact, shear banding in SSI belongs to a general phenomenon, the localized failure in granular soils, which is considered as a bifurcation for the material from the homogeneous deformation pattern to a discontinuous one. During this geometrical transformation of the material, even for experiments conducted in identical conditions, various possible patterns can be encountered, leading to various post-bifurcation mechanical responses. It has been decades since scientists paid attention to strain localization and localized failure in the granular soil. Many studies have investigated the behavior and morphology of this type of failure from experimental [3–6] and theoretical viewpoints [7–14]. Of note, much research has focused on exploring the

W.-H. Zhou and Z.-Y. Yin, *Practice of Discrete Element Method in Soil-Structure Interface Modelling*, https://doi.org/10.1007/978-981-19-0047-1_2

micro-mechanical basis of localized failure and the shear-banding process. Characteristic micro-scale properties, such as particle rotation, fabric anisotropy, and force-chains, were observed to develop in association with the localized strain [1, 2, 15–20]. The influence of some micro-mechanical parameters on the formation of the shear band was revealed [21–23]. Exploring the shear-banding process in SSI, as a typical and simple pattern of localized failure, from macro and micro viewpoints will contribute to the knowledge about the mechanism of the localized failure in soil.

Under the continuum-mechanical framework, determining the thickness of the shear band becomes important in the modeling of SSI. It is often required to define the interface zone, inside which the material will be treated with a different constitutive relation from the material outside the shear band. By adopting different criteria to detect the shear-band boundary, some researchers have drawn varying conclusions [24–27]. Especially, the study of [28], using DEM simulation, investigated the development of strain localization in densely packed granular soil sheared by interfaces of various roughness. Several characteristic points in the shear-banding process were pointed out. The influence of surface roughness on shear-band thickness was reported in the authors' study. Even though the shear-band thickness has been determined in various loading conditions and interface geometries, without understanding how the shear band emerges from the small scale strain localization and develops to its final form in SSI, the relation between shear-band thickness and its factors (the surface roughness, initial void ratio, confining stress, etc.) can be only empirical and lacks a physical background. However, scant work has engaged in this point.

This chapter presents the shear-banding process in densely and loosely packed granular soils subjected to interface shearing. The following two aspects are focused: (1) how the pattern arises from a homogeneous deformation, evolves, and gives birth to a shear band with high strain localization and how this evolution relates to the mechanical responses of SSI during shearing. Interface shear tests using DEM are introduced in Sect. 2.2 with modelling parameters, the test protocol, and some preliminary results. Methods of analyzing the kinematic field in the soil are given in Sect. 2.3. A meso-scale strain rate tensor is presented to define the local average strain inside individual polygonal cells enclosed by neighboring contact branches. This local strain rate tensor allows the strain field to form in the granular soil in a way where strain localization becomes conspicuous and can be conveniently captured and measured. We conduct comprehensive analyses on the kinematics of specimens in Sect. 2.4. A new indicator is proposed to characterize the degree of the strain localization. The evolutions of the shear velocity field, the shear velocity gradient, and the particle rotation are presented. In Sect. 2.5, evolutionary patterns and features of strain localization and their dependency on shear stress and volumetric strain are discussed for dense and loose specimens.

2.2 2D Interface Shear Test

2.2.1 *Numerical Modelling and Test Protocol*

The discrete element method (DEM) proposed by Cundall and Strack [29], coded by PFC2D 5.0 software, is employed in this chapter to solve the dynamics of the granular assembly subjected to interface shear loading. The interaction between two contacting objectives (particle–particle or particle–wall) is defined by the friction-elastic law, the combination of Hooke's law with a constant normal and tangential stiffness k_n and k_s, and the cohesionless Coulomb's law. To mimic the interlocking effect caused by the irregular shape of soil grains in natural soil, a rolling resistance is imposed onto contacts [22].

As illustrated in Fig. 2.1, the test apparatus, in width $w = 0.1$m along the x-axis and in height $h = 0.04$m along the y-axis, is constituted of a top wall and a teeth-plate on the bottom, as the latter assimilates the surface of the artificial structure, whose maximum y coordinate is set at $y = 0.0$m. The teeth-plate is regularly shaped by segments of slope alternating between 45° and 135°. Its roughness R is quantified by λ/D_{50}, where λ is the vertical span of the teeth-plate. To avoid the boundary effect along the shear direction, a periodic boundary condition is set on both sides of the apparatus. In other words, particles passing through one side will appear on the counter-side with the same velocity. It is important to note that we chose a regular surface geometry for this chapter instead of an irregular one based on the following two considerations: (1) the main interest of this chapter is the evolutionary feature of strain localization in the soil, rather than the effect of surface geometry on SSI's mechanical behavior, and (2) the surface geometry of an artificial structure is distinct in various situations; depending on the nature of the material and the way the structure is fabricated, no single pattern can be representative for all kinds of structures. Research on the mechanical behavior of SSI with various surface geometries includes

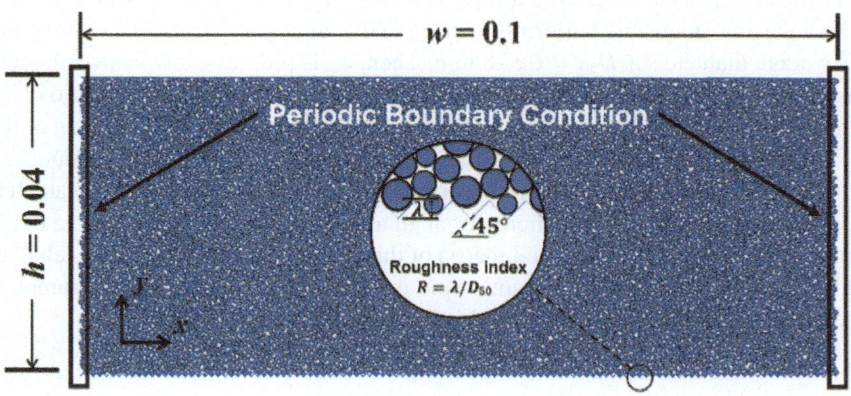

Fig. 2.1 Schematic diagram of the interface shear apparatus

Table 2.1 Parameters in the simulation

Parameters	Value
k_n	5.0×10^9 N/m
k_t/k_n	0.5
Number of particles	12,000
Particle density	2,650.0 kg/m^3
φ_g	26.57°
Particle–wall friction	41.99°
σ_n	1.0×10^5 N/m^2
v_s	2.0×10^{-3} m/min
R	0.75
Rolling resistance coefficient	0.1
Damping coefficient	0.1

works by [30–33]. Otherwise, we chose the specimen's height h in reference to the study given by [34], in which a minimum 40 D_{max} is suggested for h to sufficiently minimize the boundary effect of the apparatus to the direct shear test. As the direct shear test is normally constituted of two shear frames vertically symmetrical with each other, each frame can be basically equivalent to an interface shear box sheared by the soil contained in another frame. To this extent, a minimum 20 D_{max} is optical for h in the interface shear test. A much higher value $h = 0.04$m (around 51 D_{max}) is chosen in this chapter.

The specimen is installed by filling 12,000 particles into the test apparatus. To obtain a specimen with a relative isotropic initial fabric, the installation scheme for granular specimens proposed by Chareyre and Villard [35] is adopted by simultaneously growing the size of all particles, which are initially seeded over the apparatus's space with the diameter following a uniform distribution. This process stops once the normal stress on the top wall σ_n reaches the prescribed value. All grains in the specimen, whose radii are fixed afterwards, form a poly-dispersed granular system with the gains' diameters uniformly varying in the range [0.374 mm, 0.781 mm] and an average diameter of $D_{50} = 0.578$ mm. Then, φ_g is progressively vanished while maintaining σ_n to obtain a series of specimens with a reduced initial void ratio of e_0, wherein $e = V_{void}/V_{solid}$, as V_{void} and V_{solid} are respectively the total volume of the void and solid parts in the domain of measurement. After specimens are initialized with a certain value of e_0, a constant shear velocity v_s is applied on the teeth-plate. It is noteworthy that v_s herein is sufficiently small to maintain the quasi-static state of the granular system; namely, the total inertia of this system is negligible, as this chapter only aims to investigate the system's time-independent properties. The parameters used in numerical modeling are listed in Table 2.1.

Table 2.2 Specimens and corresponding initial void ratio

Specimens	e_0
S1	0.151
S2	0.179

2.2.2 Interface Shear Tests

The densest and loosest specimens among a series of prepared specimens, listed in Table 2.2, are chosen to carry out the interface shear test. In each monitoring strain state, the shear stress σ_s, the shear displacement d_s of the teeth-plate and the volumetric strain ε_v of the specimen are measured as $\varepsilon_v = (h_t - h_0)/h_0$, where h_0 and h_t are heights of the specimen h (the distance between the top wall and the peak of the teeth-plate) respectively at the initial state and the monitoring state. Evolutions of stress ratio σ_s/σ_n and ε_v with respect to normalized shear displacement d_s/D_{50} are shown in Fig. 2.2.

Two specimens converge into a single final value of σ_s/σ_n. S1 presents a stress peak and stress softening, while S2 does not see large-scale stress softening. In terms of volumetric evolution, S1 dilates from the beginning, reaches the highest dilatant rate at around the stress peak, and turns to a steady dilatancy when the stress residual state is reached. S2 experiences a slight contractancy at the beginning of the test and smoothly dilates afterwards. Generally, the denser specimen tends to have a higher stress peak and a larger degree of dilatancy.

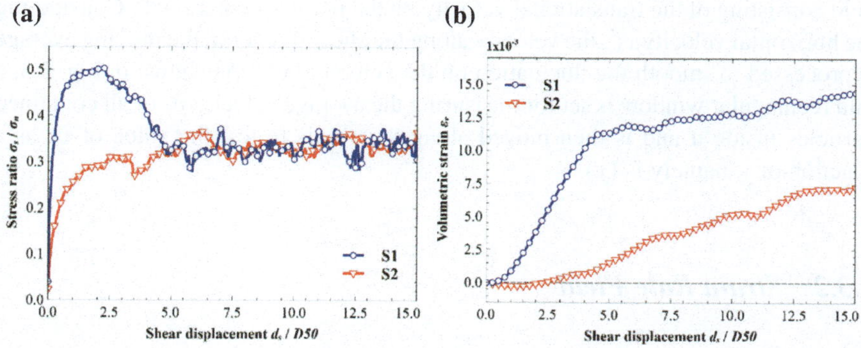

Fig. 2.2 **a** Stress ratio σ_s/σ_n and **b** volumetric strain ε_v with respect of shear displacement d_s/D_{50}

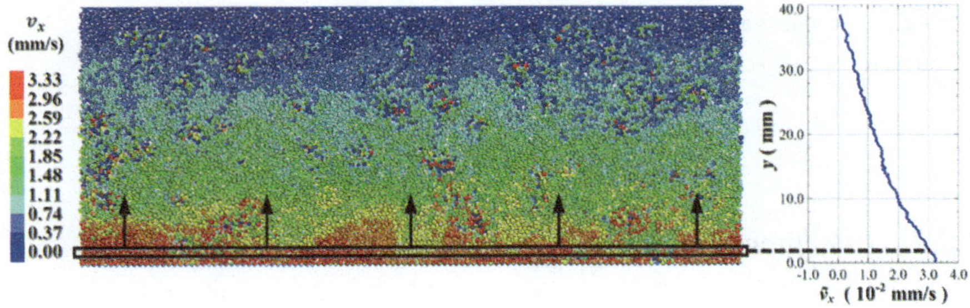

Fig. 2.3 Schematic diagram of measuring $\tilde{v}_x$ by the moving average scheme. On the left is the specimen S1 at $d_s/D_{50} = 0.5$ with particles colored according to the magnitude of v_x. On the right is the line chart of measured $\tilde{v}_x$ as function of y

2.3 Velocity Field and Its Gradient

Investigating the shear-band formation process should be based on the kinematic analysis of the granular system. The means of analyzing the velocity field and building the strain rate field will be introduced in this section.

2.3.1 Analyzing the Velocity Field

The kinematic aspect of the granular system can be largely represented by the discrete field consisting of the translational velocity on the particle centers $\overrightarrow{v}^p$. Considering the horizontal velocity v_x^p, the velocity along the shear direction, the moving average is processed to smooth the fluctuation on the velocity field. As shown in Fig. 2.3, a thin rectangular window is set for measuring the average velocity $\tilde{v}_x$ of all contained particles inside it and is then moved along the y-axis to test the value of $\tilde{v}_x$ as a function of y, namely $\tilde{v}_x(y)$.

2.3.2 Strain Rate Field

At any quasi-static state of a granular assembly, most grains are well supported by contacts with their neighbors. Contact branches, connecting the centers of two contacting grains, distribute throughout the material, forming a network, named the contact network. By the contact network, the material area can be seamlessly tessellated into polygonal loops, denoted as meso-loops, enclosed by contact branches (as shown in Fig. 2.4). The equivalent continuum is assumed inside meso-loops in this chapter. The translational velocity field on a meso-loop is defined in such a way that the velocity at each vertex is equivalent to the velocity of associated grain with it, and the velocity along the contact branch keeps linear. Based on this velocity field,

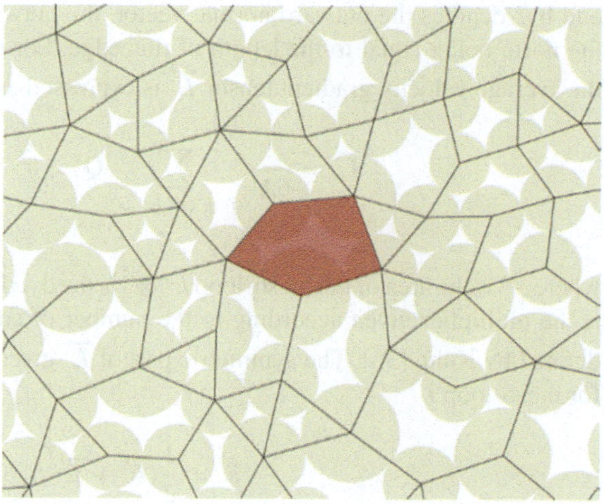

Fig. 2.4 Meso-loops enclosed by contacts represented by black branches

Kuhn [36] derived a formula to define the average strain rate field of meso-loops. We exemplify this definition with a pentagon, highlighted in red in Fig. 2.4 and illustrated with local geometrical and kinematic quantities in Fig. 2.5. Vertices and edges are respectively numbered in clockwise order. For the meso-loop i ($i = 1, 2, \cdots, N_l$), where N_l is the total number of meso-loops, $\hat{\mathbf{v}}^{i,j}$ denotes the relative velocity vector on the edge j, which is equal to $\mathbf{v}^{i,j+1} - \mathbf{v}^{i,j}$ ($\mathbf{v}^{i,j}$ is the velocity vector at vertex j),

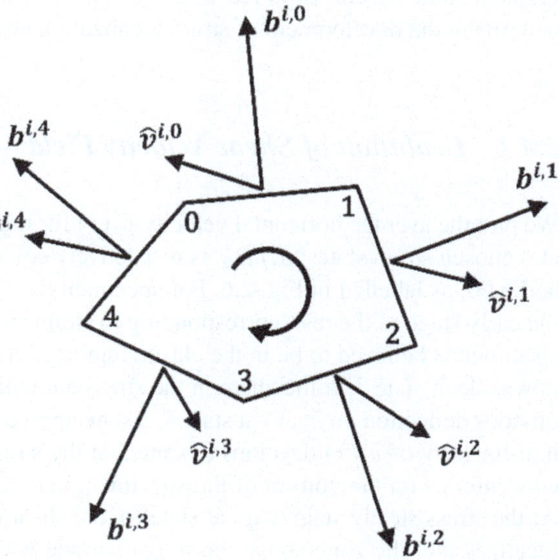

Fig. 2.5 For the edge j in the meso-loop i, the outward normal vector $b^{i,j}$ and the relative velocity vector $\hat{v}^{i,j}$

and $\mathbf{b}^{i,j}$ denotes the outward normal vector, the outward normal of the edge j with the norm being equal to the length of this edge. For the meso-loop i with s edges, the average velocity gradient tensor $\overline{\boldsymbol{L}}^i$ is expressed as below,

$$\overline{\mathbf{L}}^i = \frac{1}{6\mathbf{A}^i} \sum_{j_1, j_2 \in \{0,1,\ldots,s\}} \mathbf{Q}^s_{j_1,j_2} \hat{v}^{i,j_1} \otimes \mathbf{b}^{i,j_2} \tag{2.1}$$

where A^i is the area of the meso-loop i, "$\otimes$" is the dyadic product operator, and $Q^s_{j_1,j_2}$ is the multiplier given according to the number of edge s, which has been exactly defined by Kuhn [36]. The symmetric part of $\overline{\boldsymbol{L}}^i$ gives the average strain rate tensor for meso-loop i

$$\dot{\overline{\boldsymbol{\varepsilon}}}^i = \frac{1}{2}(\overline{\boldsymbol{L}}^i + \overline{\boldsymbol{L}}^{iT}) \tag{2.2}$$

where $\overline{\mathbf{L}}^{iT}$ is the transposition of $\overline{\mathbf{L}}^i$.

2.4 Formation of Shear Band

The shear-banding process will be studied in this section. The velocity field and its gradient field will be explored in an attempt to capture the featured evolutionary patterns in the development of strain localization and shear-band formation.

2.4.1 Evolution of Shear Velocity Field

We plot the average horizontal velocity $\tilde{v}_x(y)$ for the dense specimen S1 in Fig. 2.7, at 6 chosen strain states: d_s/D_{50} is respectively equal to 0.10, 0.50, 1.30, 2.00, 3.00 and 6.00, as labelled in Fig. 2.6. For specimen S1, $\tilde{v}_x(y)$ keeps in quasi-linearity at the early stage of the test (corresponding to strain state 1 and 2 in Fig. 2.6) when the specimen is believed to be in the elastic regime. $\tilde{v}_x(y)$ is observed to bend from the downside at state 3, somewhere in the stress hardening regime. It is followed by an obvious deflection on $\tilde{v}_x(y)$ at state 4, a state approaching the stress peak. When the non-linearity of $\tilde{v}_x(y)$ develops to state 5 at the stress softening, shear deformation concentrates on the bottom of the specimen, that is, large-scale strain localization. At the stress steady state (e.g., at state 6), the shear deformation almost completely localizes into the zone above the rough surface while vanishing outside this zone. This phenomenon signals a stable pattern in the strain localization. In other words, a steady shear band has been formed.

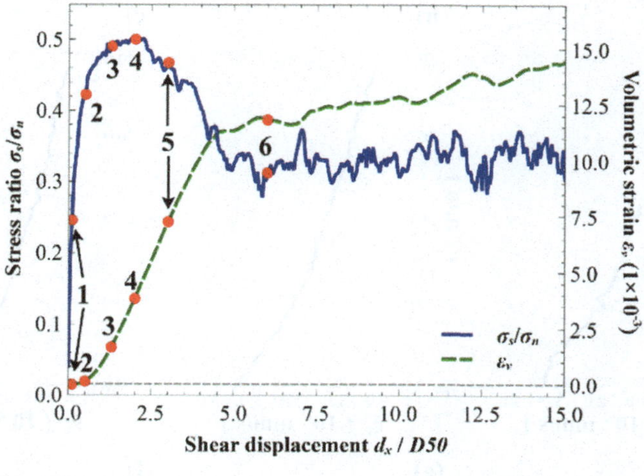

Fig. 2.6 For S1 stress ratio σ_s/σ_n and volumetric strain ε_v with respect of shear displacement d_s/D_{50}, with 6 monitoring strain states labelled by red points: d_s/D_{50} =. 0.10, 0.50, 1.30, 2.00, 3.00 and 6.00

The average horizontal velocity $\tilde{v}_x(y)$ for the loose specimen S2 is plotted in Fig. 2.9, at 6 chosen strain states: $d_s/D_{50} = 0.20, 1.00, 3.40, 6.00, 7.05$ and 9.75, as labelled in Fig. 2.8.

In the stress hardening phase of S2, the non-linearity of $\tilde{v}_x(y)$ is observed to develop from state 1 (corresponding to Fig. 2.9a) to the state 3 (Fig. 2.9c). At state 3, a pronounced deflection of $\tilde{v}_x(y)$ occurs in the lower part of the sample with a small drop in the stress. The shear zone seems to evolve to its steady form from state 3 onward. However, it is followed by a straight shaped $\tilde{v}_x(y)$ at state 4 (Fig. 2.9d). In other words, the non-affine deformational pattern recovers to an affine one. The non-linearity reappears in $\tilde{v}_x(y)$ at state 5 (Fig. 2.9e) and maintains to state 6 (Fig. 2.9f).

It is concluded that the strong strain localization has shaped in both S1 and S2, as the velocity field is stratified into two phases (as illustrated in Fig. 2.7e): the non-linear phase covering the surface and the linear phase in the remaining part of the sample. Moreover, the evolution of the velocity field in the loose sample is quite different from in the dense one. The non-linear manner of $\tilde{v}_x(y)$ can persist to the end of the test once it appears in the dense specimen but has been observed to vanish and reappear in the loose specimen.

Generally, with the help of $\tilde{v}_x(y)$, the velocity field of specimens is visualized in such a simple way that its non-affinity can be clearly manifested and easily noted. However, $\tilde{v}_x(y)$ only displays the profile of the specimen's kinematics at some individual strain states. The evolution of the kinematic field of samples should be investigated in more detail. A continuous and quantitative characterization of the strain localization is required and will be solved in the following section.

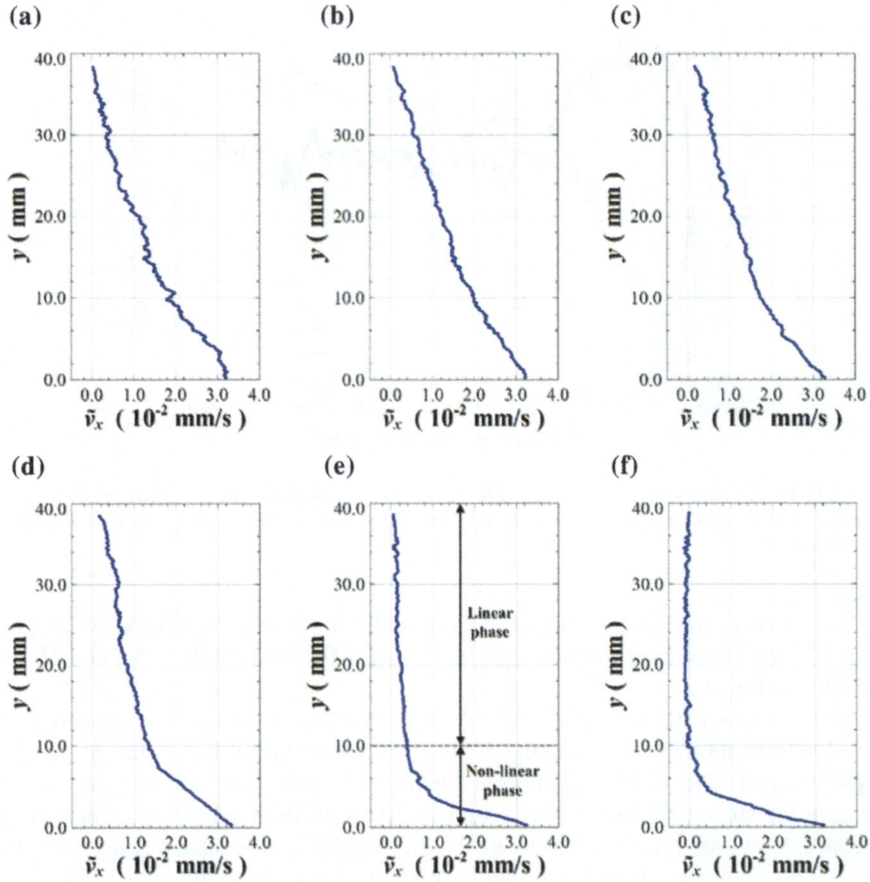

Fig. 2.7 For S1 distribution of average horizontal velocity $\tilde{v}_x$ in terms of y at strain states **a** 1, **b** 2, **c** 3, **d** 4, **e** 5 and **f** 6. In (**e**) the linear and non-linear phases of $\tilde{v}_x(y)$ are marked

2.4.2 *Development of Strain Localization*

A new indicator α is proposed to quantitatively characterize the degree of strain localization along the loading path. As illustrated in Fig. 2.10, $\tilde{v}_x(y)$ can be linearly approximated based on the boundaries' movement so that the approximation function $v_x^l(y)$ is equal to the boundaries' velocity at the depths of top and bottom boundaries: $v_x^l(0.0) = 3.33$ mm/s; $v_x^l(4.0) = 0.00$ mm/s. Of note, due to the high friction attributed to contacts between boundaries and grains, there are $\tilde{v}_x(0.0) \approx v_x^l(0.0)$ and $\tilde{v}_x(4.0) \approx v_x^l(4.0)$. An error exists in this approximation due to the non-linearity of $\tilde{v}_x(y)$. Integrating this error over y gives the total error:

$$S_{err} = \int_{0.0}^{4.0} \left| v_x^l - \tilde{v}_x \right| dy, \tag{2.3}$$

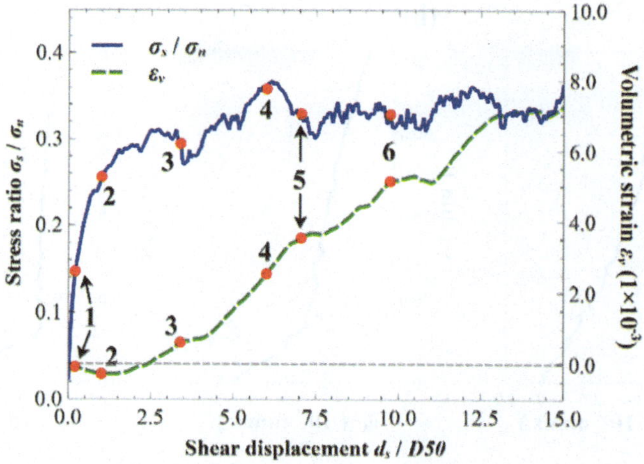

Fig. 2.8 For S2 stress ratio σ_s/σ_n and volumetric strain ε_v with respect of shear displacement d_s/D_{50}, with 6 monitoring strain states labelled by red points: $d_s/D_{50} = 0.20$, 1.00, 3.40, 6.00, 7.05 and 9.75

and α is defined as the normalized error:

$$\alpha = \frac{S_{error}}{S_{linear}} = \frac{\int_{0.0}^{4.0} |v_x^l - \tilde{v}_x| dy}{\int_{0.0}^{4.0} v_x^l dy}. \tag{2.4}$$

A larger α corresponds to a larger error in the approximation due to the non-linearity of $\tilde{v}_x(y)$ and therefore corresponds to a larger degree of strain localization. It is worth mentioning that there is always difficulty in terms of describing the strain localization through analyzing the kinematic field of the material. Most of the literature present scenes of displacement or strain fields at individual strain states, which merely provide a qualitative overview of the concentrated strain. Otherwise, it is almost impossible to use a function to fit the measured velocity field, as a normal function is insufficient to contain the enrichment of presented patterns in the strain localization of the interface shearing (such as those shown in Figs. 2.7 and 2.9). This proposed indicator allows the strain localization to be quantified into a scatter so that its evolution can be conveniently traced during shearing. In Fig. 2.11, we plot the evolution of α during the loading history for S1 and S2 and compare it with evolutions of the stress and volumetric strain. We notice that α for S2 is displayed until $d_s/D_{50} = 20.0$.

In dense specimen S1 (shown in Fig. 2.11a and b), α steadily ascends during the stress hardening regime. This scene absolutely changes with a dramatic increase of α from a characteristic point close to but before the stress peak. Strain localization promptly develops afterwards until the stress steady state is reached, and α keeps

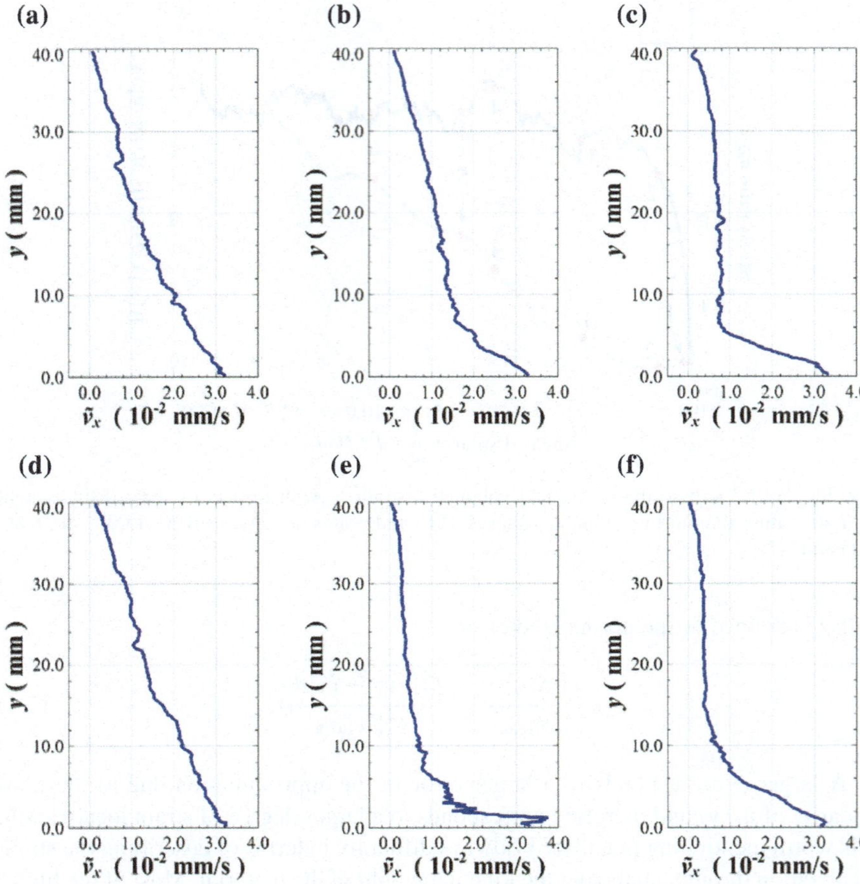

Fig. 2.9 For S2 distribution of average horizontal velocity $\tilde{v}_x$ in terms of y at strain states **a** 1, **b** 2, **c** 3, **d** 4, **e** 5 and **f** 6

fluctuating at a high level, around 0.85. This result refers to the considerable development of non-linearity in $\tilde{v}_x(y)$ from strain state 4 to 5 (shown in Fig. 2.7d and e, respectively).

In loose specimen S2 (Fig. 2.11c and d), α evolves with violent oscillation in a chaotic pattern. Both high and low values of α have temporarily emerged during the loading history. Generally viewing the curve of α, we note a big valley around in the range $d_s/D_{50} \in [5.00, 7.00]$. In this valley, α dramatically drops from a high value (close to 8.0) to vanish at strain state 4 and transiently increases afterward to touch the original level of α. This result is corroborated by the characteristic transformation of $\tilde{v}_x(y)$ along states 4, 5, and 6 (shown in Fig. 2.9c–e, respectively), that is, the transition from a highly non-linear curve of $\tilde{v}_x(y)$ to a quasi-linear one and the recovery to the original pattern. Moreover, such an unsteady pattern in the strain localization can also be captured after state 6, during $d_s/D_{50} \in [10.55, 13.15]$

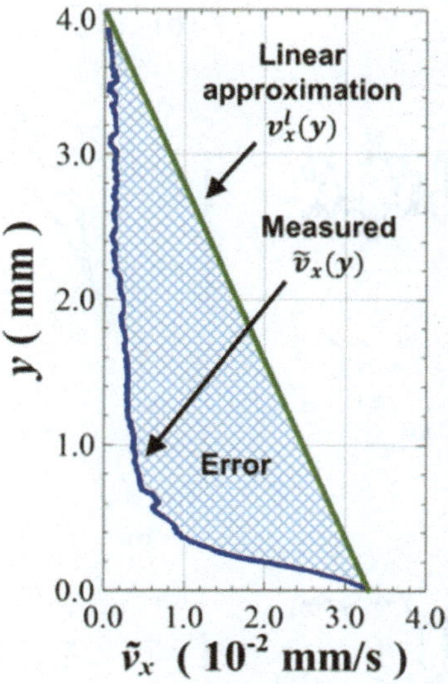

Fig. 2.10 In terms of y, average velocity $\tilde{v}_x$, its linear approximation v_x^l calculated from the boundary movement and the error $|v_x^l - \tilde{v}_x|$

and [14.30, 19.70], as highlighted by transparent blue in Fig. 2.11d. Otherwise, after the specimen reaches the stress steady state (considered as $d_s/D_{50} > 5.0$ in S2), the dilatancy rate is found to be notably relevant to α; a lighter strain localization, with a lower α, basically corresponds to a higher dilatancy rate.

2.4.3 *Evolution of Shear Velocity Gradient*

To elucidate the evolution of the spatial distribution of the shear deformation in the granular assembly, the gradient of the shear velocity $\overline{L}_{12}^i$ ($i = 1, 2, \ldots, N_l$) is plotted for S1 in Fig. 2.12 and for S2 in Fig. 2.13, at measuring strain states already labelled in Figs. 2.6 and 2.8. The negative value of $\overline{L}_{12}^i$, colored by blue (or positive value colored by red), represents the shear deformation along (or against) the subjected shear load of the specimen. It is noteworthy that instead of using the shear strain rate $\dot{\overline{e}}_{12}^i$, which is equal to $\frac{1}{2}\left(\overline{L}_{12}^i + \overline{L}_{21}^i\right)$, we chose to display $\overline{L}_{12}^i$, the gradient of the shear velocity in terms of the vertical direction. By doing so, the shear deformation along different depths of the specimen can be clearly manifested.

In the dense specimen, the deformation is more or less homogeneous at state 1 (Fig. 2.12a) in the very early stage of the test. Despite this outcome, some small-scale deformational concentrations have been found to already exist, arranging along the shear direction, preferentially at the bottom of the specimen. With the expansion and conjugation among these tiny strain concentrations, the deformation appears

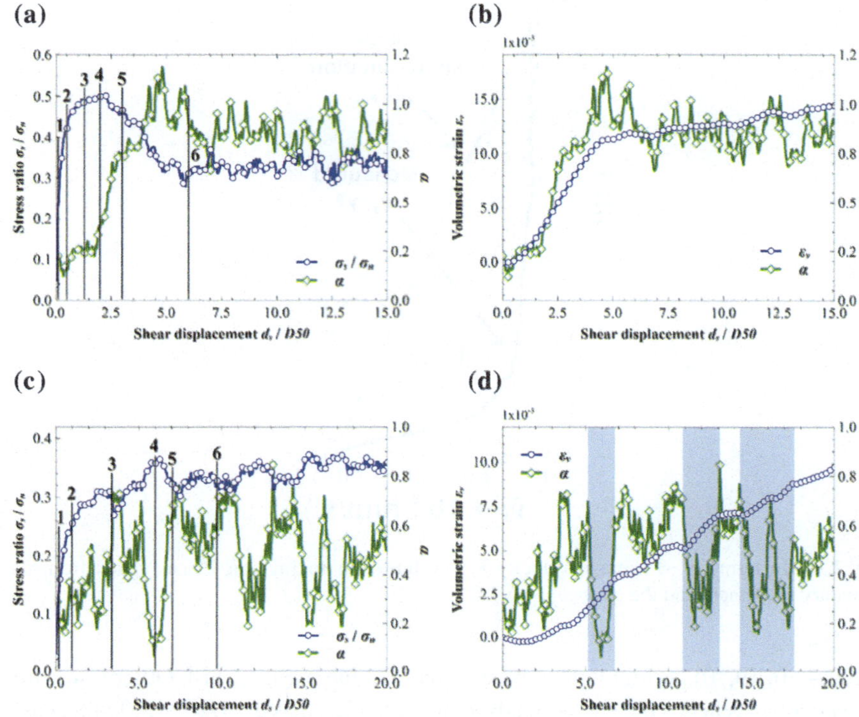

Fig. 2.11 With respect to shear displacement d_s/D_{50}, evolutions of α, compared with evolution of σ_s/σ_n for **a** S1 and **c** S2, and with evolutions of ε_v for **b** S1 and **d** S2. Monitoring strain states referred to in Figs. 2.6 and 2.8 for S1 and S2, respectively, are also labelled in **a** and **c**

to be discontinuous at the onset of stress hardening, around state 2 (Fig. 2.12b), even though no pronounced rupture exists in the material. To state 3 (Fig. 2.12c), several readily perceivable bands have formed: one clearly at the bottom and two blurrily on the top and in the middle of the sample, respectively, as marked inside the ellipses in Fig. 2.12c. The bottom band of the strain localization becomes clearer and paramount over the specimen near the stress peak, as shown in Fig. 2.12d. In the following stress softening regime, the strain acceleratingly localizes into the downside of the specimen. A notable shear band promptly shapes above the rough surface, as the shear deformation sufficiently concentrates inside the shear band and almost vanishes outside it. Meanwhile, the positive value of $\overline{L}^{i}_{12}$, which is in red, is observed somewhere outside the shear band, referring to the unloading of the material.

At state 1 (Fig. 2.13a) of the loose specimen, there is a generally homogeneous order in the distribution of $\overline{L}^{i}_{12}$, accompanied by many trivial local failures (or small scale strain localizations) diffusing over the specimen. The unloading of the material is noted in areas (of red) alongside these failures. We call this pattern hereafter the diffuse failure mode. It is noteworthy that when also believed to be generally

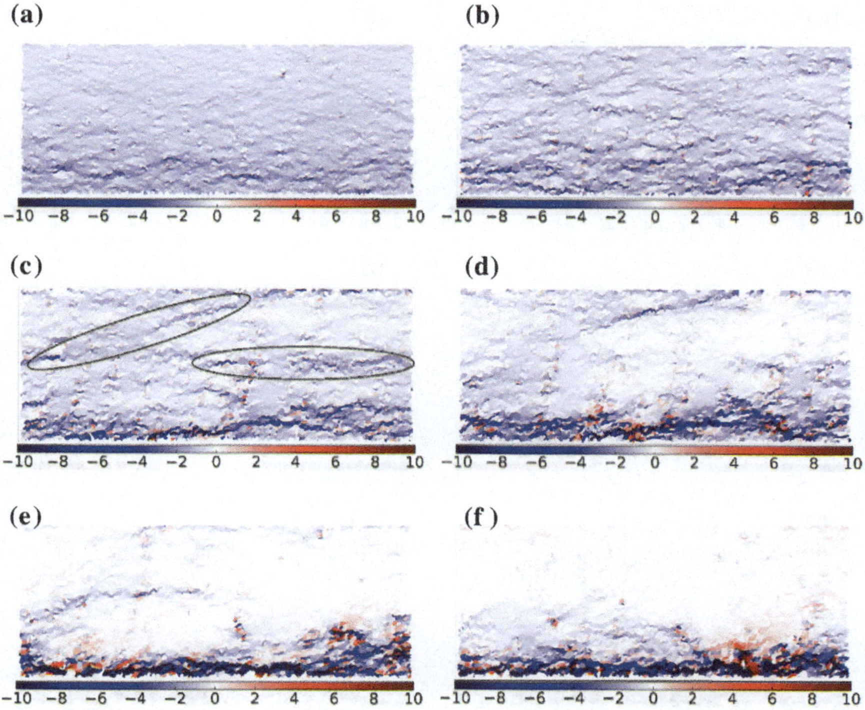

Fig. 2.12 For S1, distribution of $\overline{L}^{i}_{12}$ $(10^{-3}s^{-1})$ at strain states **a** 1, **b** 2, **c** 3, **d** 4, **e** 5, and **f** 6

homogeneous, the diffuse failure mode involves local strain localizations on a larger degree and scale than those beheld at state 1 of S1 (referring to Fig. 2.13a). Between strain states 2 and 3 (Fig. 2.13b and c), the specimen experiences a gradual strain localization into the sample's bottom and an unload of the material elsewhere. This failure mode is designated as the localized failure mode. At state 4, as previewed in the velocity field (referring to Fig. 2.9d), the velocity gradient field displays a scene reminiscent of that presented in state 1. Failures resume diffusion in the material after shortly being localized. However, this pattern still cannot be sustainable. Diffuse failures disappear afterwards and turn into an increasingly localized failure, as shown in Fig. 2.13e and f for state 5 and state 6, respectively. This is similar to what has happened from state 1 to 3, that is, the appearing and strengthening of the localized failure from a generally homogeneous deformation with diffuse failures.

2.4.4 Particle Rotation

Particle rotation plays significant role in the formation of the shear band [15, 22, 37–39]. The magnitude of rotation $|\theta|$ on particles is plotted at 4 respective monitoring

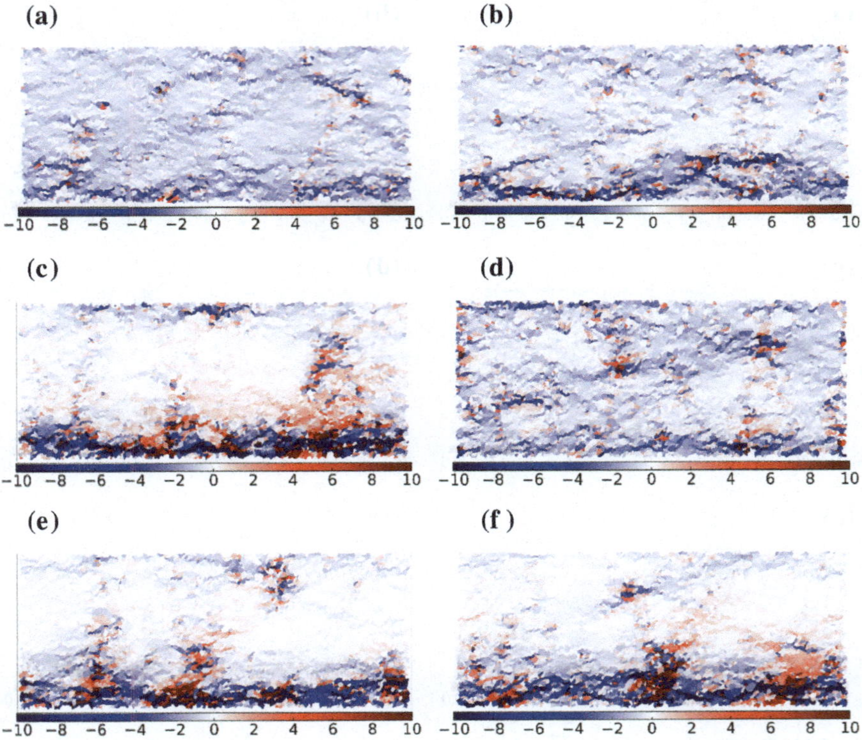

Fig. 2.13 For S2, distribution of $\overline{L}^i_{12}$ ($10^{-3}s^{-1}$) at strain states **a** 1, **b** 2, **c** 3, **d** 4, **e** 5, and **f** 6

strain states in Fig. 2.14 for S1 and in Fig. 2.15 for S2. Clock-wise and counter clock-wise rotations herein are not distinguished.

The spatial distribution of $|\theta|$ follows the way $\overline{L}^i_{12}$ evolves in both specimens, as the particle rotation concentrates in the area in which the shear deformation localizes. There is a pronounced association between particle rotation and strain localization. In S1, $|\theta|$ gradually stresses to form a distinct and continuous area with a high level of particle rotation running through the bottom of the specimen. In S2, $|\theta|$ tends to localize from state 2 to 3, resumes to be seemingly homogeneous at state 4, and localizes again at state 6. Otherwise, $|\theta|$ in S2 is generally smaller and less concentrated than that in S1.

2.5 Discussion

Generally, developments of $\tilde{v}_x(y)$, α, and $\overline{L}^i_{12}$ are consistent with each other and can be mutually supported during the loading history of both S1 and S2. In these two specimens, the shear band has been observed to be shaped on the bottom of them,

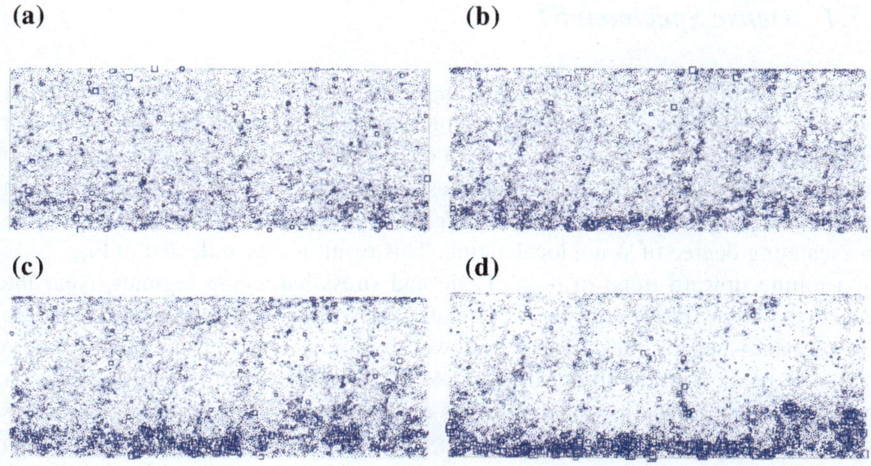

Fig. 2.14 The distribution of $|\theta|$ for S1 at strain states **a** 2 **b** 3 **c** 4 and **d** 5. The square's size is proportional to the magnitude of the particle rotation

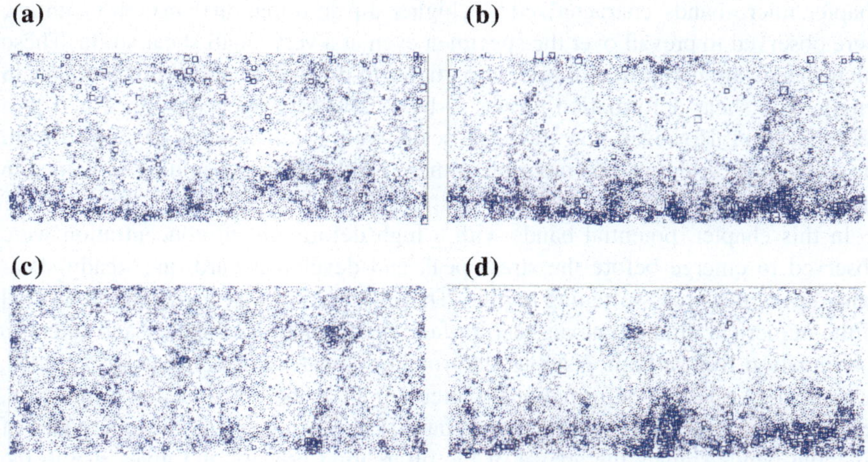

Fig. 2.15 The distribution of $|\theta|$ for S2 at strain states **a** 2 **b** 3 **c** 4 and **d** 6. The square's size is proportional to the magnitude of the particle rotation

featured by considerable and sufficient strain localizations into a band-like area. In particular, several characteristic trends were presented, which will be discussed in the following paragraphs.

2.5.1 Dense Specimen S1

As shown in Fig. 2.12a, tiny deformational concentrations have already existed at the fairly small strain, especially in the soil downside the specimen, indicating that the deformation was structured even from the very early stage of the shearing. With the progress of the shear deformation, these local deformational heterogeneities in the small scale grow, elongate, and conjugate with each other afterward, leading to an escalating degree of strain localization. This result is also reflected in Fig. 2.11a by a jolting upward trend of α in elastic and stress hardening regimes. Near the end of the stress hardening, the deformation has been sufficiently localized into several band-shaped areas. This is followed by a characteristic moment, prior to the stress peak, with an abrupt uprising of α, when the strain largely localizes at the bottom of the specimen and vanishes elsewhere. During the subsequent stress softening regime, a distinct pattern appears and is promptly strengthened, along with a mounting deformational concentration, indicating the formation of the shear band, which can persist to the end of the test.

Tiny concentrations of the shear deformation found at the very early stage of the test is reminiscent of what was noticed by Kuhn [36] in the drained biaxial test. In the chapter, micro-bands, characterized by a higher slip deformation than their vicinities, were observed to prevail over the specimen even at a very small shear strain. These micro-bands are composed of several particles in thickness and tens of particles in length. The shear deformation proves to be locally heterogeneous since it is at a very low level. However, how the large-scale strain localization is developed from the expanding, bridging, and merging of these local patterns should be explained in more detail.

In this chapter, potential bands with a high deformational concentration were observed to emerge before the stress peak and develop toward one steady shear band during the stress-softening regime. This observation was also reported by [17] in the numerical drained biaxial test. In fact, the formation of shear band refers to a bifurcation from a homogeneous deformation toward a discontinuous one [40]. A number of possible deformational patterns hold at the onset of this bifurcation, resulting in diversified post-bifurcation states [41]. To this extent, the emergence of potential bands signals the arrival of the bifurcation process and gives rise to the diverse mechanical performance of SSI in the post-bifurcation section.

2.5.2 Loose Specimen S2

The spatial distribution of the shear deformation in the loose specimen evolves in a significantly different manner from that in the dense specimen, even though the localized pattern was beheld in both of them. Failures in localized and diffuse modes are observed to appear alternatively during the loading history. As far as the end of the test ($d_s/D_{50} = 20.0$), the granular system of S2 has not deformed into a

Table 2.3 Two complementary loose specimens and corresponding initial void ratios

Specimens	e_0
S21	0.179
S22	0.181

steady localized pattern. All experienced large-scale localized deformation can be only temporary and variable, contrary to the shear band in the dense specimen, which can persist to the end of the test once it appears. For those who may argue against the representativeness of S2's mechanical preformation, we add two complementary loose specimens listed in Table 2.3, which are also installed and sheared according to parameters listed in Table 2.1. Evolutions of σ_s/σ_n, ε_v and α for these two additional specimen S21 and S22 are demonstrated in Fig. 2.16. Similar to what is observed for S2, these two additional specimens present chaotic variations of α; Transitions between the highly localized pattern and the generally homogeneous one, like those

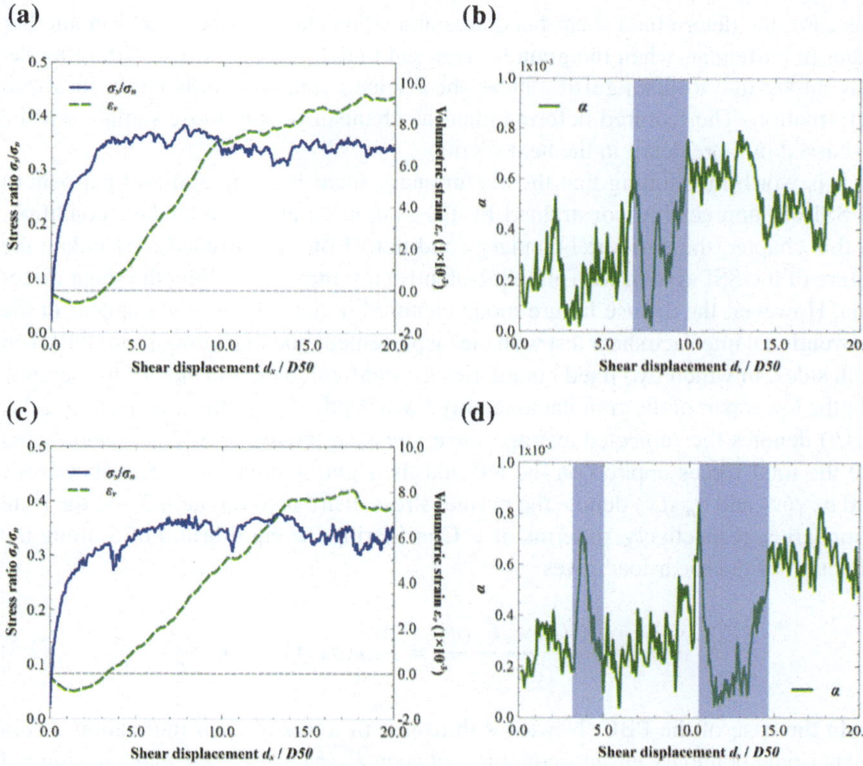

Fig. 2.16 With respect to shear displacement d_s/D_{50}, evolutions of σ_s/σ_n, and ε_v for **a** S21 and **c** S22 and evolutions of α for **b** S21 and **d** S22

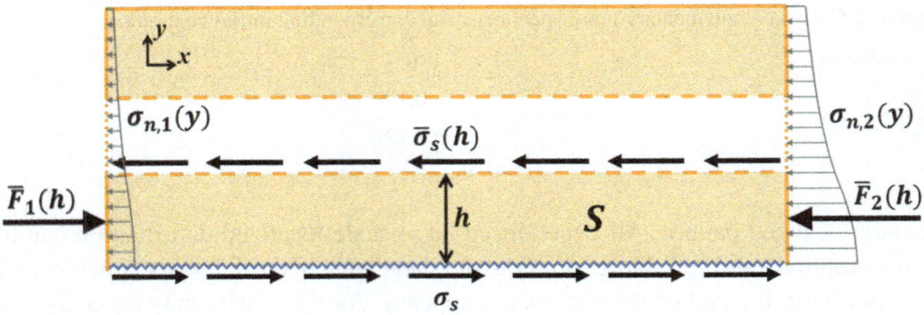

Fig. 2.17 Subjected forces and the equilibrium of a lower part of the granular assembly S along the shear direction

observed in S2, also exist in specimens S21 and S22, at the sections highlighted in the transparent blue in Fig. 2.16b and d.

It is concluded that a steady shear band may not arise at the large strain in the loosely assembled granular soil subjected to the interface shearing. Considering this case, the determined shear-band area at a strain state will be invalid in another state. In particular, when the granular material undergoes the diffuse failure mode, it is impossible to distinguish a clear shear zone containing sufficiently localized deformation. The featured deformational mechanism for the loose sample will be discussed in more detail in the next section.

It is worth mentioning that the evolutionary shear banding and its final pattern in SSI are appreciably constrained by the system's intrinsic boundary condition. In this chapter, the periodic boundary condition (PBC) is assigned to simulate the nature of the SSI system with an (equivalently) infinite length along the shear direction. However, the diffuse failure mode captured in the PBC will not appear in the conventional interface shear test with the displacement boundary condition (DBC) on both sides, in which two fixed boundaries are configured at both side of the sample. For the lower part of the granular assembly S at a depth of h, as illustrated in Fig. 2.17, $\overline{\sigma}_s(h)$ denotes the subjected average shear stress on the top of S, $\overline{F}_1(h)$ and $\overline{F}_2(h)$ are the total forces applied on the left and the right boundaries of S, respectively, and $\sigma_{n,1}(y)$ and $\sigma_{n,2}(y)$ denote the normal stress distributed on the left and the right boundaries, respectively, in terms of y. Considering the equilibrium of S along the direction of the shear load gives.

$$\overline{\sigma}_s(h) - \sigma_s = \frac{\overline{F}_1(h) - \overline{F}_2(h)}{w} = \frac{1}{w}\int_0^h (\sigma_{n,1}(y) - \sigma_{n,2}(y))dy \qquad (2.5)$$

In the case of the PBC, Newton's third law in terms of each interaction across the periodic boundary ensures equaling between $\overline{F}_1(h)$ and $\overline{F}_2(h)$, and vanishing of the item $\frac{\overline{F}_1(h)-\overline{F}_2(h)}{w}$. $\overline{\sigma}_s(h)$ with PBC is therefore constant, being equal to σ_s. With the DBC of fixed side boundaries, due to the moving trend of mass toward the right boundary, arguably, $\sigma_{n,2}(y)$ is larger than $\sigma_{n,1}(y)$ for various y. This results in an increasing negativity of $\overline{\sigma}_s - \sigma_s$ with the increasing h, that is, the reduction of $\overline{\sigma}_s$ with increasing depth. In consequence, the shear strain can concentrate nowhere but on the downside of the specimen. To this extent, the interface shear test with fix side boundaries is probably unable to produce a sensible stress and deformational mechanisms for SSI with an equivalently infinite length in the shear direction.

2.5.3 Volumetric Evolution Versus Strain Localization

Figure 2.11d shows that after the stress steady state is reached in S2 (more or less at $d_s/D_{50} = 5.0$), the strain localization is highly related to the dilatancy of the material. Evolutions of the global volumetric strain rate $\dot{\varepsilon}_v$ and α of S2 are plotted in Fig. 2.18, showing a generally reverse trend between them. A lower degree of the strain localization corresponds to a higher dilatancy rate during the stress steady state. This result is supported by a highly negative correlation between $\dot{\varepsilon}_v$ and α, measured by Pearson coefficient r ($r = -0.717$, as shown in Fig. 2.19b). The Pearson coefficient r, ranged in $[-1.0, 1.0]$, and its confident interval (CI) are given by the Pearson test to evaluate the linear correlation between two variables [42]. Reducing r from 1, through 0, to -1 refers to changing from the perfectly positive correlation, to no correlation, to the perfectly negative correlation. S1 also weakly exhibits this trend during the stress steady state, with $r = -0.577$ (Fig. 2.19a).

Previous research has reported that in a well-developed shear band, a singular and constant material fabric is developed with characteristic micro- and meso-structures [1, 2, 19], leading to a constant volume and shear strength of the material. Thus, there must finally be dynamical equilibrium between the local contractancy and local dilatancy inside the shear-band, which has well-developed during the stress steady state. The global volumetric evolution during the stress steady state, to this extent, turns to attribute to the shear-dilatancy nature of the material outside the shear band. A degradation of the strain localization, corresponding to an increasing shear

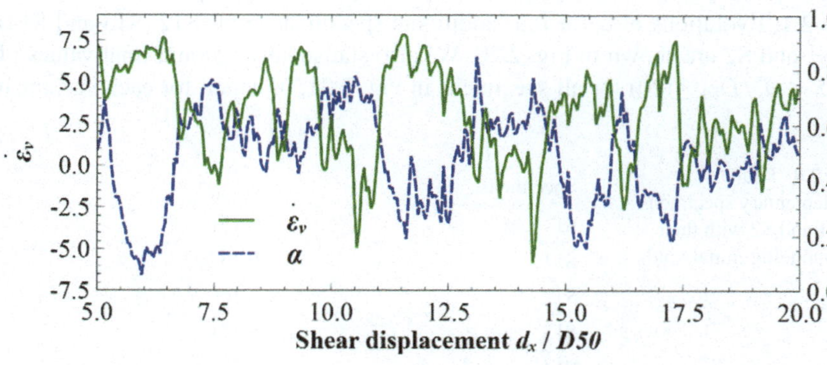

Fig. 2.18 $\dot{\varepsilon}_v$ and α with respect to d_s/D_{50} in S2 during $d_s/D_{50} \in [5.0, 20.0]$

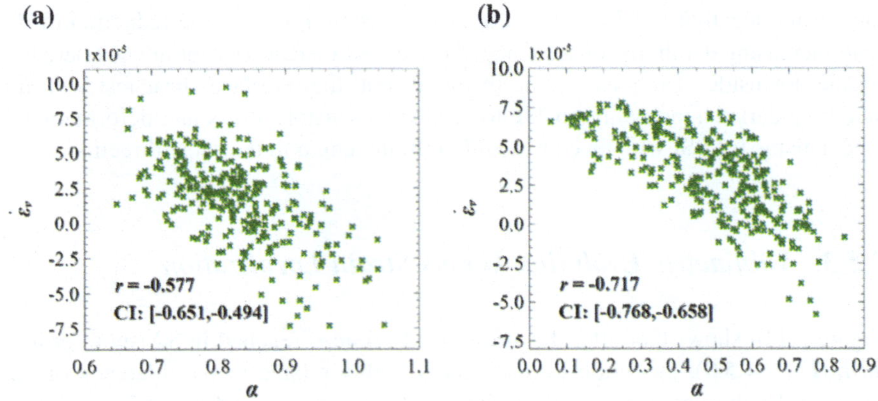

Fig. 2.19 $\dot{\varepsilon}_v$ versus α (**a**) for S1 during $d_s/D_{50} \in [6.0, 20.0]$ and (**b**) for S2 during $d_s/D_{50} \in [5.0, 20.0]$. Pearson correlation r and its confidence interval CI with 95% confidence level

strain imposed on the outside shear-band area, will lead to a higher dilatancy rate of the soil. This is likely the reason why there is a negative correlation between the degree of the strain localization and the dilatancy rate.

2.5.4 Patterns Transition Between Dense and Loose Specimens

The strain localization presents distinct pattern in loose specimen from that in the dense specimen. Why the pattern differ according to the initial void ratio remains not answered. Finding transitional patterns between these two distinct ones might shed light on understanding the underlying mechanism. Four complementary specimens with intermediate initial void ratios between the dense and loose specimens (listed and labelled in Table 2.4) are prepared and loaded according to parameters listed in Table 2.1. Evolutions of α for four additional specimens (S11, S12, S13 and S14) and S1 and S2 are shown in Fig. 2.20. We also statistically account final values of α ($7.5 \leq d_s/D_{50} \leq 20$) for all specimens in Fig. 2.21, in which for each specimen

Table 2.4 Four complementary specimens and S1 and S2 with their corresponding initial void ratios

Specimens	e_0
S1	0.151
S11	0.157
S12	0.166
S13	0.171
S14	0.175
S2	0.181

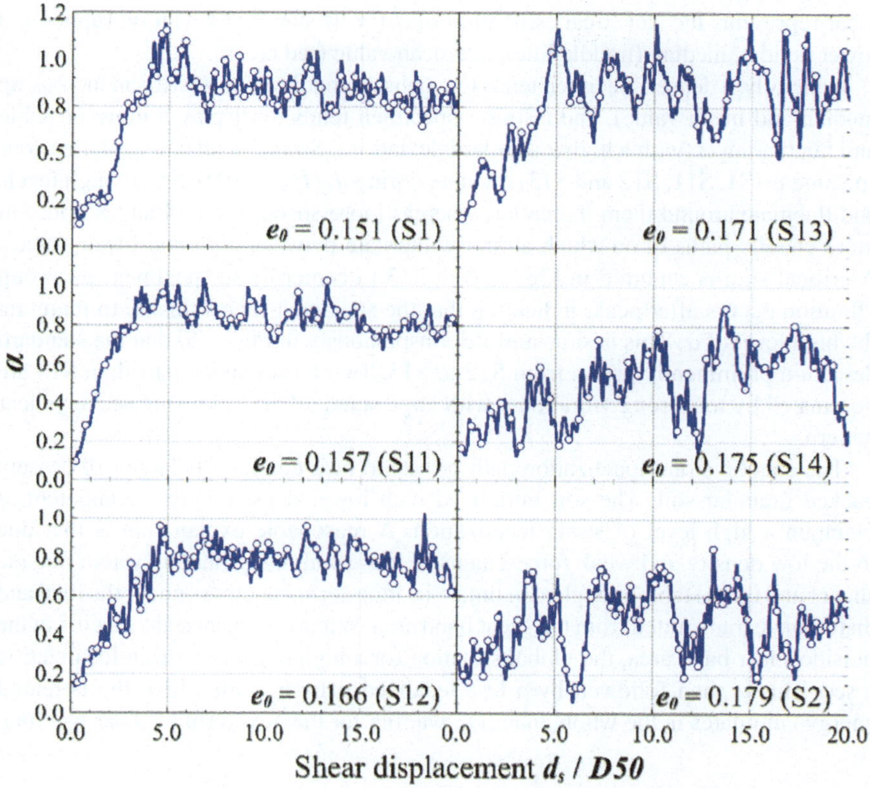

Fig. 2.20 Evolutions of α for S1 (dense specimen), S11, S12, S13, S14 and S2 (loose specimen) with respect to shear displacement d_s/D_{50}

Fig. 2.21 Statistical quantities of α during $7.5 \leq d_s/D_{50} \leq 20$ for S1, S11, S12, S13, S14 and S2: standard deviation (upper and lower bonds), median (middle line) and mean value (red cross)

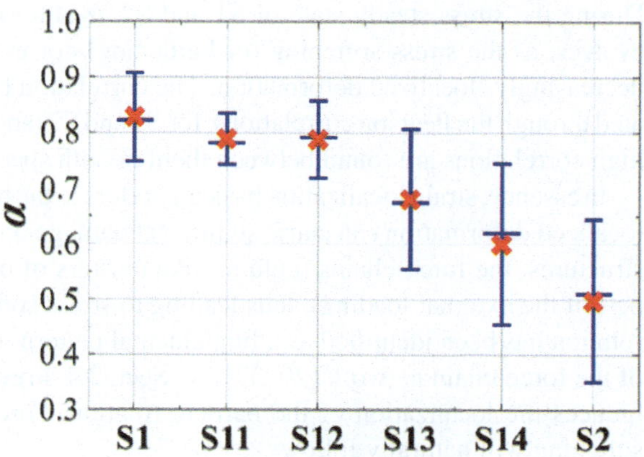

a bar represents the statistical estimation of α for its standard deviation (upper and lower bonds), median (middle line) and mean value (red cross).

Generally, a denser specimen tends to reach a higher level of α (with an increasing median and mean value), and a looser specimen tends to display a more unstable and fluctuating α (with a higher standard deviation). Strain localizations in first four specimens (S1, S11, S12 and S13) escalate during $d_s/D_{50} = [0.0, 5.0]$ to high levels and fluctuate around them thereafter, whereas loose specimens S14 and S2 present more chaotic paths of α, which abruptly alternate between high and lower values. A critical sign is captured in Fig. 2.20 in S13 (a center-loose specimen) as strong vibration occurs after peak, indicating that the specimen seems unable to maintain the high level of α. This is shown more conspicuously in Fig. 2.20 that the standard deviation prominently grows from S12 to S13. Two looser specimens than S13 are dominated by too strong vibrations (with large standard deviations) to see any clear pattern.

The chaotic strain localization path proves to be a common behavior of loosely packed granular soil. The soil initialized with lower density is less competent to maintain a high level of strain localization. A reasonable explanation is that due to the low density and weak force-chains in loose soil, the granular system outside shear band is not stable enough to maintain its long-term intactness under the load and disturbance transmitted from the shear band area. When divergence also occurs in the outside shear band area, the global situation for a high degree of strain localization loses. This is then followed by a less localized pattern, with which the potential energy cumulates in the whole material, waiting for the next term of shear banding.

2.5.5 Shear Strength Versus Strain Localization

During the stress steady state of S1 and S2, α almost reverses the ways the stress evolves, as the stress softening (or hardening) corresponds to the increasingly (or decreasingly) localized deformation. The correlation between σ_s/σ_n and α is evaluated through the Pearson correlation r for S1 and S2 (shown in Fig. 2.22). Remarkably high correlations are found between them in both specimens.

In essence, strain localization basically refers to the material's failure. The strongly localized deformation can cause significant damage on the pre-existed force-bearing structures, the force-chains (column-like clusters of particles, which carry a major part of the external loading), thus leading to stress softening. Of note, the particles' rotation has been identified as a fundamental pattern in the movement and buckling of the force-chain network [20, 37]. As Sect. 2.4.4 revealed, strain localization also induces the localization of the particle rotation. Force-chains buckling and stress softening will naturally follow.

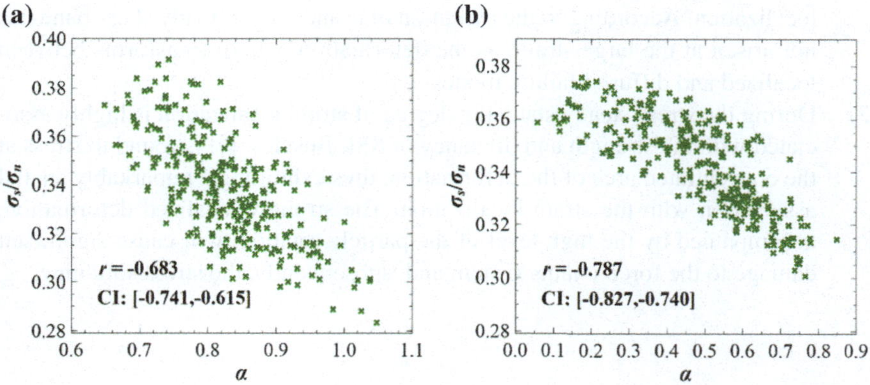

Fig. 2.22 σ_s/σ_n versus α (**a**) for S1 during $d_s/D_{50} \in [6.0, 20.0]$ and (**b**) for S2 during $d_s/D_{50} \in [5.0, 20.0]$. Pearson correlation r and its 95% confidence interval (CI)

2.6 Conclusions

The development of strain localization and the shear-banding process have been studied in densely and loosely packed granular soil subjected to interface shearing. By analyzing the kinematical field in specimens, how the strain localization evolves and how this evolution is related to the mechanical shearing responses of SSI have been explored on the basis of DEM simulation. Based on the velocity field given by grains' translational and rotational velocities, several kinematic quantities are tested at the monitoring strain states of both specimens. They are the average shear velocity $\tilde{v}_x(y)$, the gradient of the shear velocity $\overline{L}^i_{12}$, and the spatial distribution of particle rotation $|\theta|$. By measuring the non-linearity of $\tilde{v}_x(y)$, an indicator α is proposed to quantify the degree of the strain localization along the loading path. The main conclusions are summarized as follows:

(1) According to $\tilde{v}_x(y)$ in dense and loose specimens, the velocity field at strain states with a strong strain localization has structurized into two phases: the non-linear phase covering the rough surface and the linear phase in the remaining part. The linear phase vanishes after the localized deformation has been sufficiently developed. The SSI system then can be essentially seen as the combination of the rough surface and the soil nearby with a highly non-linear velocity field.

(2) For the dense specimen, tiny deformational concentrations have been observed at the very early stage of the shearing. The level of strain localization steadily ascends during the stress-hardening regime. To a strain state prior to the stress peak, shear deformation has largely concentrated at several areas, preferentially at the specimen's bottom. During the stress-softening phase, one of the former shear concentrated areas becomes dominant, while the remainders gradually disappear. For the loose specimen, there is a chaotic evolution to strain

localization. According to the evolution of α and $\overline{L}^{i}_{12}$, a steady shear band has not arisen at the large strain, as the deformation pattern transforms between localized and diffused failure modes.

(3) During the stress steady state, the degree of strain localization is highly associated with the strength and dilatancy of SSI. Besides, $|\theta|$ is found to stress at the concentrated area of the deformation, thus exhibiting a remarkably spatial association with the strain localization. The strongly localized deformation, accompanied by the high level of the particle rotation, will cause significant damage to the force-chains system and, subsequently, the stress softening.

References

1. Fu P, Dafalias YF (2011) Fabric evolution within shear bands of granular materials and its relation to critical state theory:1918–1948. https://doi.org/10.1002/nag
2. Zhu HX, Nguyen HNG, Nicot F, Darve F (2016) On a common critical state in localized and diffuse failure modes. J Mech Phys Solids 95:112–131
3. Desrues J, Chambon R, Mokni M, Mazerolle F (1996) Void ratio evolution inside shear bands in triaxial sand specimens studied by computed tomography. Geotechnique 46:529–546. https://doi.org/10.1680/geot.1996.46.3.529
4. Desrues J, Lanier J, Stutz P (1985) Localization of the deformation in tests on sand sample. Eng Fract Mech 21:909–921. https://doi.org/10.1016/0013-7944(85)90097-9
5. Finno RJ, Blackburn JT, Roboski JF (2007) Three-dimensional effects for supported excavations in clay. J Geotech Geoenvironmental Eng 133:30–36. https://doi.org/10.1061/(ASCE)1090-0241(2007)133:1(30)
6. Finno RJ, Harris WW, Mooney MA, Viggiani G (1997) Shear bands in plane strain compression of loose sand. Geotechnique 47:149–165. https://doi.org/10.1680/geot.1997.47.1.149
7. Vardoulakis I, Goldscheider M, Gudehus G (1978) Formation of shear bands in sand bodies as a bifurcation problem. Int J Numer Anal Methods Geomech 2:99–128
8. Petryk H (1993) Theory of bifurcation and instability in time-independent plasticity. Springer. https://doi.org/10.4028/www.scientific.net/AMM.224.362
9. Tejchman J, Wu W (1996) Numerical simulation of shear band formation with a hypoplastic constitutive model. Comput Geotech 18:71–84
10. Chambon R, Caillerie D (1999) Existence and uniqueness theorems for boundaryvalue problems involving incrementally non linear models. Int J Solids Struct 36:5089–5099
11. Bigoni D (2000) Bifurcation and instability of non-associative elastoplastic solids. In: Material instabilities in elastic and plastic solids. Springer, pp 1–52
12. Tejchman J, Górski J (2008) Deterministic and statistical size effect during shearing of granular layer within a micro-polar hypoplasticity. Int J Numer Anal Methods Geomech 32:81–107
13. Lin J, Wu W, Borja RI (2015) Micropolar hypoplasticity for persistent shear band in heterogeneous granular materials. Comput Methods Appl Mech Eng 289:24–43
14. Sulem J, Vardoulakis IG (1995) Bifurcation analysis in geomechanics. CRC Press
15. Bardet JP, Proubet J (1991) A numerical investigation of the structure of persistent shear bands in granular media. Géotechnique 41:599–613. https://doi.org/10.1680/geot.1991.41.4.599
16. Wang Y-H, Leung S-C (2008) A particulate-scale investigation of cemented sand behavior. Can Geotech J 45:29–44. https://doi.org/10.1139/T07-070
17. Gu X, Huang M, Qian J (2014) Discrete element modeling of shear band in granular materials. Theor Appl Fract Mech 72:37–49. https://doi.org/10.1016/j.tafmec.2014.06.008
18. Tordesillas A, Walsh DCS (2002) Incorporating rolling resistance and contact anisotropy in micromechanical models of granular media 124:106- 111

19. Zhu H, Veylon G, Nicot F, Darve F (2017) On the mechanics of meso-scale structures in two-dimensional granular materials. Eur J Environ Civ Eng 21:912–935
20. Tordesillas A (2007) Force chain buckling, unjamming transitions and shear banding in dense granular assemblies. Philos Mag 87:4987–5016
21. Vardoulakis I (1980) Shear band inclination and shear modulus of sand in biaxial tests. Int J Numer Anal methods Geomech 4:103–119
22. Iwashita K, Oda M (1998) Rolling resistance at contacts in simulation of shear band development by DEM. J Eng Mech 124:285–292. https://doi.org/10.1061/(ASCE)0733-9399(1998)124:3(285)
23. Mohamed A, Gutierrez M (2010) Comprehensive study of the effects of rolling resistance on the stress–strain and strain localization behavior of granular materials. Granul Matter 12:527–541
24. Dejong JT, White D, Randolph MF (2006) Microscale observation and modeling of soil-structure interface behavior using particle image. Soils Found 46:15–28. https://doi.org/10.3208/sandf.46.15
25. DeJong JT, Westgate ZJ (2005) Role of overconsolidation on sand-geomembrane interface response and material damage evolution. Geotext Geomembranes 23:486–512. https://doi.org/10.1016/j.geotexmem.2005.04.001
26. Zhang G, Liang DF, Zhang JM (2006) Image analysis measurement of soil particle movement during a soil-structure interface test. Comput Geotech 33:248–259
27. Wang J, Gutierrez MS, Dove JE (2005) Strain localization during shear of an idealized interphase system. In: Proceedings of the 2005 Joint ASME/ASCE/SES conference on mech, Mat
28. Wang J, Gutierrez MS, Dove JE (2007) Numerical studies of shear banding in interface shear tests using a new strain calculation method. Int J Numer Anal Methods Geomech 31:1349–1366
29. Cundall PA, Strack ODL (1979) A discrete numerical model for granular assemblies. Geotechnique 29:47–65
30. Dove J, Jarrett J (2002) Behavior of dilative sand interfaces in a geotribology framework. J Geotech Geoenvironmental 128:25–37. https://doi.org/10.1061/(ASCE)1090-0241(2002)128:1(25)
31. Wang J, Dove JE, Asce M, Gutierrez MS (2007) Anisotropy-Based Failure Criterion for Interphase Systems. J Geotech Geoenvironmental Eng 133:599–608. https://doi.org/10.1061/(ASCE)1090-0241(2007)133:5(599)
32. Wang J, Dove JE, Gutierrez MS (2007) Determining particulate–solid interphase strength using shear-induced anisotropy. Granul Matter 9:231–240. https://doi.org/10.1007/s10035-006-0031-1
33. Wang J, Jiang M (2011) Unified soil behavior of interface shear test and direct shear test under the influence of lower moving boundaries. Granul Matter 13:631–641
34. Wang J, Gutierrez MS (2010) Discrete element simulations of direct shear specimen scale effects. Géotechnique 60:395–409
35. Chareyre B, Villard P (2017) Discrete element modeling of curved geosynthetic anchorages with known macro-properties. In: Numerical modeling in micromechanics via particle methods. Routledge, pp 197–203
36. Kuhn MR (1995) A flexible boundary for three-dimensional DEM particle assemblies. Eng Comput 12:175–183. https://doi.org/10.1108/02644409510799541
37. Iwashita K, Oda M (2000) Micro-deformation mechanism of shear banding process based on modified distinct element method. Powder Technol 109:192–205
38. Oda M, Kazama H (1998) Microstructure of shear bands and its relation to the mechanisms of dilatancy and failure of dense granular soils. Geotechnique 48:465–481
39. Bardet JP (1994) Observations on the effects of particle rotations on the failure of idealized granular materials. Mech Mater 18:159–182

40. Nicot F, Darve F (2011) Diffuse and localized failure modes: two competing mechanisms. Int J Numer Anal Methods Geomech 35:586–601
41. Ikeda K, Yamakawa Y, Desrues J, Murota K (2008) Bifurcations to diversify geometrical patterns of shear bands on granular material. Phys Rev Lett 100:198001
42. Pearson K (1895) Notes on regression and inheritance in the case of two parents. Proc R Soc Lond 58:240–242

Chapter 3
Fabric Evolution at the Soil–Structure Interface

3.1 Introduction

One commonality between soil subjected to anisotropic compression and that subjected to interface shearing is that both are able to asymptotically reach a steady state—referred to as the critical state—whereby shear strain develops without any stress or volumetric change. Furthermore, this state does not depend on a soil's initial state but rather on its effective confining pressure. Since the critical state of soil was found and mathematically modeled by Roscoe et al. [1] and Schofield and Wroth [2] in the so-called Critical State Soil Mechanics (CSSM), it has constituted a fundamental aspect of modeling the constitutive behavior of soil in both phenomenological and micromechanical approaches [3–9]. The concept has also been introduced in reference to the SSI as the basis for formulating its constitutive behaviour [10–12]. Despite critical state theory having served as the cornerstone of soil mechanics for several decades, the physics underlying the critical state phenomenon, especially in reference to the SSI, is still not fully understood.

It is widely accepted that the mechanical behaviors of soils at the macroscale are significantly dependent on the evolutionary state of the soil fabric at the microscale [13–18]. The development of photographic and radiographic techniques and the emergence of the discrete element method (DEM) have created opportunities to test soil's micromechanical properties. Renewed interest has recently brought about a rethinking of the formation of the critical state and its physical mechanism via the following question: is there any state variable (especially for the microscale) other than the void ratio that is a common characteristic of the critical state? The uniqueness of fabric morphologies has been revealed through explorations of the soil fabrics' evolution to their final steady state [19–22]. Some constitutive correlations have also been developed based on the fabric uniqueness observed at the critical state [23–25] while considering quantified fabric anisotropy as the additional state parameter in addition to the void ratio in classical critical state theory and micromechanical plasticity [26, 27]. Although some developments have furthered our knowledge of the critical state physical mechanism, more micromechanical insight into this topic is

W.-H. Zhou and Z.-Y. Yin, *Practice of Discrete Element Method in Soil-Structure Interface Modelling*, https://doi.org/10.1007/978-981-19-0047-1_3

required. Studies of the micromechanical basis of the critical state of the SSI system will contribute to the knowledge of the SSI's asymptotic behaviour.

Another key feature of the SSI is its strain localization zone. Once the strain localization has been progressively developed, the granular system is gradually structured into two domains: (1) the domain with highly localized strain as the weak and most deformable layer of the material, and the remaining domain, in which material acts in a quasi-elastic manner over a relatively constant displacement gradient. The domain within the strain localization domain is the main area from which the nonelastic behaviour of the SSI is derived. In other words, the SSI is essentially the interaction between the strain localization domain of soil and the rough surfaces of artificial structures. Based on this, the SSI's behavior at both the macro- and microscales inside the strain localization domain should be discussed in depth.

The aim of this chapter is to present the evolution of the soil fabric in the SSI, with a focus on whether the soil's microstructural rearrangement will ultimately lead to the development of a common soil fabric in the critical state, how this critical state soil fabric is achieved in an evolutionary sense, and how it is related to macroscale critical state behavior. To this end, a series of numerical interface shear tests are first described and then validated for specimens installed under different initial states. Then, the thickness of the strong strain localization zone is determined for each specimen based on an analysis of the shear displacement field. Void ratios inside and outside of said zones are then followed to check the achievement of the critical state void ratio. The evolution of the anisotropy and the rotation of the soil contact (also referred to as the material fabric) and internal force fabric (also referred to as the force fabric) distribution are investigated in a separate section. By monitoring characteristic quantities of the material and force fabrics, the evolution of various aspects of the soil fabric is exploited, and the presence of a common fabric during the final state of interface shear tests is elucidated. All the research described in this chapter was conducted using two-dimensional (2D) simulations via DEM.

3.2 2D Interface Shear Test

3.2.1 Numerical Modeling and Test Protocol

The DEM coded by PFC2D 5.0 software (Itasca Coop.) was, in this chapter, used to simulate the behavior of the soil subjected to interface shear loading. As proposed by Cundall and Strack [28], DEM was used as a fundamental tool to measure granular dynamics and to explore their physics at the macro- to microscales. The 2D DEM features low computation costs and generates concise results that can be easily processed and interpreted, thus allowing us to conveniently conduct micromechanical analyses of granular soil.

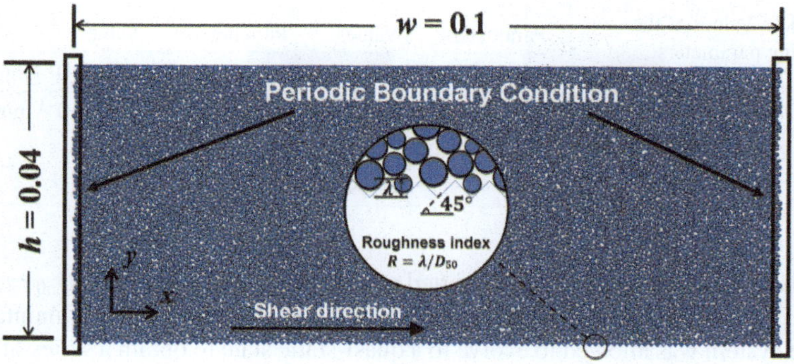

Fig. 3.1 Schematic diagram of the interface shear apparatus

Circles in the 2D simulation box succinctly represent spheres of centroids constrained on a plane; in other words, motions perpendicular to the box and rotations within its plane are fixed. The interaction between two spheres is described by the Hertz-Mindlin contact law [29], where only shear modulus G_s and Poisson's ratio v are required as material parameters. A cohesionless friction is also attached, parameterized by the internal friction angle φ_g for particle–particle contacts and φ_w for particle–wall contacts.

As illustrated in Fig. 3.1, the test apparatus of $w = 0.1$m in width along the x-axis and $h = 0.04$m in height along the y-axis was composed of an upper wall and teeth-plate on the base because the latter assimilated the surface of the artificial structure for which the maximum value of y was set at $y = 0.0$m. The teeth-plate was shaped by segments alternately inclining at 45° and 135° and linking to a sequence. Its roughness was quantified by $R = \lambda/D_{50}$, where λ is the vertical span of the teeth-plate and D_{50} is the average particle diameter. To prevent the occurrence of boundary effects along the shear direction, a periodic boundary condition was set on both sides of the apparatus to allow particles to pass through one side and appear on the counter-side at the same velocity. It is important to note that we used a regular surface geometry in this chapter instead of an irregular one based on the following two considerations: (1) the main focus of this chapter is the evolutionary features of soil's microscale properties' interplay with an artificial structure rather than the effect of surface geometries on the mechanical behaviors of the SSI; and the geometry of an artificial structure varies depending on the nature of the material of which a structure is composed and the way in which it is fabricated, and no pattern can be representative of all surface geometries. Other studies on the effect of surface geometries on the SSI have been conducted [30, 31].

The specimen was installed by adding 12,000 particles to the test apparatus. To obtain specimens with a relative isotropic initial fabric, they were prepared by growing the sizes of all the particles, which were initially dispersed across the apparatus with a diameter that followed a uniform distribution. This process was halted once the normal stress on the top wall σ_n reached the prescribed value. Then all grains in the specimen, whose radii were thereafter fixed, formed a poly-dispersed

Table 3.1 Values of the modelling parameters

Parameters	Value	Parameters	Value
G_s	29 GPa	σ_n	1.0×10^5 N/m2
v	0.5	v_s	2.0×10^{-3} m/min
φ_g	26.57°	R	0.75
φ_w	41.99°	Damping	0.1

granular system of a diameter that varied at [0.65 D_{50}, 1.35 D_{50}] with $D_{50} = 0.578$ mm. φ_g was then reduced to zero, step by step. After each step, while maintaining σ_n, the system was allowed to evolve to a quasi-static state to obtain a series of specimens of reducing initial void ratios e_0, because void ratio $e = V_v/V_s$, V_v and V_s are, respectively, the volume of void and solid phase inside the considered domain Ω. Take note that, as will be shown in the later section, it was expected that void ratio would be measured in different areas of the sample. We thus adopted a special method of defining Ω for different selections, which will be specified in Sect. 3.2. After the specimens were initialized with certain values of e_0 and σ_n, a constant shear velocity v_s was applied to the teeth-plate. We noted that v_s must be sufficiently small to maintain the quasi-static state of the granular system because we only aimed to investigate the time-independent properties of the system. Parameters used for the numerical modeling are listed in Table 3.1.

3.2.2 *Interface Shear Tests and Results*

We carried out interface shear tests on specimens with different initial void ratios e_0, as listed in ascending order of e_0 in Table 3.2. Under each monitoring strain state, the shear stress σ_s and shear displacement d_s of the teeth-plate and the volumetric strain ε_v of the specimen were measured as $\varepsilon_v = (h_t - h_0)/h_0$, where h_0 and h_t are the heights of specimen h (the distance between the upper wall and the teeth-plate's peak) at the initial and monitoring states, respectively. The evolution of the stress ratio σ_s/σ_n and ε_v, with respect to the normalized shear displacement d_s/D_{50}, is shown in Fig. 3.2.

In terms of stress responses, before σ_s/σ_n reached their final values, all specimens entered a hardening phase. S1, S2, and S3 showed clear indications of stress-softening

Table 3.2 Specimens and corresponding initial void ratio

Specimens	e_0
S1	0.174
S2	0.183
S3	0.196
S4	0.209
S5	0.213

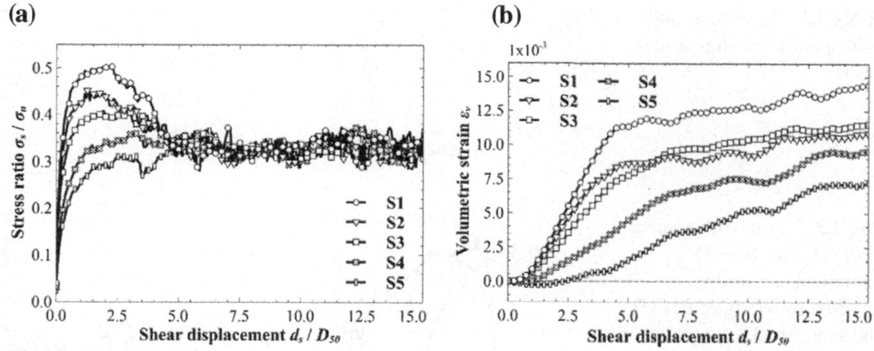

Fig. 3.2 **a** Stress ratio σ_s/σ_n and **b** volumetric strain ε_v with respect to shear displacement d_s/D_{50}

after stress levels peaked at $d_s/D_{50} = 2.0$. The specimen that was initialized more densely held a higher stress peak. All specimens eventually converged to a final value of σ_s/σ_n. Curves of ε_v were similar in shape. All the specimens ultimately dilated, while S1, S2, S3, and S4 barely underwent contraction. The degree of dilation decreased with increasing e_0 values.

3.2.3 Model Validation

Before further exploring the mechanical properties of the numerically modeled SSI, we validated the model by comparing its stress response and displacement field with those derived from numerical and laboratory interface shear tests conducted in the following studies. In Westgate and DeJong's [32] laboratory interface shear sand–steel test (hereafter referred to as T1), the specimen used was composed of soil of $D_{50} = 0.74$mm and a sample size of 101.6mm × 63.4mm × 25.4mm. The 2D numerical interface test undertaken by Wang et al. [33] (hereafter referred to as T2) was conducted in a 128.0mm × 28.0mm apparatus with $D_{50} = 0.70$mm soil. For the two aforementioned interface shear tests, two quantities were used to represent the profile roughness of different types of interfaces: (1) the average slope $\overline{\alpha}$, defined as $\tan(\overline{\alpha}) = (1/L)\int_0^L |dz/dx| dx$, where $z(x)$ is the height of the interface profile and L is the length of the interface; and the normalized average deviation R_a/D_{50}, where $R_a = (1/L)\int_0^L |z - \overline{z}| dx$ and the mean level $\overline{z} = (1/L)\int_0^L z dx$. Roughness quantities for tests conducted in the validation section are listed in Table 3.3. In a practical sense, difficulties are involved in quantitatively matching the lab experiment to the 2D simulation. In particular, producing the equivalent 3D stress according to the 2D response must entail an assumption of the definition of the 3D stress and lead to a mismatch between 2D and the lab stress–displacement correlation. Reasonable comparisons could be based on the shear ratio σ_s/σ_n and the vertical distribution of d_x. σ_s/σ_n is plotted in Fig. 3.3 in terms of the normalized shear displacement d_s/D_{50}

Table 3.3 Specimens and corresponding initial void ratios

Test	$\overline{\alpha}$	R_a/D_{50}
S1	45°	0.19
T1 [32]	30.2°	0.18
T2 [33]	33.7°	0.25

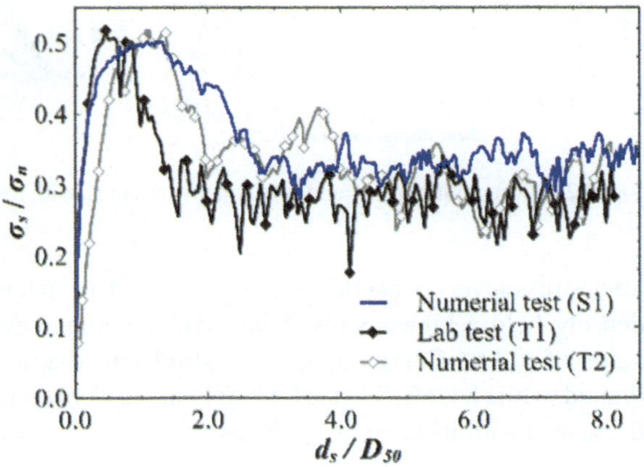

Fig. 3.3 σ_s/σ_n with respect to d_s/D_{50} for S1, 2D numerical test results reported by Wang et al. [33] and sand–steel test results reported by Westgate and DeJong [32]

for specimens S1, T1, and T2. For the interface's displacement, d_s = 7mm for the three tests, the shear displacement d_x in terms of depth y is shown in Fig. 3.4. d_x for lab test T1 was obtained via image processing, and the same value for numerical tests S1 and T2 was obtained by measuring the shear displacement of a column of labeled particles at the center of each specimen.

All three tests produced almost the same level of peak shear stress, as well as shear displacement distributions $d_x(y)$ of similar patterns, as shown in Figs. 3.3 and 3.4, respectively. All showed prominently nonlinear $d_x(y)$, because d_x rose exponentially

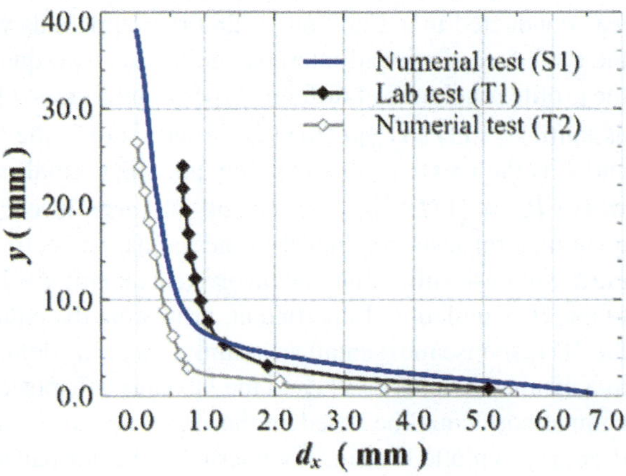

Fig. 3.4 Shear displacement d_x in terms of the depth y to the interface for S1, 2D numerical test results reported by Wang et al. [33] and sand–steel test results reported by Westgate and DeJong [32]

when approaching the interface. In other words, a zone of strong strain localization existed above the interface (hereafter referred to as the localized zone). Maximum d_x values occurred on the interface, and S1 performed larger soil shear displacement on the interface (about 6.54 mm) than did the corresponding values of T1 and T2 (both around 5.2 mm). Because the interface had sheared 7 mm, the soil's lowest layer did not keep pace with the moving interface, and a relative displacement δ_s was seen between the structure and the soil, which will be explored in Sect. 3.2. In fact, a considerable portion of the structure's movement did not exert a force on the soil body, and this portion varied depending on the soil's mechanical properties, surface roughness levels, and loading conditions [34]. Because we adopted a periodic boundary in this chapter, no additional lateral force was imposed to the soil body to handicap its horizontal movement. This was the reason why the soil adjacent to the interface of S1 traveled further than its counterparts.

3.3 Strain Localization and Associated Void Ratio

3.3.1 Measuring the Scale of the Localized Zone

In this section, the kinematic field is analyzed to determine the thickness of the strong strain localization zone, i.e., the localized zone. For a considered strain specimen state, the velocity field $\vec{v}(x, y)$ of the granular system is represented by velocities measured at particle centers. The average horizontal velocity measured at layers of various depths $\tilde{v}_x(y)$ can be plotted through a moving average process.

Essentially, once the stress steady state is reached, $\tilde{v}_x(y)$ has likely developed two distinct domains (as shown, for example, in Fig. 3.5): domain-I, the domain with a highly nonlinear velocity field, which corresponds to the layer of grains near the interface of a certain thickness; and domain-II, the remaining quasi-linear domain, which contains a relatively linear velocity field. In this chapter, domain-I is defined as the localized zone. The characteristic depth at which the velocity field changes from quasi-linear to nonlinear gives the thickness of the localized zone t_s. In this chapter, t_s was determined (at $d_s/D_{50} = 9.95$, when a steady localization is considered to have developed) by the point at which the second-order derivative of $40.0 - \hat{v}_x(y)$ first exceeded a given threshold a, i.e., $-\hat{v}''_x(y) > a$, where $\hat{v}_x(y)$ is the smoothed $\tilde{v}_x(y)$. When t_s of each specimen was measured and listed in Table 3.4, the localized zone thickness appeared to increase as the initial void ratio e_0 increased. This result is reminiscent of the common observation in triaxial and biaxial tests that the localized zone thickness tends to be wider in a loosely initialized than in a densely initialized specimen.

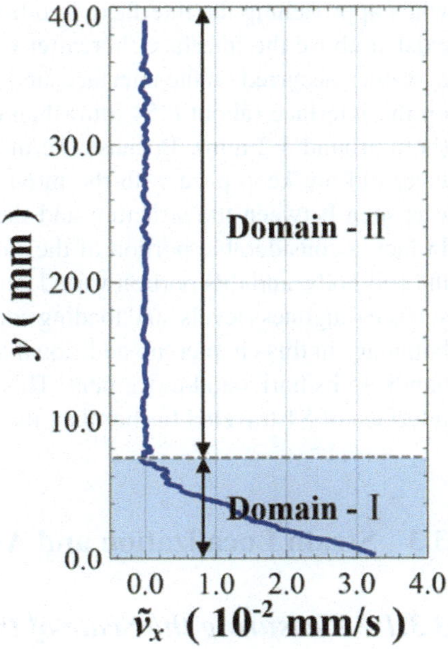

Fig. 3.5 Average horizontal velocity $\tilde{v}_x$ in terms of y in S1 at strain state $d_s/D_{50} = 9.00$

Table 3.4 Specimens and their determined localized zone thicknesses

Specimens	t_s(mm)
S1	7.11
S2	10.07
S3	11.82
S4	12.30
S5	15.05

3.3.2 Sliding on the Interface

As mentioned in Sect. 3.2.3, uneven shear displacement was present between the soil and the structure, which caused sliding on the interface. The distance of such sliding δ_s, denoting the soil–structure relative displacement, will be explained in this subsection. In terms of d_s/D_{50}, evolutions of normalized sliding distance δ_s/D_{50} and δ_s/d_s for S1, S2, and S3 are demonstrated in Fig. 3.6. δ_s of three samples is shown to linearly increase against d_s, thus standing at relatively constant proportions in the total shear displacement during loading (only around 3–6%). In particular, larger sliding would occur in a denser sample because more sufficiently packed soil presents greater resistance to being driven by the shearing structure, limiting the performance of larger nonsynchrony on the interface.

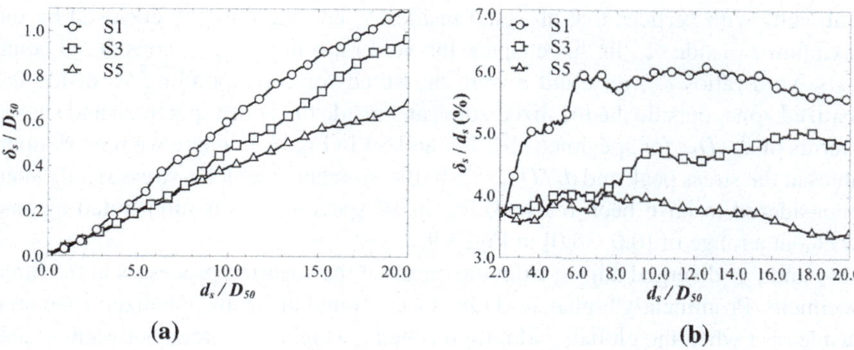

Fig. 3.6 **a** Sliding on the interface δ_s / D_{50} and **b** the proportion of sliding to the total shear displacement δ_s / d_s with respect to shear displacement d_s / D_{50}

3.3.3 Void Ratio e

The void ratio of the soil e is an important quantity for the behavior of soil and the SSI. To obtain a relatively accurate average void ratio for a selected area, the minimum cell enclosed by contact branches was taken as the basis for calculating the void ratio. We show the solution in Fig. 3.7 by finding the void ratio within a polygonal domain Ω enclosed by L. Ω here includes five cells numbered as i, with each cell including grain portions highlighted in light blue with an area a_s^i and the void portion shown in white with an area of a_v^i. The average void ratio is given by.

$$e_\Omega = \frac{\sum_{i=0}^{N_c} a_v^i}{\sum_{i=0}^{N_c} a_s^i}, \tag{3.1}$$

where N_c is the number of cells in the considered area. For any selection domain Ω', the considered polygonal domain Ω for assessing the void ratio is the assembly

Fig. 3.7 Illustration of the solid portion (in blue) and the void portion (in white) in a polygonal domain Ω enclosed by line L and selected by Ω'

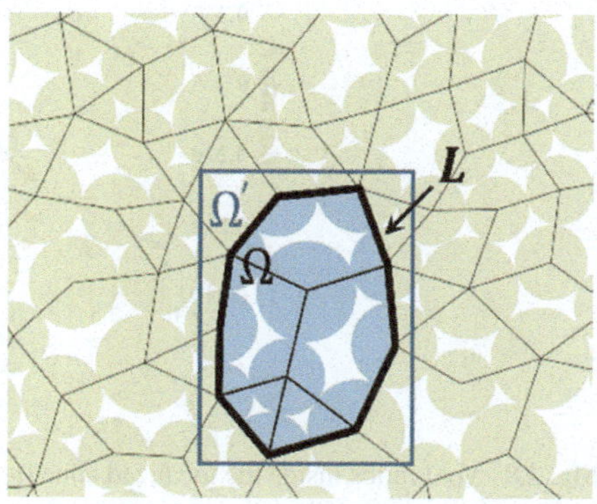

of all cells with vertices that all stand inside Ω', i.e., the polygon enclosed by the maximum L inside Ω'. In this chapter, the void ratio in Ω' $e_{\Omega'}$ is considered equal to e_{Ω}. Void ratios e_{lz}, e_{olz}, and e were measured for corresponding Ω' inside the localized zone, outside the localized zone, and inside the entire specimen and shown in terms of d_s/D_{50} for specimens S1, S2, and S3 in Fig. 3.8, where we remark upon values at the stress peak and $d_s/D_{50} = 5.0$ (the moment at which a stress steady state is considered to have been reached). e_{lz} in all specimens is demonstrated against d_s/D_{50} at a range of [0.0, 35.0] in Fig. 3.9.

e_{lz} and e_{olz} diverged almost from the outset of the shearing processes in the three specimens. Prominently higher void ratios were found inside the localized zone than outside of it when the global void ratio reached a weighted average between e_{lz} and e_{olz}. Similar evolutionary patterns were observed between the localized zone void ratio e_{lz} and the global value e. In S1 and S3, e_{olz} became stationary and barely rose after the stress peak, and yet e_{lz} reached its maximum increasing rate at the peak, deflected off upon reaching stress steadiness, and then increased at a fairly slow pace thereafter. All three kinds of void ratios presented steady ascendances in S5.

At the start of the test, dense and intermediate specimens dilated preferentially near the interface but also in the remaining area. Differentiation between the dilation

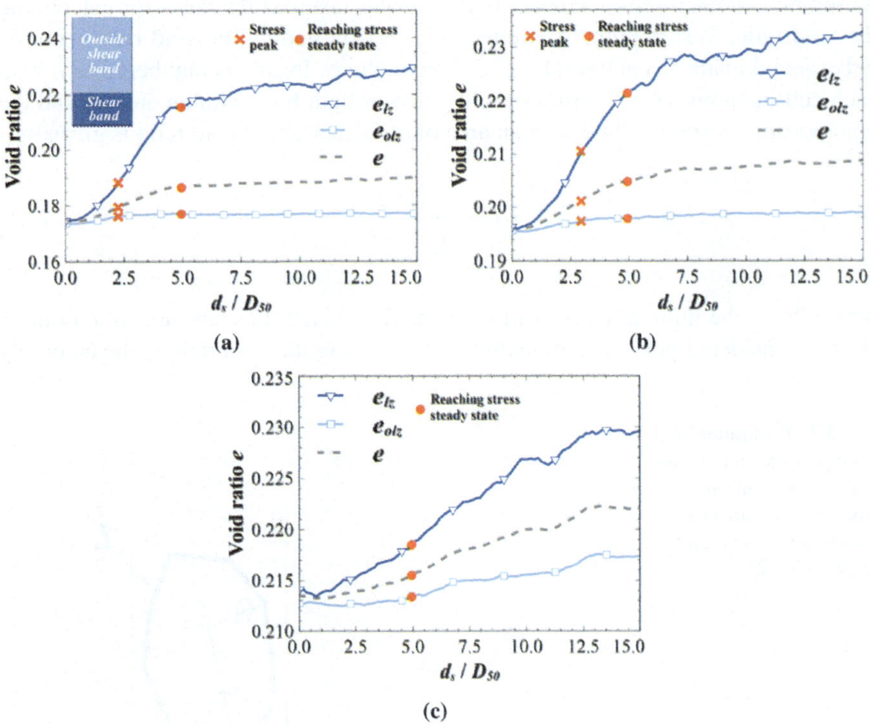

Fig. 3.8 Void ratios inside the localized zone (e_{lz}), outside the localized zone (e_{olz}), and within the entire specimen (e) with respect to d_s/D_{50} for specimens **a** S1, **b** S3, and **c** S5

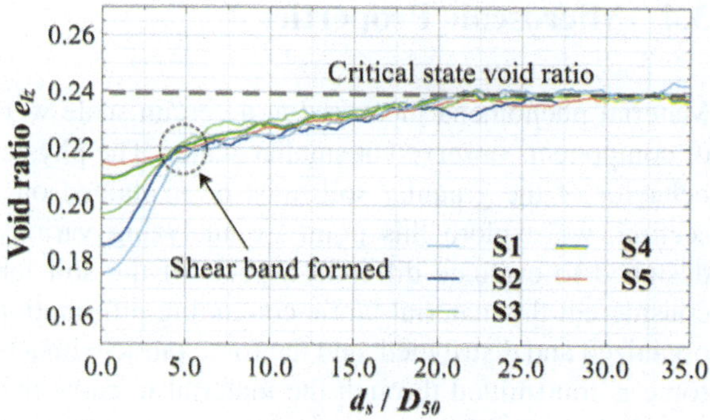

Fig. 3.9 Void ratios inside the localized zone (e_{lz}) with respect to d_s/D_{50} for specimens S1, S2, S3, S4, and S5

of lower and upper layers of the soil appeared as the stress peak was approached, i.e., in the hardening phase as dilation accelerated near the interface. At roughly the peak stress, in concurrence with the formation of the localized zone, the soil near the interface reached a maximum dilation rate, while the remaining soil started to stabilize in terms of volume. After the localized zone was fully shaped, two phases structured the material (the localized zone phase and the remaining phase) because the former involved a much lower density level than the latter. The shear-dilatancy near the interface accounted for the largest fraction of global dilation, thereby determining the volumetric evolution of the SSI system.

In the loose specimen, the shear stress-dilatancy rose both inside and outside the localized zone, steadily driving the global value upward. Notably, this trend persisted even after the system reached a stress steady state. The cause of this was alluded to by, Zhu et al. [35] who studied the strain localization process in loose specimens. The failure zone in the loose soil presented a so-called diffuse mode, indicating that the SSI system had not developed a stable localized zone that could persist in a finite but large deformation. This shows that the localized zone area in the loose specimen indicated on the velocity field at $d_s/D_{50} = 9.95$ may have changed before or could change later. This changing boundary of the localized zone must have caused shear-dilatancy to occur alternately in the two considered areas, i.e., the determined localized zone area and the remaining section.

Most importantly, all the specimens converged to a common final void ratio in the large deformation. The critical state void ratio was found inside the localized zone rather than in the entire specimen. Determination of the localized zone thickness is thus crucial to achieving a reliable modeling of SSI as a boundary value problem, especially for computations involving continuum-mechanically based methods such as the Finite Element Method (FEM) and Finite Difference Method (FDM). Otherwise, we note that once the curve inflection point is reached, at which point the boundary of the specimen's localized zone is considered to be formed, specimens still require time to reach the critical state void ratio. The specimens appear to have undergone a process that made the localized zone area fully sheared and dilated prior to reaching the final state.

3.4 Microscale Properties

Material phenomena measured at a certain scale were derived from the behaviors of components observed at smaller scales. The physical essence of the critical state behavior of the granular soil must be explained on a particle-scale basis. In this section, we explore this point by analyzing various micromechanical quantities designed to evaluate different aspects of the soil fabric. Two kinds of fabric are considered: the material fabric containing information on how material entities are organized and distributed, and the force fabric with a twin focus: on how the internal force is transmitted through the material to carry the external loading; and on the morphology of this force network.

3.4.1 Soil Fabric

By describing the distribution of contacts in the granular system, we will gain valuable information on how grains are organized and connected to form a granular assembly. Oda et al. [36] presented a simple way of statistically characterizing the soil fabric in a granular system that involves calculating the fabric tensor $\overline{\overline{F}}$, defined as

$$F_{ij} = \frac{1}{N_c}\sum_{c=1}^{N_c} n_i^c n_j^c = \frac{1}{2\pi}\int_0^{2\pi} E(\theta) n_i n_j d\theta \tag{3.2}$$

which averages, over all N_c contacts, the self-dyadic product of the contact norm vector $\overrightarrow{n}^c$ at contact c; $E(\theta)$ denotes the probability density distribution (PDD) in terms of the contact direction θ as the contact norm $\overrightarrow{n} = (cos\theta, sin\theta)$. The deviator of the soil fabric D^F denotes the degree of its anisotropy, defined as.

$$D^F = \frac{F_1 - F_2}{F_1 + F_2}, \tag{3.3}$$

where F_1 and F_2 are, respectively, the major and minor principals of $\overline{\overline{F}}$. Fabric deviators inside D_{lz}^F and outside the localized zone D_{olz}^F, compared to σ_s/σ_n, are plotted against d_s/D_{50} for S1, S3, and S5 in Fig. 3.10. The evolution of D_{lz}^F in specimens S1, S2, S3, S4, and S5 is shown in Fig. 3.11. Reference lines were drawn and sequentially numbered for S1 for five shear states: $d_s/D_{50} = 0.20$, 1.30, 2.25, 4.45, and 12.00. For these states (plus the initial state) of the inner localized zone, we plotted the PDD of the contacts' orientations $E(\theta)$ and its fitted probability density function (PDF) $\widehat{E}(\theta)$ as a truncated Fourier series of the second order:

$$\hat{E}(\theta) = 1 + a\, cos2(\theta - \theta_a), \tag{3.4}$$

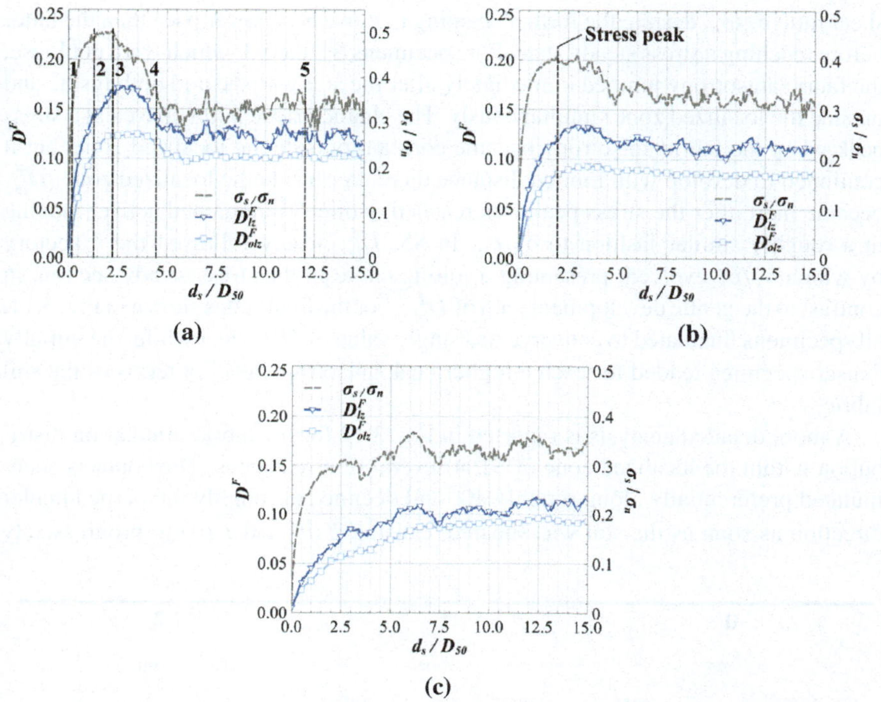

Fig. 3.10 Fabric deviator inside D^F_{lz} and outside the localized zone D^F_{olz} and the stress ratio σ_s/σ_n with respect to d_s/D_{50} for **a** S1, **b** S3, and **c** S5. For S1, five shear states, $d_s/D_{50} = 0.20$, 1.30, 2.25, 4.45, and 12.00, are sequentially labeled with vertical reference lines, and state 3 corresponds to the stress peak

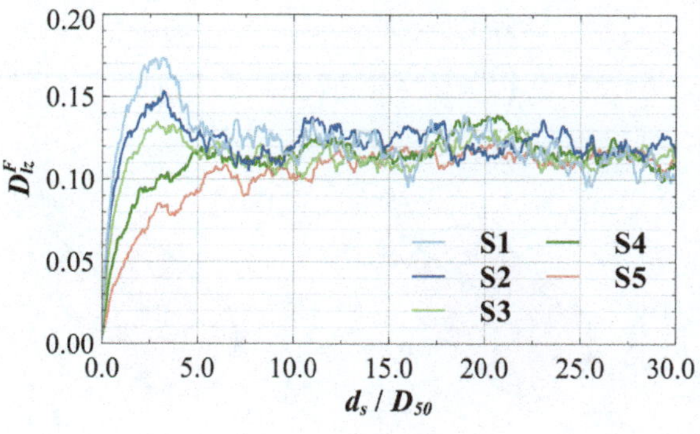

Fig. 3.11 Evolution of the fabric deviator inside the localized zone D^F_{lz} in terms of d_s/D_{50} for S1, S2, S3, S4, and S5

where parameter a also denotes the fabric anisotropy and θ_a gives the orientation preferences of the contacts.

A significantly higher degree of fabric anisotropy is shown in all three specimens, especially in the stress-hardening and -softening sections of dense and intermediate specimens S1 and S3. D^F_{lz} in the three specimens presents an analogous trend to the shear ratio σ_s/σ_n, despite the former seeming to move at a slower pace than the latter before reaching a stress steady state. For specimens S1 and S3, which were not loose, the fabric anisotropy jumped immediately after the soil was sheared both inside and outside the localized zone simultaneously. The characteristic point arrived at a stress peak when the fabric anisotropy became constant outside the localized zone, but it continued to develop with limited distance thereafter inside the localized zone. D_{olz}^F became inert after the stress peak was reached, while D_{lz}^F remained active, moving in a roughly similar fashion to σ_s/σ_n. In S5, D_{lz}^F closely followed the trajectory by which σ_s/σ_n evolved, presenting a jolting curve in the stress steady section, in contrast to the gentle development path of D_{olz}^F. For the final states shown in Fig. 3.11, all specimens fluctuated to converge to a single value of D_{lz}^F. Meanwhile, the initially denser specimen tended to reach a higher peak anisotropy level in terms of the soil fabric.

A more detailed analysis is reported in Fig. 3.12 for the fabric orientation distribution within the localized zone of S1 at several referred states. The contacts accumulated preferentially along roughly 40° and declined at roughly the perpendicular direction as soon as the soil was sheared, causing $E(\theta)$ and $\widehat{E}(\theta)$ to progressively

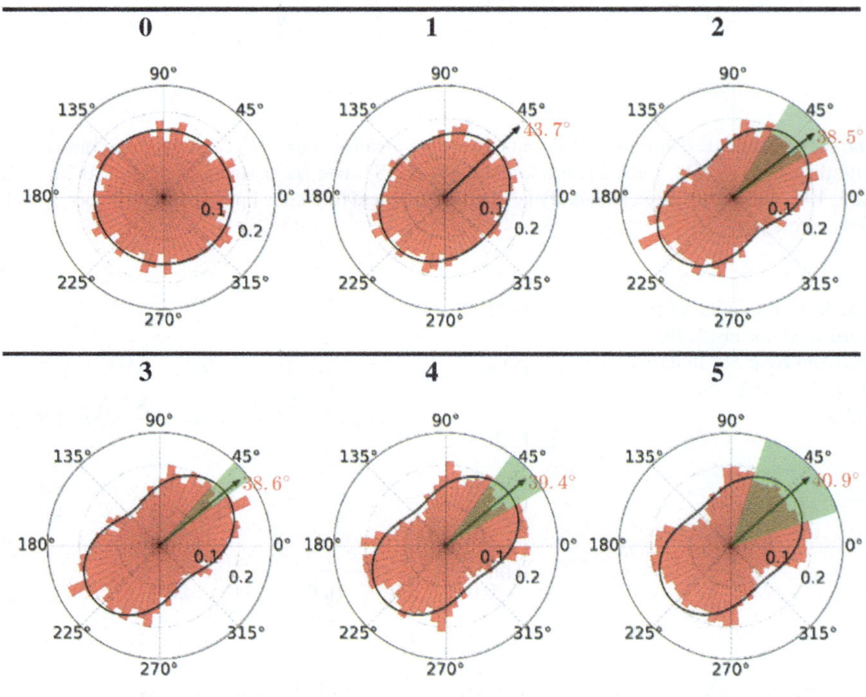

Fig. 3.12 Probability density distribution (PDD) of contact orientation $E(\theta)$ and its fitted probability density function (PDF) $\hat{E}(\theta)$ under localized zone of S1 at six shear states $d_s/D_{50} = 0.00$, 0.20, 1.30, 2.25, 4.45, and 12.00 for the fabric anisotropy direction

elongate from states 0 to 3. In the two subsequent states, 4 (somewhere after the stress peak) and 5 (at the stress steady state), the fabric appeared to drift back to a low anisotropy level. A notable phenomenon was that the fabric rearranged in favor of building contacts along a preferential direction (herein referred to as the major direction). However, we saw a slight rise in the contact population in a section fairly close to the main direction, namely a "stagnant zone" in terms of redistributing contacts (referred to as α-zone hereafter). Furthermore, this angular trough appeared to spread after the stress peak was reached, as is qualitatively shown in Fig. 3.12, gradually extending the green-lighted sector from state 3 to state 5. We then plotted $E(\theta)$ evolving in space E-θ-d_s/D_{50} and the contour plot from the project of E onto the θ-d_s/D_{50} plane shown in Fig. 3.13. We found that a prominent valley narrowly formed at the beginning of the shearing period from roughly $\theta = 40°$ before traversing

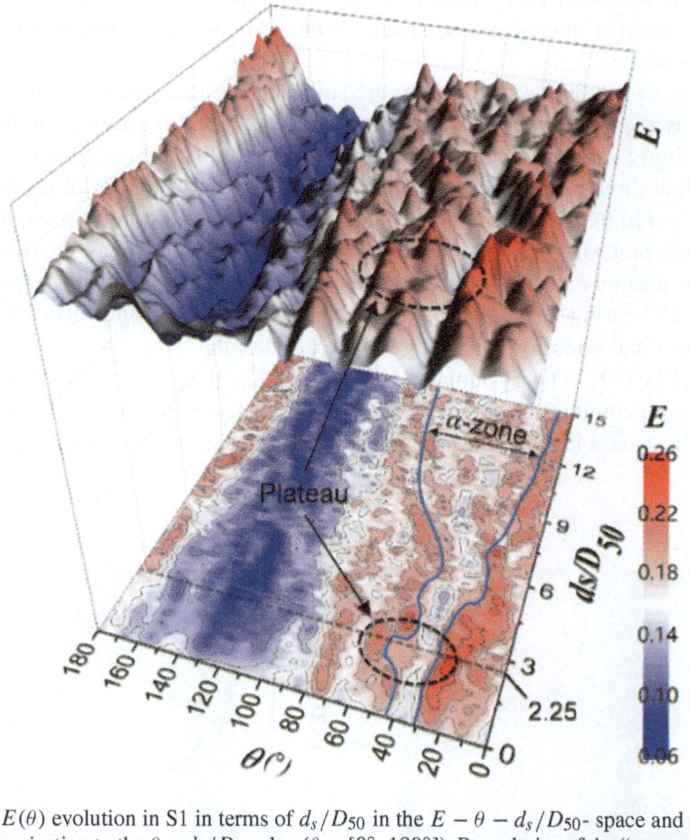

Fig. 3.13 $E(\theta)$ evolution in S1 in terms of d_s/D_{50} in the $E - \theta - d_s/D_{50}$- space and its contour plot as the projection to the $\theta - d_s/D_{50}$ plan ($\theta \in [0°, 180°]$). Boundaries of the "stagnant zone" in terms of redistributing contacts are approximately outlined by blue curves, in which contacts exhibit slow growth. A plateau on the $E(\theta) - d_s/D_{50}$- surface is labeled near the stress peak ($d_s/D_{50} = 2.25$)

through the θ-d_s/D_{50} plan in an increasingly aggressive manner, forming a horn-shaped trough along the loading history. This zone included a significantly lower E value than at its boundaries and was indeed, therefore, the aforementioned α-zone. In fact, more or less after d_s/D_{50}= 6.0, E became so chaotic inside the α-zone that the zone's boundaries could only be roughly sketched. Notably, the α-zone underwent an abrupt dilation near the stress peak ($d_s/D_{50} = 6.0$) and resumed thereafter, resulting in the formation of an observed plateau on the E surface.

The highly heterogeneous shear deformation pattern spurred the formation of remarkably non-uniform soil, which was reflected not only in the higher void ratio e but also in the higher degree of fabric anisotropy within the localized zone rather than outside of it. A common fabric anisotropy was eventually found inside the localized zone in the large shear deformation. This echoes the findings of previous studies on unique textures at the microscale [25, 37] and mesoscale [38] under common critical state soil behaviors. These findings and related works invite further research to review, from a microscale perspective, the critical state phenomena and their physical essence.

In terms of the fabric's evolution, the soil fabric rearranged itself in a principal direction as a result of the imposed shear (anisotropic) deformation, evolving toward an increasingly anisotropic layout. Even though the redistributing contacts preferred the principal direction, they were barely able to undergo substantial growth in this direction and in the surrounding area, i.e., the α-zone. To explain the existence and extension of said zone, we must mention the concept of force chains, the fundamental force transmission elements of the granular system. Force chains, which are column-like clusters bearing a major proportion of the external loading, distribute weight along a principal direction following the direction of the fabric anisotropy, as shown in Fig. 3.14a [39]. The systematic buckling and collapse of force chains inhibit contacts from cumulating in the α-zone, and the deterioration in the stability of the force chain system leads to the extension of the α-zone. In Fig. 3.14b, we plot force

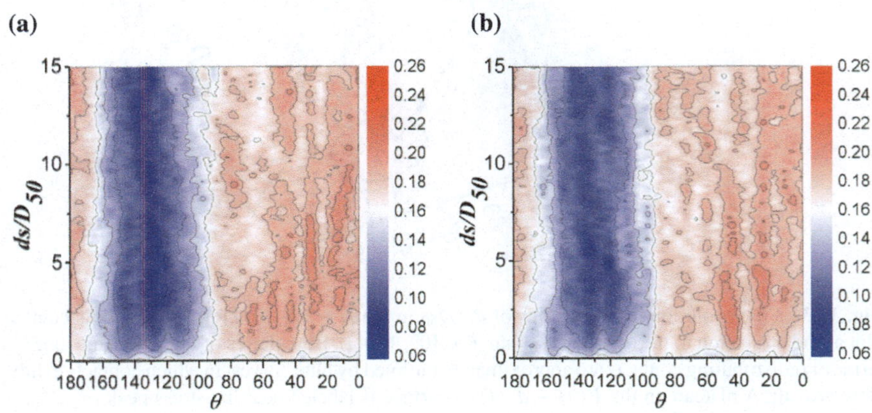

Fig. 3.14 Contour plot of $E(\theta)$ on $\theta - d_s/D_{50}$ plan ($\theta \in [0°, 180°]$) for **a** S12 and **b** S13

chains (composed of particles with a magnitude of transmitted force that is higher than the average) and buckling points (measured based on force chain kinematics) at the stress peak $d_s/D_{50}= 2.25$, following a method for detecting these chains and their buckling patterns proposed by Peters et al. [40]. Significant levels of force chain collapse were found in the forming localized zone at the stress peak in conjunction with the first and sudden extension of the α-zone, causing the development of a short plateau around the stress peak along the $E(\theta, d_s/D_{50})$ surface, as shown in Fig. 3.13. To determine whether the α-zone in the localized zone of S1 was representative of a densely packed SSI, two additional samples (S12 and S13) were tested. Both S12 ($e_0 = 0.153$) and S13 ($e_0 = 0.152$) were in the same dimension as S1 and contained 50,000 grains of the physical and mechanical properties listed in Table 3.1. From the contour plots of E for the localized zone areas of S12 and S13 shown in the θ-d_s/D_{50} plan in Fig. 3.14, similar patterns to those of S1 appeared as two troughs that traversed the $E(\theta, d_s/D_{50})$ surfaces and somehow became extended. We note that the normalized localized zone thickness t_s/D_{50} in S12 and S13 did not significantly differ from that of S1, whereas more grains were involved within localized zones as the normalized width of the shear box w/D_{50} increased.

In a granular assembly, media among the force chains are structures composed of low-stress soil grains and encompassing voids. These structures are sheltered by the arching and ceiling effects of force chains but have confining effects to prevent the chains from becoming damaged by external loading. This confining quality is dependent on the properties of force chains surrounding structures and, in particular, on their density, because a higher density provides a better confinement for force chain stability. As a result, the soil density substantially drives the behavior of force chains and the soil fabric by influencing force chain confinement conditions [38, 41]. As dilation develops after the stress peak is reached in the soil, especially within the localized zone, more and more force chains show dysfunction in the principal and neighboring directions based on a lack of reliable supporters. This rationally results in chronic α-zone extension. In its path toward a sufficiently large shear deformation (namely the critical state), the soil fabric is gradually trapped into a state characterized by a well-developed α-zone. Because of the poor confinement of force chains, which range along this widespread α-zone, the distribution of contact orientation remains largely unchanged while contacts build, buckle, and collapse frequently to create a dynamic invariant.

3.4.2 *Internal Force Anisotropy*

In response to a soil's shear deformation, in a similar manner to the contact fabric, transmitting internal forces also become anisotropic. It is, without doubt, this internal force anisotropy that gives rise to the stress anisotropy and shear resistance. A conventional way of characterizing this force transmission anisotropy, which imitates the method for describing contact anisotropy, involves expressing the average normal and shear contact force onto a polar plan based on contact orientation θ, because

$\overline{f}_n(\theta)$ and $\overline{f}_t(\theta)$, respectively, denote the average normal and shear force acting on contacts following direction θ. For a 2D granular assembly or part of one, Rothenburg [42] proposed using the orientation distribution of the average contact force to regress toward a truncated Fourier series in the second order for the normal force:

$$\frac{\overline{f}_n(\theta)}{\overline{f}_0} \approx 1 + a_n \cos(\theta - \theta_n), \tag{3.5}$$

and in the fourth order for the shear force:

$$\frac{\overline{f}_t(\theta)}{\overline{f}_0} \approx -a_s \cos 2(\theta - \theta_s) - a_{s1} \sin 4(\theta - \theta_{s1}), \tag{3.6}$$

where $\overline{f}_0$ is the average normal stress of the material obtained as:

$$\overline{f}_0 = \frac{1}{2\pi} \int_0^{2\pi} \overline{f}_n(\theta) d\theta. \tag{3.7}$$

$\overline{f}_n(\theta)/\overline{f}_0$ and $\overline{f}_t(\theta)/\overline{f}_0$ at the stress peak ($d_s/D_{50} = 2.25$) of the interior localized zone of S1 are plotted with their fitting curves in Fig. 3.15. The normal forces form a dumbbell-shaped distribution that basically follows formula (3.5) because a_n and θ_n denote the degree and direction of the internal force anisotropy, respectively. Meanwhile, the shear forces form a fan-like distribution because a_s (if ignoring the disturbance from the fourth-order item) gives the maximum average shear force for the fitted curve, and θ_s points in the zero-shear direction. Otherwise, $\overline{f}_t(\theta)/\overline{f}_0$ exhibits a severely non-orthogonal configuration when two pairs of "fans" are not

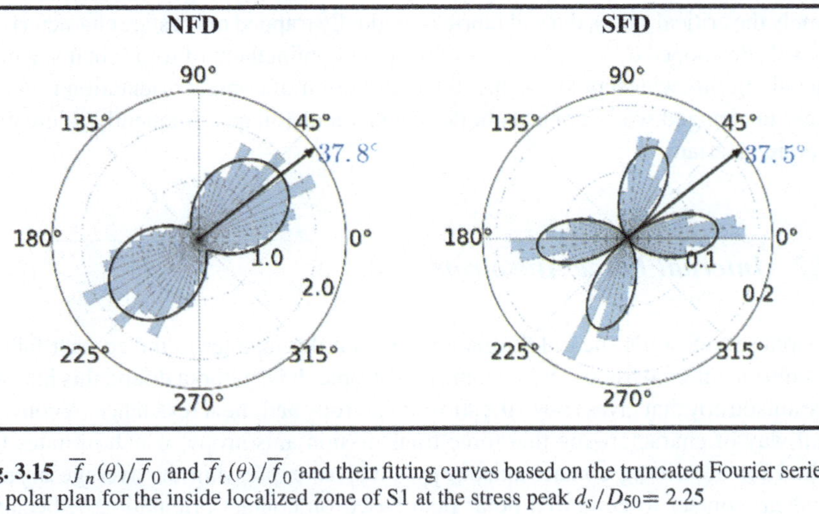

Fig. 3.15 $\overline{f}_n(\theta)/\overline{f}_0$ and $\overline{f}_t(\theta)/\overline{f}_0$ and their fitting curves based on the truncated Fourier series of the polar plan for the inside localized zone of S1 at the stress peak d_s/D_{50}= 2.25

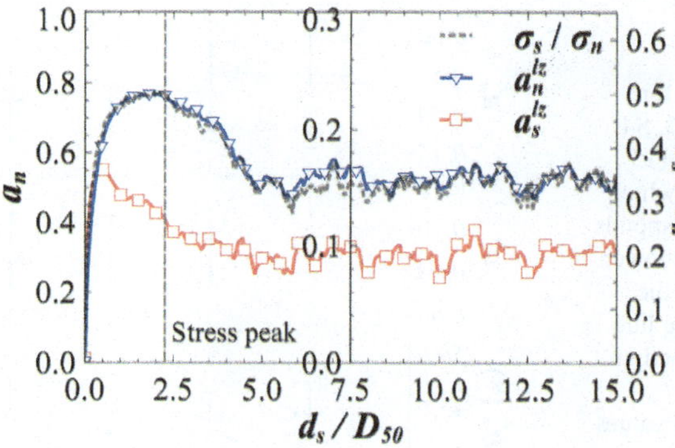

Fig. 3.16 a_n^{lz} and a_s^{lz} compared to σ_s/σ_n with respect to d_s/D_{50} for inside the localized zone of S1

perpendicular to one another. The fourth-order item of the Fourier series is, therefore, appended at the end of formula (3.7) as a correction for the orthogonal second-order item.

For d_s/D_{50} of S1, a_n^{lz} and a_s^{lz} are plotted inside the localized zone (denoted by superscript "*lz*") in Fig. 3.16 relative to the stress ratio σ_s/σ_n. Anisotropy directions inside the localized zone for the normal force θ_n^{lz}, shear force θ_s^{lz}, and contact fabric θ_a^{lz} are plotted for specimens S1, S3, and S5 in Fig. 3.17. We found a prominent analogy between the evolving a_n^{lz} and σ_s/σ_n across the loading history shown. a_s^{lz} reached its maximum value far before the stress peak at roughly the start of stress hardening before dropping dramatically over a long period until the arrival of the stress steady state. Figure 3.17 shows that in the three presented specimens, three characteristic angles of internal forces generally took similar values inside the localized zone, particularly for the strong tie between θ_n^{lz} and θ_a^{lz}.

The final values ($15 \leq d_s/D_{50} \leq 30$) of the four parameters, in terms of the internal force anisotropy inside the localized zone (a_n^{lz}, a_s^{lz}, θ_n^{lz} and θ_s^{lz}), are statistically accounted for in Fig. 3.18, in which each bar represents the estimated tendency of the corresponding parameter toward its maximum (upper bond), minimum (lower

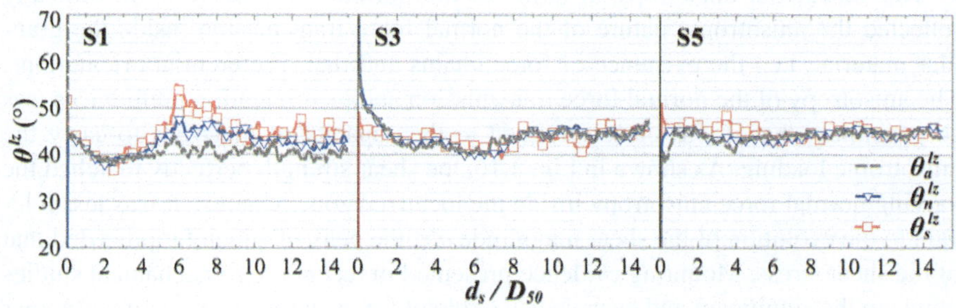

Fig. 3.17 θ_n^{lz}, θ_n^{lz}, and θ_a^{lz} inside the localized zone with respect to d_s/D_{50} for specimens S1, S3, and S5

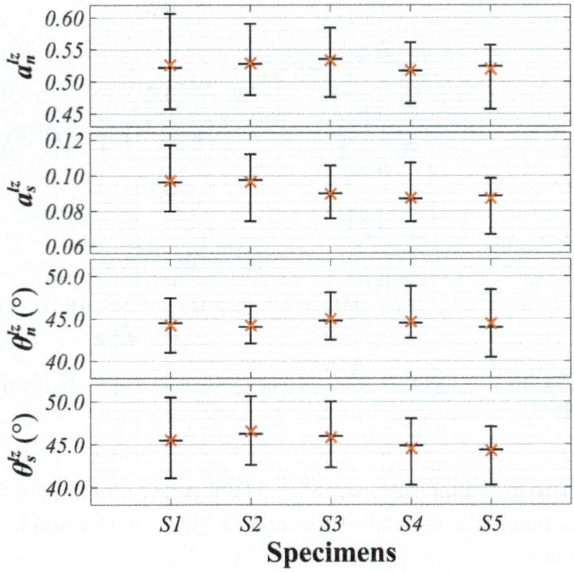

Fig. 3.18 Statistical quantities of a_n^{lz}, a_s^{lz}, θ_n^{lz}, and θ_s^{lz} for specimens S1, S2, S3, S4, and S5 at final states ($15 \leq d_s/D_{50} \leq 30$), in which the upper bond corresponds to the maximum, the lower bond corresponds to the minimum, the middle line corresponds to the median, and the red cross corresponds to mean values

bond), median (middle line), and mean values (red cross). For all four internal force parameters, slight differences can be found between the final median and mean values among the five specimens. Disregarding minor fluctuations, a_n^{lz} and a_s^{lz} for the five specimens stabilized roughly at 0.53 and 0.09 and for θ_n^{lz} and θ_s^{lz} at approximately 45°.

A common and dynamically stable force fabric clearly formed in the large shear deformations of localized zone areas for all the tested specimens. Together with the former common contact fabric, the large deformed material exhibited micromechanical uniqueness in both the way the material was organized (soil fabric) and the way internal forces were transmitted (force fabric). A unified and constant fabric embedded in the soil acted as a fundamental micromechanical basis for attributing constant stress and volumetric responses to the soil in large and continuous shear deformations.

The substantial anisotropy in terms of the normal force distribution basically reflected the anisotropic nature of the normal force transmission inside the granular material, i.e., the existence of force chains and their preferential orientations. The anisotropy of the normal force transmission and its orientation preferences was so prominent that it was closely related to the capacity of materials to carry the anisotropic loading. As shown in Fig. 3.16, the shear strength perfectly matched the varying normal force anisotropy inside the localized zone, whereas it was less relevant to the evolution of the shear magnitude a_s^{lz}, the peak of which far preceded that of the shear stress. Mounting evidence presented in recent micromechanical studies points to the significant and even decisive role of force chains in protecting granular systems against shearing and in giving rise to the shear strength of granular materials [43–46]. In other words, variations in shear stress can be largely interpreted based on the creation, strengthening, buckling, and collapse of entities in the force chain system. Meanwhile, under a large normal force, force chains are characterized by remarkable grain rotation during movement [47–51]. These phenomena illustrate the basic mechanism of granular soil failure to be reviewed from a microscale perspective because frictional sliding between grains is likely not the only derivation of shear failure and deformation. In other words, the weight of grain rolling under shear failure should be considered and estimated.

Otherwise, bonding occurs between the orientations of material and force fabrics inside the localized zone because θ_n^{lz}, θ_s^{lz}, and θ_a^{lz} are generally coincident with one another, according to Fig. 3.17. Bathurst and Rothenburg [52] once assumed equality between θ_n^{lz}, θ_s^{lz}, and θ_a^{lz} when deriving the global stress ratio of 2D granular material $(\sigma_1 - \sigma_2)/(\sigma_1 + \sigma_2)$ based on microscale quantities (σ_1 and σ_2 denote the major and minor stress principals, respectively). The observations illustrated in Fig. 3.17 support this assumption.

3.4.3 Critical State and Its Micromechanical Basis in the SSI

The evidence described in the previous section points to the formation of a unique state in the soil phase of the SSI in both macro- and microscale aspects after a sufficient shear deformation, a state that existed not in the entire specimen but in the localized shear zone. This result supports, in the condition of the SSI, the opinion offered by some recent researchers [19, 25, 37, 38] that, in simple terms, behind the critical state void ratio lies a convergent micro- (or meso-) structure (as shown in this chapter by, for instance, the identical force and fabric anisotropy). The classic critical state theory only formulates structureless scalars, i.e., the void ratio and mean effective stress, to describe soil's hardening. Many other authors, however, have gradually come to an important consensus on the following points: (1) the shear-induced fabric anisotropy is a prevailing phenomenon accompanied by shear deformation [19, 37, 42, 53, 54]; and the initial fabric anisotropy plays an important role in the mechanical performance of the soil [13, 55, 56]. It is, therefore, necessary for the geotechnical community to contain fabric-related quantities within the form of constitutive relations to account for the role of the fabric anisotropy and its evolution. The existence of the convergent fabric at large shear deformation, as confirmed by this chapter, provides an opportunity to keep the constitutive formulation of the critical state fabric in a simple form because it is independent from the initial fabric. The stress hardening, under the elasto-plastic framework, could then be referred to as the initial fabric being gradually driven toward the single critical state fabric. In this way, the fabric could act as the core parameter for soil's plasticity in place of the void ratio so that the fabric evolution and the soil behaviors relating to it could be explained.

Otherwise, the formation of the critical state fabric and how it relates to the loading condition still lacks a micromechanical basis. The observation on the occurrence and opening of the α-zone could be a clue to revealing the underlying mechanism. The mode by which the evolution of the α-zone is related to the build and collapse of force chains may merit further research.

3.5 Conclusions

In this chapter, the evolution of fabric (the contact and internal force fabric) in soil subjected to interface shearing was presented with a focus on whether a soil's micromechanical rearrangement ultimately leads to the development of a common soil fabric in the critical state, how this critical state soil fabric is attained in an evolutionary sense, and how it is related to macroscale critical state behaviour. Employing DEM, I conducted 2D numerical interface shear tests with periodic boundary conditions along the shear direction for specimens of different initial density levels, which were later validated by stress and displacement data derived from the numerical and laboratory interface shear test results of previous studies. After studying these numerical tests at the macro- and microscales, I present the following conclusions.

(1) Heterogeneity emerged in the soil from the start of the tests and accumulated thereafter until the localized zone had formed. The area inside the localized zone was characterized by a significantly higher level of fabric anisotropy than the area outside the zone.
(2) Inside the localized zones of all specimens, a common critical state microstructure was found in large shear deformations, with generally the same measured void ratio and orientation distributions in terms of contact fabric and internal force transmission.
(3) Force chain behaviours inside the localized zone, denoted by the anisotropy of the average normal contact force, were in perfect agreement with the SSI's shear stress. From the systematic forming, buckling, and collapse of the force chains, a so-called α-zone formed in the main direction of the soil contact fabric within the localized zone, in which the contact density was slow to increase and decrease. The gradual deterioration of the force chain system's stability, as a result of an increasing void ratio, gave rise to the α-zone's extension and eventually led to the formation of a large α-zone at the critical state. This large α-zone singlehandedly maintained the dynamic invariability of the critical state soil fabric, which acted as the basis of constant macroscopic critical state behaviors.

References

1. Roscoe KH, Schofield A, Wroth CP (1958) On the yielding of soils. Geotechnique 8:22-53
2. Schofield AN, Wroth P (1968) Critical state soil mechanics. McGraw-hill London
3. Yin ZY, Hicher PY, Dano C, Jin YF (2016) Modeling mechanical behavior of very coarse granular materials. J Eng Mech C4016006
4. Gajo A, Wood M (1999) Severn-Trent sand: a kinematic-hardening constitutive model: the q–p formulation. Géotechnique 49:595–614
5. Jefferies MG (1993) Nor-Sand: a simle critical state model for sand. Géotechnique 43:91–103
6. Taiebat M, Dafalias YF (2008) SANISAND: Simple anisotropic sand plasticity model. Int J Numer Anal Methods Geomech 32:915–948
7. Zhou WH, Yin JH, Hong CY (2011) Finite element modelling of pullout testing on a soil nail in a pullout box under different overburden and grouting pressures. Can Geotech J 48:557–567
8. Yin ZY, Zhao J, Hicher PY (2014) A micromechanics-based model for sand-silt mixtures. Int J Solids Struct 51:1350–1363. https://doi.org/10.1016/j.ijsolstr.2013.12.027
9. Yin ZY, Jin YF, Shen JS, Hicher PY (2018) Optimization techniques for identifying soil parameters in geotechnical engineering: Comparative study and enhancement. Int J Numer Anal Methods Geomech 42:70–94. https://doi.org/10.1002/nag.2714
10. Hu L, Pu JL (2003) Application of damage model for soil-structure interface. Comput Geotech 30:165–183. https://doi.org/10.1016/S0266-352X(02)00059-9
11. Hu L, Pu J (2004) Testing and Modeling of Soil-Structure Interface. J Geotech Geoenvironmental Eng 130:851–860. https://doi.org/10.1061/(ASCE)1090-0241(2004)130
12. Boulon M, Nova R (1990) Modelling of soil-structure interface behaviour a comparison between elastoplastic and rate type laws. Comput Geotech 9:21–46
13. Tantono JTÆEBÆSF (2007) Influence of initial density of cohesionless soil on evolution of passive earth pressure 53–63. https://doi.org/10.1007/s11440-007-0022-3
14. Yin ZY, Chang CS (2013) Stress–dilatancy behavior for sand under loading and unloading conditions. Int J Numer Anal Methods Geomech 37:855–870
15. Yin JH, Zhou WH (2009) Influence of grouting pressure and overburden stress on the interface resistance of a soil nail. J Geotech Geoenvironmental Eng 135:1198–1208
16. Li X, Yu H-S (2014) Fabric, force and strength anisotropies in granular materials: a micromechanical insight. Acta Mech 225:2345–2362
17. Subhash G, Nemat-Nasser S, Mehrabadi MM, Shodj HM (1991) Experimental investigation of fabric-stress relations in granular materials. Mech Mater 11:87–106
18. Kumar N, Luding S, Magnanimo V (2014) Macroscopic model with anisotropy based on micro–macro information. Acta Mech 225:2319–2343
19. Fu P, Dafalias YF (2011) Fabric evolution within shear bands of granular materials and its relation to critical state theory 1918–1948. https://doi.org/10.1002/nag
20. Liu J, Zou D, Kong X (2014) A three-dimensional state-dependent model of soil-structure interface for monotonic and cyclic loadings. Comput Geotech 61:166–177. https://doi.org/10.1016/j.compgeo.2014.05.012
21. Guo N, Zhao J (2013) Computers and Geotechnics The signature of shear-induced anisotropy in granular media. Comput Geotech 47:1–15. https://doi.org/10.1016/j.compgeo.2012.07.002
22. Kruyt NP, Rothenburg L (2014) On micromechanical characteristics of the critical state of two-dimensional granular materials. Acta Mech 225:2301–2318
23. Gao Z, Zhao J (2012) Constitutive modeling of artificially cemented sand by considering fabric anisotropy. Comput Geotech 41:57–69
24. Gao Z, Zhao J, Li X, Dafalias YF (2014) A critical state sand plasticity model accounting for fabric evolution. Int J Numer Anal methods Geomech 38:370–390
25. Li XS, Dafalias YF (2012) Anisotropic critical state theory: role of fabric. J Eng Mech 138:263–275
26. Chang CS, Yin Z-Y (2010) Micromechanical Modeling for Inherent Anisotropy in Granular Materials. J Eng Mech 136:830–839. https://doi.org/10.1061/(asce)em.1943-7889.0000125
27. Yin ZY, Chang CS, Hicher PY (2010) Micromechanical modelling for effect of inherent anisotropy on cyclic behaviour of sand. Int J Solids Struct 47:1933–1951

28. Cundall PA, Strack ODL (1979) A discrete numerical model for granular assemblies. Geotechnique 29:47–65
29. Mindlin RD, Deresiewicz H (1953) Elastic spheres in contact under varying oblique forces. J Appl Mech 20:327–344
30. Dove J, Jarrett J (2002) Behavior of dilative sand interfaces in a geotribology framework. J Geotech Geoenvironmental 128:25–37. https://doi.org/10.1061/(ASCE)1090-0241(2002)128:1(25)
31. Wang J, Jiang M (2011) Unified soil behavior of interface shear test and direct shear test under the influence of lower moving boundaries. Granul Matter 13:631–641
32. Westgate ZJ, DeJong JT (2006) Evolution of sand-structure interface response during monotonic shear using particle image velocimetry. In: GeoCongress 2006: geotechnical engineering in the information technology age, pp 1–6
33. Wang J, Gutierrez MS, Dove JE (2007) Numerical studies of shear banding in interface shear tests using a new strain calculation method. Int J Numer Anal Methods Geomech 31:1349–1366. https://doi.org/10.1002/nag.589
34. Zhang G, Zhang J (2009) State of the art : Mechanical behavior of soil – structure interface. Prog Nat Sci 19:1187–1196. https://doi.org/10.1016/j.pnsc.2008.09.012
35. Zhu HX, Zhou WH, Yin ZY (2017) Deformation mechanism of strain localization in 2D numerical interface tests. Acta Geotech 1–17
36. Oda M, Konishi J, Nemat-Nasser S (1982) Experimental micromechanical evaluation of strength of granular materials: effects of particle rolling. Mech Mater 1:269–283
37. Zhao J, Guo N (2013) Unique critical state characteristics in granular media considering fabric anisotropy. Géotechnique 63:695–704
38. Zhu HX, Nguyen HNG, Nicot F, Darve F (2016) On a common critical state in localized and diffuse failure modes. J Mech Phys Solids 95:112–131
39. Cambou B, Jean M, Radjaï F (2013) Micromechanics of granular materials. John Wiley & Sons
40. Peters JF, Muthuswamy M, Wibowo J, Tordesillas A (2005) Characterization of force chains in granular material. Phys Rev E 72:41307
41. Zhu H, Veylon G, Nicot F, Darve F (2017) On the mechanics of meso-scale structures in two-dimensional granular materials. Eur J Environ Civ Eng 21:912–935
42. Rothenburg L (1981) Micromechanics of idealized granular systems
43. Tordesillas A (2007) Force chain buckling, unjamming transitions and shear banding in dense granular assemblies. Philos Mag 87:4987–5016
44. Mueggenburg NW, Jaeger HM, Nagel SR (2002) Stress transmission through three-dimensional ordered granular arrays. Phys Rev E 66:31304
45. Radjai F, Roux S, Moreau JJ (1999) Contact forces in a granular packing. Chaos Interdiscip J Nonlinear Sci 9:544–550
46. Tordesillas A, Walker DM, Lin Q (2010) Force cycles and force chains. Phys Rev E 81:11302
47. Bardet JP (1994) Observations on the effects of particle rotations on the failure of idealized granular materials. Mech Mater 18:159–182
48. Bardet JP, Proubet J (1991) A numerical investigation of the structure of persistent shear bands in granular media. Géotechnique 41:599–613. https://doi.org/10.1680/geot.1991.41.4.599
49. Iwashita K, Oda M (1998) Rolling Resistance at Contacts in Simulation of Shear Band Development by DEM. J Eng Mech 124:285–292. https://doi.org/10.1061/(ASCE)0733-9399(1998)124:3(285)
50. Oda M, Kazama H (1998) Microstructure of shear bands and its relation to the mechanisms of dilatancy and failure of dense granular soils. Geotechnique 48:465–481
51. Tordesillas A, Shi J, Tshaikiwsky T (2011) Stress–dilatancy and force chain evolution. Int J Numer Anal Methods Geomech 35:264–292
52. Rothenburg L, Bathurst RJ (1989) Analytical study of induced anisotropy in idealized granular materials. Géotechnique 39:601–614
53. Calvetti F, Combe G, Lanier J (1997) Experimental micromechanical analysis of a 2D granular material: Relation between structure evolution and loading path. Mech Cohesive-Frictional Mater 2:121–163

54. Kuhn MR (1999) Structured deformation in granular materials. Mech Mater 31:407–429
55. Oda M (1972) The mechanism of fabric changes during compressional deformation of sand. Soils Found 12:1–18
56. Wan RG, Guo PJ (2004) Stress dilatancy and fabric dependencies on sand behavior. J Eng Mech 130:635–645

Chapter 4
Soil-Structure Interface Under Constant Normal Stiffness

4.1 Introduction

In the early research on the soil-structure interface, specimens were usually sheared under the conditions of constant normal stress or constant volume. The condition of constant normal stress is widely used because a constant load is applied at the top of the specimen in traditional experiments such as direct and simple shear tests [1–5]. The constant volume condition was also adopted in some studies to study the effect of fully restrained volume change. However, in many practical problems, neither the normal stress nor the soil volume remains constant, such as the interface between a pile shaft and the surrounding soil, where the volume change at the interface is constrained by the soil beyond this zone [6]. The normal stiffness, defined by the incremental normal stress over the incremental normal displacement, is a more stable value in such conditions. And the constant normal stress and constant volume conditions are two limiting cases with the normal stiffness respectively equal to zero and infinity [7]. Therefore, by modifying traditional shear devices, the effect of normal stiffness is studied with both monotonic and cyclic shear tests, and the macroscopic interface behaviors and microscopic observations based on particle image velocimetry are presented [6, 8–11]. However, previous experimental studies on the soil-structure interface with constant normal stiffness mainly focus on macroscopic behaviors and observations, while ignores the micro-mechanical analysis and micro-mechanism, such as the evolutions of local void ratio, force chains and contact fabric. In terms of numerical analysis with DEM, although the soil-structure interface has been simulated at both 2D and 3D, the general condition of constant normal stiffness has not been investigated. Therefore, in this study, a series of interface shear tests are simulated with DEM at the constant normal stiffness condition. An effective algorithm to achieve the condition of constant normal stiffness is developed. Both loose and dense specimens are generated for simulations. The normal stiffness has a wide range representing different interface conditions. The macroscopic interfacial behaviors are investigated first, and the microscopic properties, including shear

W.-H. Zhou and Z.-Y. Yin, *Practice of Discrete Element Method in Soil-Structure Interface Modelling*, https://doi.org/10.1007/978-981-19-0047-1_4

zone, global and localized void ratio, coordination number, and soil fabric are also discussed.

It is important to mention that numerical tests in this study are performed using two-dimensional (2D) DEM code PFC2D because a large number of particles raise the problem of calculation efficiency in 3D, particularly in this model with an iterative process. In a previous study, by comparing with the results of 3D simulations, 2D models have been confirmed to be able to predict the qualitative nature of soil particles, although quantitative difference still exists [12]. Some successful 2D DEM simulations can also be found in [13–16]. Therefore, although the 2D DEM model was not quantitatively calibrated against experimental results, it is validated through similar evolution trends of shear stress, normal stress and stress ratio at the macroscale. In addition, a 3D model with the same model parameters is presented in Sect. 4.3 and yields realistic stress ratios, which indirectly validates the 2D model.

4.2 DEM Simulation

4.2.1 Sample Preparation and Soil Properties

In order to investigate the behavior of soil-structure interface under constant normal stiffness condition, a series of interface shear tests are simulated using DEM. The DEM was first introduced by Cundall for modeling jointed rock masses in the 1970s and is now widely used in simulating the behavior of soils and rocks [17–19].

The schematic diagram of the interface shear test model is shown in Fig. 4.1. The length (L) and height (H) of the model are respectively 100 and 40 mm. The periodic boundary condition is adopted on both sides of the model. Under the periodic boundary condition, when the centroid of a particle falls outside of one side of the boundary, it will be translated back to the opposite side with the same velocity. Therefore, the side walls and the associated boundary effect are eliminated in the

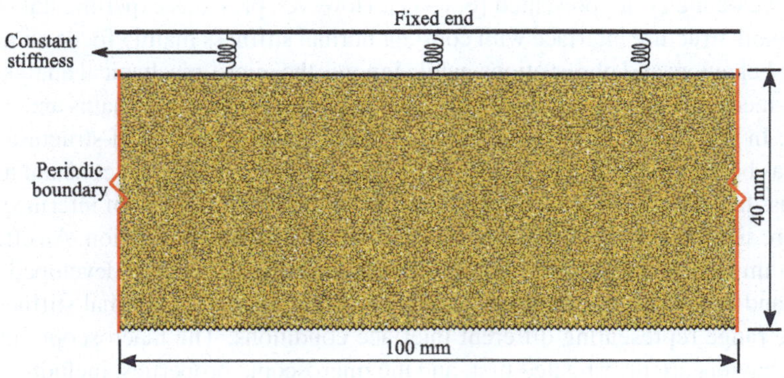

Fig. 4.1 Schematic diagram of the DEM model

simulations. Another advantage of using the periodic boundary condition is the low computational cost which is achieved by reducing the shearing length and number of particles. A sawtooth plate is placed at the bottom of the soil sample, which is made of inclined saw teeth with inclination alternatively equal to 45° and 135°. The surface roughness of the bottom plate is quantified by the parameterR_n, which is defined as $R50_{max}$, where R_{max} is the absolute vertical distance between the highest peak and lowest valley along with the surface profile over a length equal toD_{50}. To transfer shear localization into the adjacent soil packings [20], a rough surface with R_n equal to 0.7 is used for all simulations. It was reported by Wang and Gutierrez [21] that a minimum value of 40 for H/D_{max} should be used as optimal for direct shear testing. Therefore, the H/D_{max} for interface shear test should be greater than 20 to eliminate the box scale effect. In this study, the height of the sample is chosen as 40 mm so that the value of H/D_{max} is around 50. It is also worthwhile to mention that in order to focus on the effect of normal stiffness, dry sand is used in this study, and the pore pressure and capillary cohesion in unsaturated soils are not considered [22].

Both loose and dense specimens with relative densities ($D_r = \frac{(e_{max}-e)}{(e_{max}-e_{min})}$) of 20 and 80% are simulated in this study, where e_{max}, e_{min} and e are respectively the maximum, minimum and current void ratios of a soil sample. The method proposed by Wood and Maeda [23] is used to determine the e_{max} and e_{min}, in which the loosest and densest samples are generated by setting initial particle friction coefficient and void ratio respectively as 1.0 and 0.3 to get e_{max} and 0.0 and 0.1 for e_{min}. Then with a given relative density, the soil sample is created as follows: (1) the top and sawtooth walls are created with a dimension of 100 × 40 mm; (2) the number of soil particles is calculated based on the particle size distribution (PSD) and relative density, and then non-overlapped particles are generated within the domain using the Multi-layer with Undercompaction Method (UCM) [24]; (3) the particle assembly is subjected to a compressive stress of σ_n by moving the top wall towards soil particles. Note that because particles are in disc shape in 2D DEM, it is difficult to prepare a very loose specimen. Therefore, the loose specimen with a relative density of 20% may also experience some dilation during shearing. The minimum and maximum particle diameters are respectively 0.374 and 0.781 mm. During the particle generation process, the size range of particles is divided into ten intervals, and the number of particles in each interval is calculated based on the corresponding mean particle size of the interval. Therefore, a quasi-linear PSD is obtained with the average diameter, D_{50}, equal to 0.578 mm. The density of particles is 2650 kg/m^3. The particle interactions are governed by the linear contact model in which the linear elastic frictional behavior for both normal and shear directions is provided. The normal stiffness of both walls and particles is set as 5.0×10^7 N/m according to previous DEM studies [25, 26], which ensures small overlaps (less than 1% of average particle diameter) between contacting particles. The ratio of normal and shear stiffness is 2.0, which is within the recommended range for soil particles [27, 28]. An empirical value of friction coefficient 0.5 is used for particle–particle contacts according to previous studies [29, 30], and a higher value of 0.9 is adopted for particle–wall contacts to ensure that shear localization occurs within the soil zone.

Table 4.1 Parameters in DEM simulations

Parameter	Value
Density ρ (kg/m^3)	2650
Normal stiffness of particle k_n^p (N/m)	5.0×10^7
Shear stiffness of particle k_s^p (N/m)	2.5×10^7
Normal stiffness of wall k_n^w (N/m)	5.0×10^7
Shear stiffness of wall k_s^w (N/m)	2.5×10^7
Particle–particle friction coefficient μ…	0.5
Particle–wall friction coefficient μ_w…	0.9
Rolling resistance coefficient μ_r…	0.1
Compressive stress σ_n (kPa)	100
Damping coefficient	0.1

In order to numerically account for the effect of particle shape, the rolling resistance method is applied. In this method, an artificial rotational torque is provided at the contact, and its direction is against the direction of the relative rotational increment [31–33]. This method is able to reproduce both macro- and micro-scopic behaviors of slightly elongated granular material when the rolling resistance coefficient is smaller than 0.3 [34–36]. In this study, a low rolling resistance coefficient of 0.1 is adopted to simulate rounded particles with slight elongation according to a series of biaxial tests with DEM conducted by the authors [37]. The model parameters are summarized in Table 4.1, which were validated in previous DEM studies by the authors [14, 38, 39].

4.2.2 Development of Constant Stiffness Control of Interface Shearing

As discussed in several studies [6, 8, 9], the condition of constant normal stiffness is likely to prevail in many geotechnical practices such as pile foundations, because any volume change at the interface is constrained by the soils beyond this zone. Therefore, to better understand the soil-structure interfacial behavior under the condition of constant stiffness, a series of interface shear tests are conducted on specimens prepared in the previous section.

It was proposed by Boulon and Foray that the normal stiffness of the soil surrounding a pile is given as $k = 4G/d$, where G is the shear modulus of soil and d is the pile diameter. Therefore to consider soils of different types and modulus, the values of constant stiffness are chosen as 0 kPa/mm, 100 kPa/mm, 1000 kPa/mm and 10,000 kPa/mm. The shear tests are conducted incrementally in the DEM model, and the following relationship between normal stress increment, $\Delta\sigma_n$, and vertical displacement increment, Δu, is fulfilled in each increment through a back-calculation process.

$$k = \frac{\Delta\sigma_n}{\Delta u} \tag{4.1}$$

For a given $\Delta\sigma_n$, the determination of Δu needs special attention because any vertical movement of the loading platen will change the normal stress as well as the normal stiffness of the soil body k_{soil}. As a result, Eq. (4.1) no longer holds. In order to solve this issue, an iterative process is adopted to calculate Δu in each shear increment, and the flowchart is shown in Fig. 4.2. At the beginning of each shear increment, a constant shear velocity of 0.1 m/s is applied to the sawtooth wall, and the sample is sheared with a displacement Δd_s equal to 0.01 D_{50}. An increment of normal stress, $\Delta\sigma_n^i$, is generated due to the shear process in this increment. Then the shear is paused and the iterative process to adjust the position of loading platen is initiated. By envisioning the soil sample as a spring with stiffness of k_{soil} working in parallel with the loading spring with a constant stiffness k, the displacement of the loading platen Δu^i is given by

$$\Delta u^i = \frac{\Delta\sigma_n^i}{k + k_{soil}} \tag{4.2}$$

And the desired normal stress after the vertical movement is $\sigma_n^i + k\Delta u^i$. However, because k_{soil} is not a constant and affected by particle arrangement, the true normal

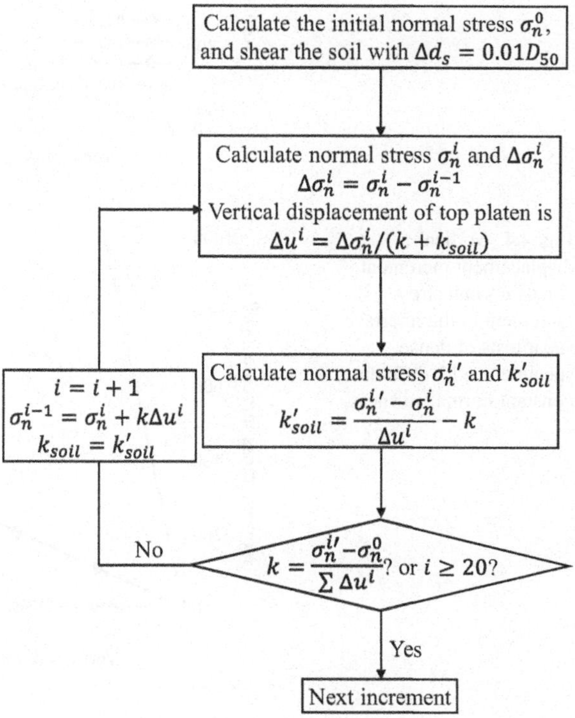

Fig. 4.2 Flowchart of the shear increment to achieve constant normal stiffness

stress after movement, $\sigma_n^{i'}$, may not equal to the desired normal stress. If the convergence criterion of constant stiffness is satisfied or the total iterative trials is greater than 20, the model continues to the next shear increment. Otherwise, the iteration is repeated with updated values of σ_n^{i-1} and k_{soil}. The interface shear test stops when the shear displacement reaches 20 D_{50}. It takes about 40 h on a workstation (Inter Xeon E5-2690A CPU and 128 GB RAM) to simulate one test. The relationships of Δu and $\Delta\sigma_n$ in the interface tests for both loose and dense specimens with various k are shown in Figs. 4.3 and 4.4. As shown in Figs. 4.3 and 4.4, the condition of constant stiffness is successfully maintained in all the interface shear tests. The slight fluctuations in loose specimens with k equal to 0 kPa/mm may come from the convergence issue in a few shear increments, except which the condition of constant normal stiffness is satisfied reasonably well.

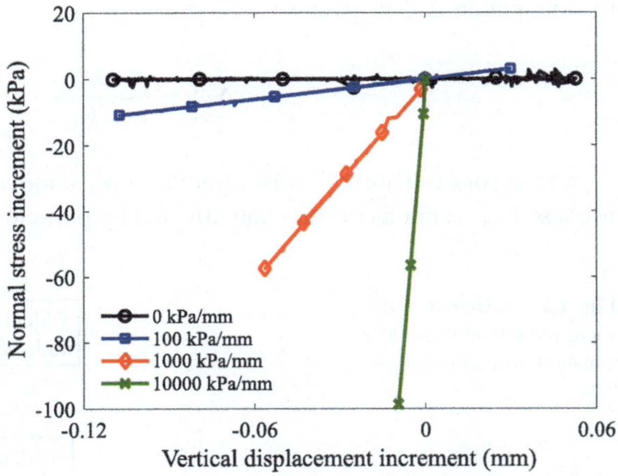

Fig. 4.3 Vertical displacement increment versus normal stress increment in the interface shear tests of loose specimens with different constant normal stiffness

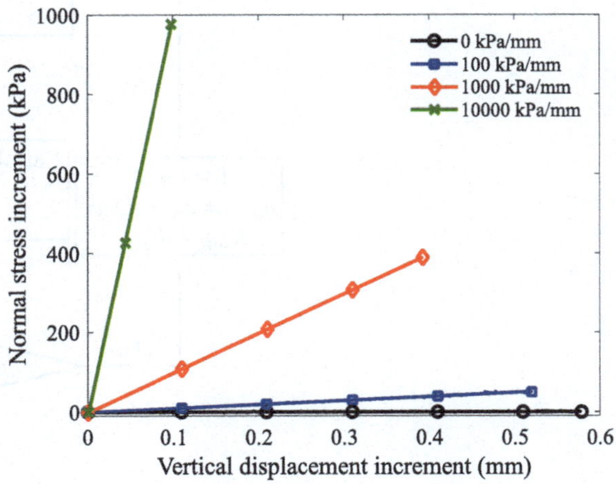

Fig. 4.4 Vertical displacement increment versus normal stress increment in the interface shear tests of dense specimens with different constant normal stiffness

4.3 Results and Discussion of Interface Tests

4.3.1 Macroscopic Behaviors

Figure 4.5 shows the shear stress versus shear displacement obtained from DEM simulations and experiments performed on both loose and dense specimens at different normal stiffness. For loose specimens in Fig. 4.5a, the increase of the normal stiffness causes a significant reduction of the shear stress. Particularly, the shear stress in the test with $k = 10000$ kPa/mm decreases to zero after a shear displacement of 16 D_{50}, which is due to the detachment of the specimen and the loading platen. Contrary to the loose specimens, the shear stress increases with the increase of the normal stiffness for the dense specimens as shown in Fig. 4.5b. When the normal stiffness is low (100 kPa/mm), the shear stress is very close to that in the constant normal stress condition ($k = 0$ kPa/mm). As k continues to increase, the shear stress increases significantly. At the condition of $k = 10000$ kPa/mm, the simulation is

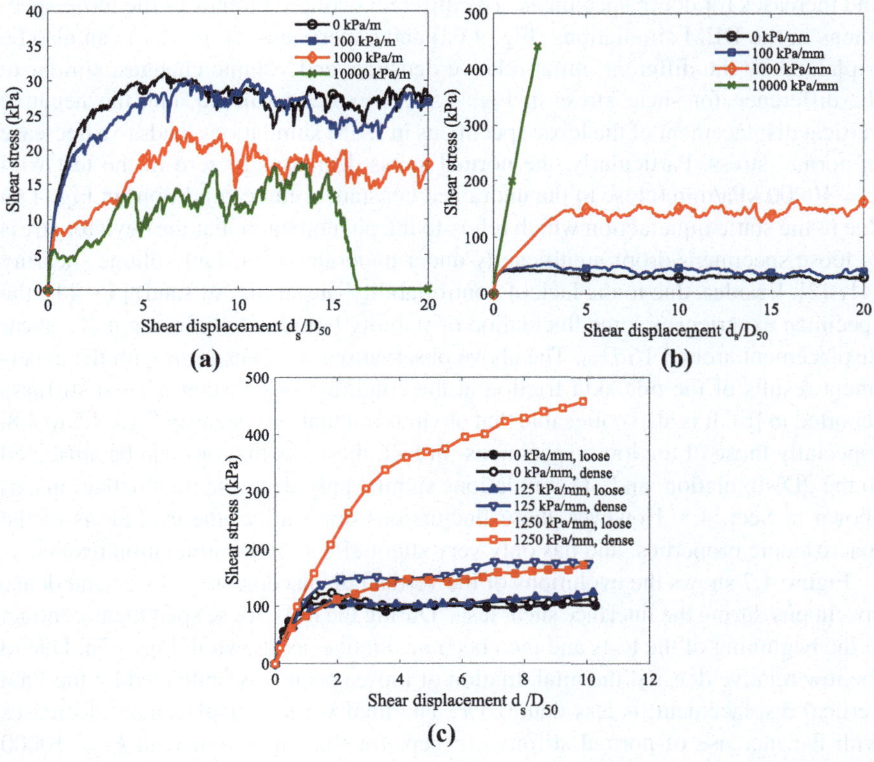

Fig. 4.5 Shear stress versus shear displacement of interface shear tests with different normal stiffness: **a** DEM simulations with loose specimens; **b** DEM simulations with dense specimens; and **c** experiments [7]

stopped much earlier because of the large overlaps induced by the high normal stress at the particle contacts. These large overlaps contradict the small-overlap assumption of DEM and make the simulation results no longer reliable [40]. Compared with the experimental results at similar conditions ($R_n = 0.95$, $k = 0$–1250 kPa/mm) from DeJong and Westgate [7] in Fig. 4.5c, the effect of normal stiffness on the shear strength of dense specimens is qualitatively similar, i.e., the shear stress increases with normal stiffness. However, the shear stress also increases with normal stiffness for loose specimens in the experiments, compared with a decreased stress ratio in the DEM simulations (Fig. 4.5a). The seemly opposite results can be explained by the different relative densities in the DEM simulations and experiments. A lower relative density of 20%, compared with 35% in experiments, is adopted in the DEM simulations, which leads to an overall contraction or very slight dilation of the soil specimens. As a result, shear stress decreases with the increase of normal stiffness. In another experimental study with very loose specimens, the interface shear stress is also found to decrease with the increase of normal stiffness [6].

The evolutions of normal stress are presented in Fig. 4.6. In general, the normal stress decreases with the increase of normal stiffness in tests with loose specimens, and increases for dense specimens. The different evolution trends of the loose specimens in the DEM simulations (Fig. 4.6a) and experiments (Fig. 4.6c) can also be explained by the different initial relative densities and volume changes, similar to the difference for shear stress in Fig. 4.5. As expressed in Eq. (1), the negative vertical displacement of the loose specimens in DEM simulations leads to a decrease in normal stress. Particularly, the normal stress decreases to zero in the test with $k = 10000$ kPa/mm (close to the undrained constant volume condition) in Fig. 4.6a due to the static liquefaction which refers to the phenomenon that the deviator stress of loose specimens drops significantly under underained/constant volume shearing [41, 42]. Besides, due to the lack of controllability (instability of sand) [43, 44], the specimen experiences some fluctuation of stability before totally losing it at a shear displacement around 16 D_{50}. The above observations are consistent with the experiment results of the pile skin friction at the condition of constant normal stiffness reported in [6]. It is also noticeable that obvious fluctuations occur in Figs. 4.5 to 4.8, especially those of the loose specimens. In fact, these fluctuations can be attributed to the 2D simulation, and 3D simulations significantly decrease the fluctuations, as shown in Sect. 4.3. However, these fluctuations don't affect the evolutions of the macroscopic properties, and has only very slight effect on the simulation results.

Figure 4.7 shows the evolutions of the vertical displacements of loose and dense specimens during the interface shear tests. During the tests, loose specimens contract at the beginning of the tests and then become dilative, as shown in Fig. 4.7a. Due to the low relative density, the total dilation of loose specimens, indicated by the final vertical displacement, is less than 0.5%. The final vertical displacement decreases with the increase of normal stiffness, except for the simulation with $k = 10000$ kPa/mm, which is due to the collapse of the specimen and detachment of the soil from the loading platen. For the dense specimens in Fig. 4.7b, soils are always dilative and the magnitudes of dilation are much larger than those of the loose ones. For example, the final vertical displacement is 0.5 mm at a dense state compared to 0.04 mm at

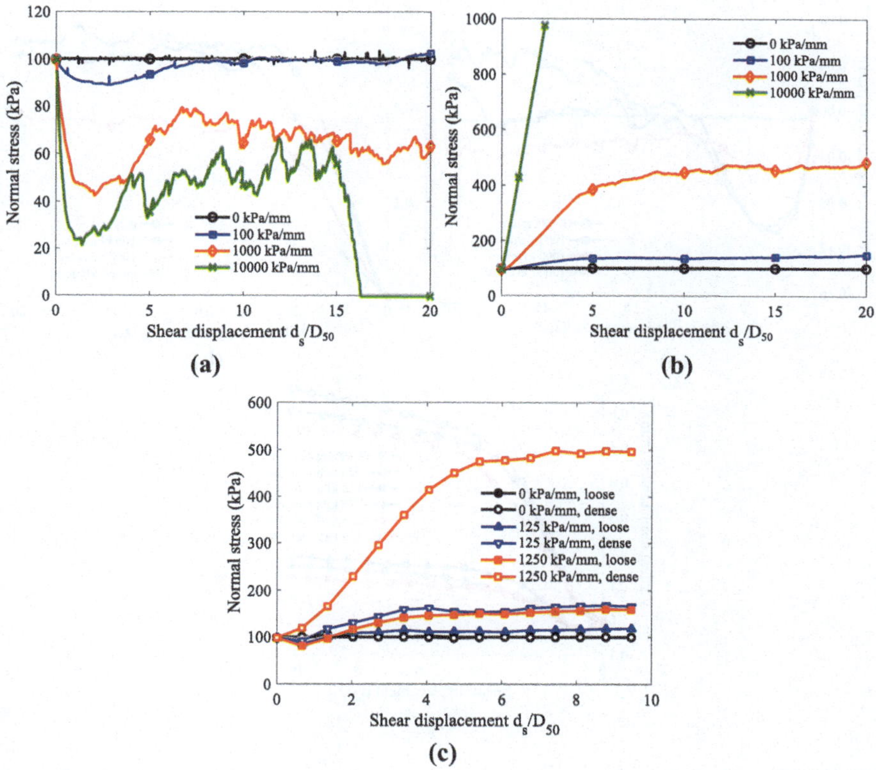

Fig. 4.6 Normal stress versus shear displacement of interface shear tests with different normal stiffness: **a** DEM simulations with loose specimens; **b** DEM simulations with dense specimens; and **c** experiments [7]

a loose state when $k = 100$ kPa/mm. And the vertical displacement decreases with stiffness for dense specimens. The evolution pattern of vertical displacement and the effect of normal stiffness for dense specimens are similar to those in the experiments in Fig. 4.7c [7].

Evolutions of stress ratio (shear stress/normal stress) with respect to normalized shear displacement are shown in Fig. 4.8. In simulations with loose specimens in Fig. 4.8a, the stress ratio increases with shear displacement up to a stable state, similar to the loose specimens in experiments in Fig. 4.8c. For the dense specimens, the stress ratio increases much faster than the loose ones until reaching the peak values, after which a softening process can be observed. Results from Fig. 4.8 also indicate that the evolution of stress ratio is mainly determined by the relative density of soil specimens, and the effect of normal stiffness is very slight. This observation is consistent with the experimental results that the peak stress ratio of uncrushable granular materials is insensitive to normal stiffness [9, 45]. In the tests by DeJong and Westgate [7], however, it is found that the stress ratio decreases with the increase of normal stiffness for dense specimens, as shown in Fig. 4.8c. The exact reason for

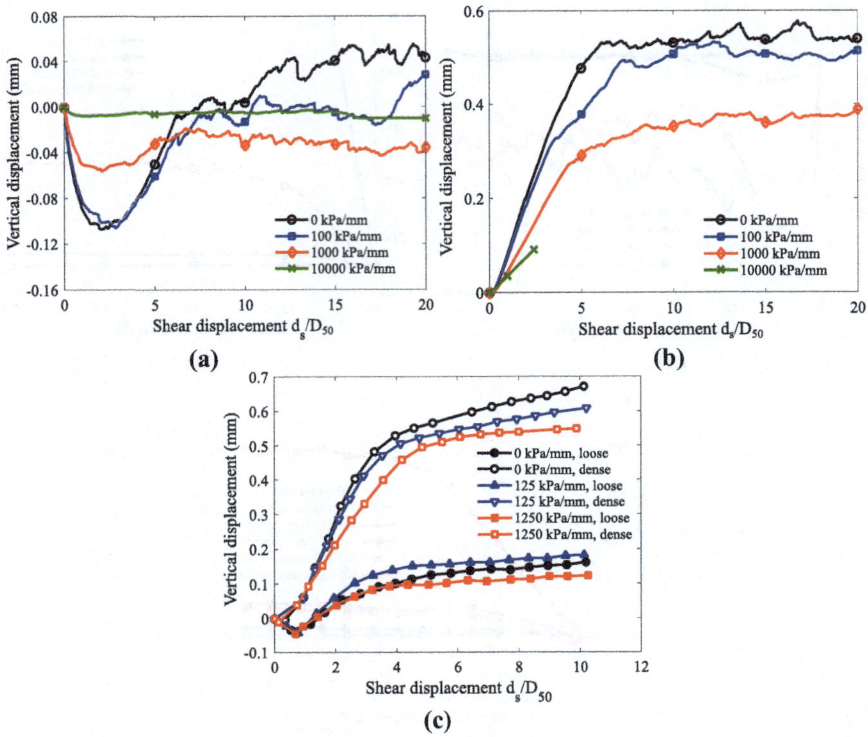

Fig. 4.7 Vertical displacement versus shear displacement of interface shear tests with different normal stiffness: **a** DEM simulations with loose specimens; **b** DEM simulations with dense specimens; and **c** experiments [7]

different observations in the above experimental tests is unclear and beyond the scope of this study, but may be explained by the different amount of particle breakage in experiments, which significantly decreases the shear resistance of granular material. This assumption is confirmed by the results of interface shear tests of the sand-woven geotextile interface [46], in which the peak stress ratio of uncrushable glass beads is insensitive to the normal stress, while the ratio of crushable sand decreases with the increase of normal stress. The values of stress ratios at the end of the simulations (residual stress ratios) are around 0.3 for both loose and dense specimens, which can be explained by the constant slope of the failure envelope in the critical state soil mechanics. Using the Mohr–Coulomb equation, the peak and residual friction angles are respectively given as 23.8° and 17.2°. Note that because particles have a lower average coordination number in 2D than that in 3D, both of the two values are smaller than the tested results of sands [7].

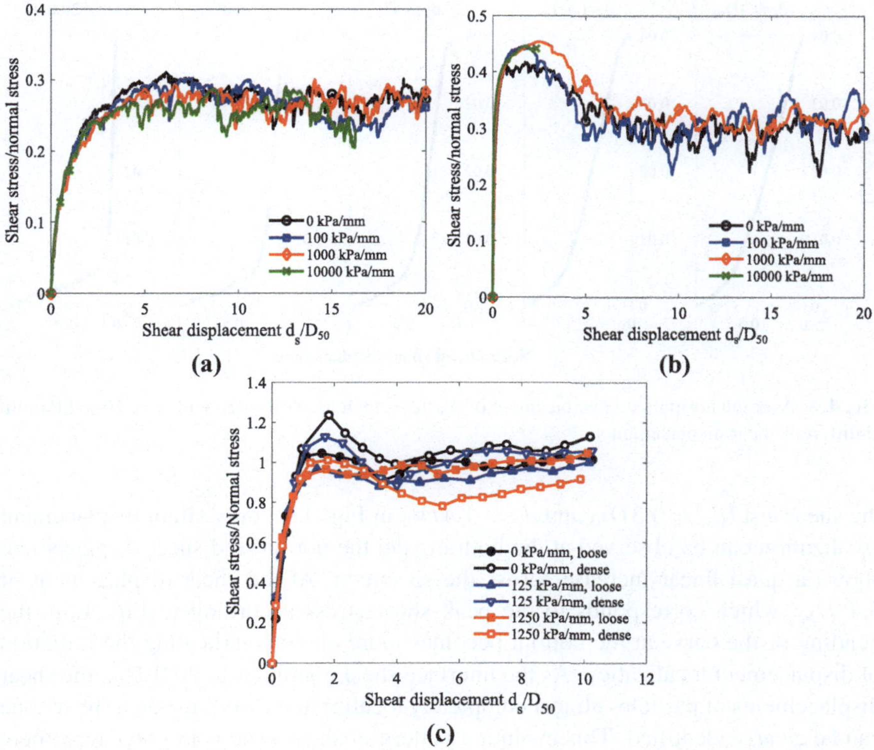

Fig. 4.8 Shear stress/normal stress versus shear displacement of interface shear tests with different normal stiffness: **a** DEM simulations with loose specimens; **b** DEM simulations with dense specimens; and **c** experiments [7]

4.3.2 Thickness of the Shear and Dilation Zones

In the interface test, the strain localization arises at the rough surface and gradually develops within the area a few times D_{50} from the surface [47, 48]. In the following context, this area will be called the shear zone. Accurate measurement of the thickness of the shear zone is important for theoretical models and continuum-based numerical simulations, in which the thickness of the shear zone should be given as a priori value. In this study, the formation and development of the shear zone are investigated by monitoring the horizontal displacement of particles at different heights. And the thickness is obtained by separating a discrete zone at the interface in which particles have strong displacement localization [49].

In order to investigate the development of displacement localization, the evolution of average normalized shear displacement of soil particles in the interface test with the dense specimen at $k = 1000$ kPa/mm is presented in Fig. 4.9. The normalized shear displacement is defined as the ratio of particle displacement in the shear direction and the shear displacement of the bottom rough surface (d_s). At the early stages of

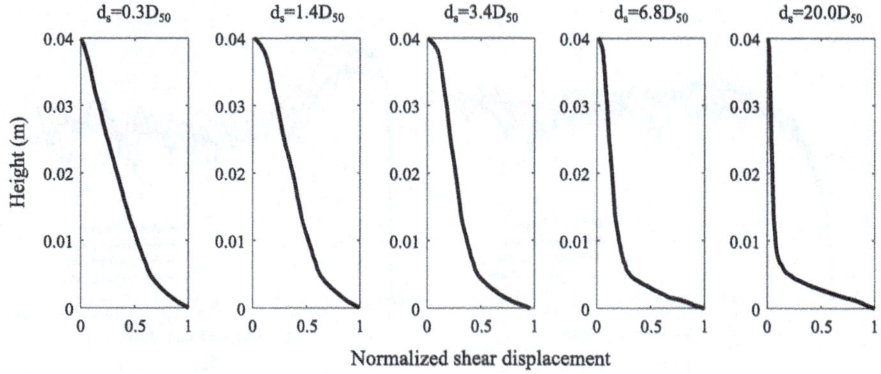

Fig. 4.9 Average normalized displacement of particles in dense specimen with $k = 1000$ kPa/mm at different shear displacements

the shear test ($d_s = 0.3D_{50}$ and $d_s = 1.4D_{50}$ in Fig. 4.9), only slight displacement localization can be observed at the bottom, and the normalized shear displacement shows a quasi-linear increase across the specimen. At the shear displacement of 3.4 D_{50}, which corresponds to the peak shear stress according to Fig. 4.5b, the bending of the curve at the bottom becomes more obvious indicating the initiation of displacement localization. As the interface shear continues to 20.0 D_{50}, the shear displacements of particles almost completely localize at the bottom and a shear zone can be clearly identified. This evolution pattern of shear zone is in good agreement with the experimental observations [49]. Noth that the developments of displacement localization in other tests are similar and therefore not presented here.

The horizontal displacements of particles at the end of interface shear tests are presented in Fig. 4.10. Particles at the bottom show much larger displacements than the upper ones, and a shear zone with strong strain localization is well-developed in

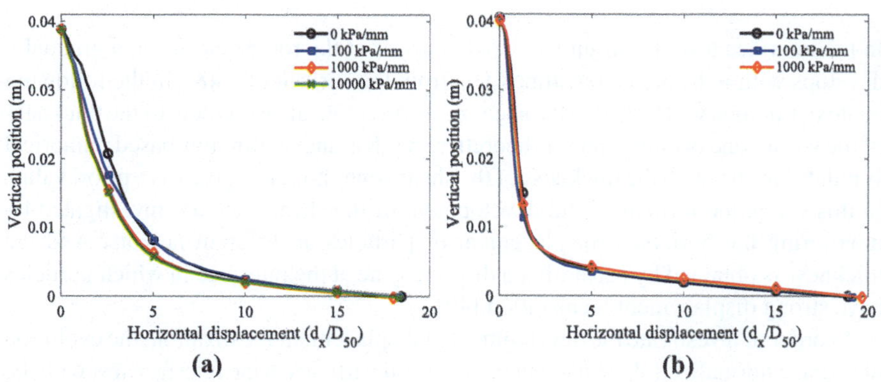

Fig. 4.10 Horizontal displacement of particles from simulations of interface shear tests with different constant stiffness: **a** loose specimen and **b** dense specimen

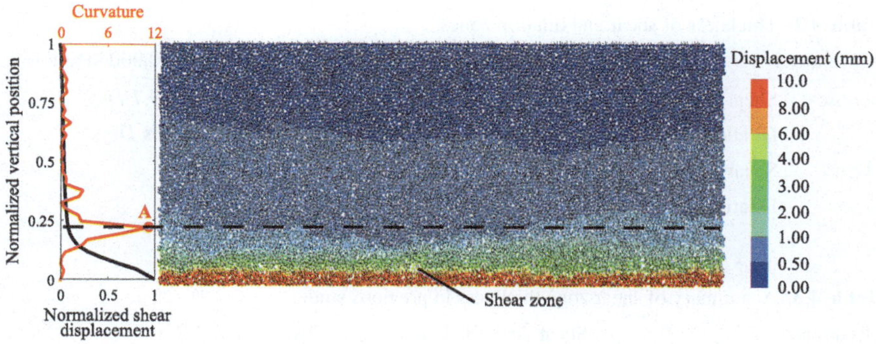

Fig. 4.11 Determination of the shear zone thickness in the interface shear test

all loose and dense specimens. Above the shear zone, the horizontal displacement gradually decreases to zero as the vertical position increases. The loose specimens have lower strain localization, indicated by larger horizontal displacements of particles above the shear zone. This can be explained by the weaker interlocking among particles in the loose specimens, which provides less contraints no the movement of particles. Similar results were also reported in [50, 51]. Slightly different and almost identical horizontal displacements are observed respectively for loose and dense specimens at different normal stiffness in Fig. 4.10, indicating the low effect of normal stiffness and normal stress on the horizontal displacements of particles. In addition, particles at the bottom have a similar magnitude of horizontal displacements with the shearing platen, i.e. very slight interface sliding. This observation is consistent with previous interface studies with rough surfaces, in which particles are trapped by the roughness and the interface failure is induced by the internal failure of soil particles [7, 51].

A simple method to obtain the thickness of the shear zone is proposed and shown in Fig. 4.11. This method is based on the idea that the boundary of the localization zone is marked by an abrupt change in the displacement gradient. After the interface test, the curvature of the normalized shear displacement versus normalized vertical position curve is calculated, as the red line shown in Fig. 4.11. The point with the maximum curvature (point A in Fig. 4.11) and the vertical position of point A can be obtained. The shear zone is defined as the region below this vertical position. As shown in Fig. 4.11, the shear zone thickness in the test with the dense specimen at $k = 1000$ kPa/mm is 8.29 mm, which is equal to 14.2 D_{50}. The thicknesses of the shear zones in the tests with different relative densities and stiffness are summarized in Table 4.2. For the loose specimens, the thickness of shear zone ranges from 18.5 D_{50} to 20.2 D_{50}, not showing a dependence on the magnitude of normal stiffness. For the dense specimens, the thickness slightly decreases with normal stiffness from 15.0 D_{50} for $k = 0$ kPa/mm to 14.2 D_{50} for $k = 1000$ kPa/mm. In addition, the shear zone thickness decreases as the relative density increases, which is consistent with the observation in Fig. 4.10 and in previous experimental results [52]. In previous studies, the thickness of shear zone, summarized in Table 4.3, varies from 1 D_{50} to 25

Table 4.2 Thickness of shear and dilation zones

Specimens		0 kPa/mm	100 kPa/mm	1000 kPa/mm	10,000 kPa/mm
Loose	Shear zone	19.8 D_{50}	18.5 D_{50}	20.2 D_{50}	19.7 D_{50}
	Dilation zone	4.4 D_{50}	4.4 D_{50}	6.6 D_{50}	8.8 D_{50}
Dense	Shear zone	15.0 D_{50}	14.4 D_{50}	14.2 D_{50}	–
	Dilation zone	22.3 D_{50}	20.3 D_{50}	17.1 D_{50}	–

Table 4.3 A summary of shear zone thickness in previous studies

Reference	Shear zone thickness/D_{50}	Test method
Sadrekarimi and Olson [53]	10–14	Interface ring shear test
Iwashita and Oda[54]	10	2D DEM simulation
Uesugi et al. [55]	3–4	Sand-steel interface tests
Gu et al. [50]	8–10	2D DEM simulation
Grabowski et al. [52]	1–14	3D DEM simulation
Rui et al. [56]	1–14	Interface ring shear test
Chen et al. [57]	4.0–4.2	3D DEM simulation
Ho et al. [58]	4–13	Sand-steel interface test
Zhao et al. [59]	8–14	Sand-steel interface test
Vangla and Latha [60]	5–25	Sand-geomembrane interface test

D_{50} and is affected by many factors such as surface roughness and relative density. The shear zone thickness in this study is within the range of previous studies, and close to the upper bound values at high interface roughness.

The vertical displacements of particles in the simulations are presented in Fig. 4.12. For the loose specimens, particles close to the interface tend to dilate

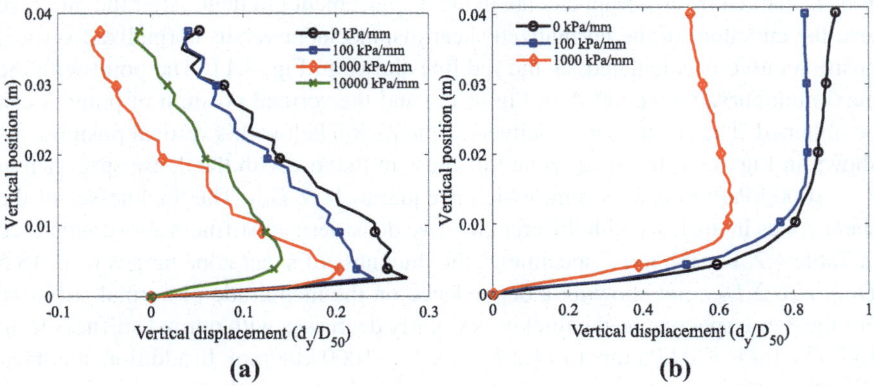

Fig. 4.12 Vertical displacement of particles from simulations of interface shear tests with different constant stiffness: **a** loose specimen and **b** dense specimen

and reach a maximum vertical displacement at the position of 4 to 9 D_{50} depending on the normal stiffness. The position with maximum vertical displacement divides the soil into two parts, i.e., the dilation zone below this position and the contraction zone above it. Because of the initial loose state, the thickness of the dilation zone is much smaller than that of the contraction zone. For the dense specimens in Fig. 4.12b, although the dilation still concentrates near the interface, the thickness of the dilation zone is much larger than the corresponding loose ones. Specifically, the whole specimen dilates in the simulation with $k = 0$ kPa/mm (constant normal stress condition). A similar method based on the maximum curvature is used to determine the dilation zone for the dense specimens. The thickness of the dilation zone is summarized in Table 4.2. The initial relative density plays a more significant role in the thickness of the dilation zone, i.e., the thickness increases with the increase of relative density. The normal stiffness indirectly affects the thickness of dilation zone by changing the normal stress, and the thickness is negatively correlated with the normal stress.

4.3.3 Void Ratio

During the interface shear test, the volume of the specimen changes (Fig. 4.7), which can be attributed to the rearrangement of soil particles. At the same time, the void ratio also varies according to the volume change. In order to understand the mechanism of volume change from the microscopic perspective, the localized void ratio, which is the void ratio in a meso-loop, is investigated. The meso-loops are the polygonal loops that are enclosed by contact branches, as shown in Fig. 4.13. For each meso-

Fig. 4.13 Polygonal meso-loops enclosed by contact branches

loop, the volume of solid particles, V_S, is obtained by the summation of the areas of circle sectors within the meso-loop. And the volume of the meso-loop, V_T, can be calculated by knowing the positions of all vertices. Therefore, the localized void ratio of the meso-loop is given as $(V_T - V_S)/V_S$.

Figure 4.14 shows the localized void ratios of the meso-loops before and after the interface shear test in the dense specimen at $k = 1000$ kPa/mm. In Fig. 4.14, each circle represents a meso-loop whose center is the same as the sphere. The color and size of a sphere respectively indicate the void ratio and void size of the corresponding meso-loop. Before the interface shear test, the voids are randomly distributed both inside and outside of the shear zone, and no void concentration can be observed (Fig. 4.14a). Although the localized void ratio ranges from 0.1 to over 1.0 in different meso-loops, the void sizes are relatively small. The random void distribution indicates that the specimen is homogeneous. The localized void ratio after the shear test is presented in Fig. 4.14b. Outside of the shear zone, small voids are randomly distributed, which is in a similar condition before the shear test. Within the shear zone, however, an obvious concentration of large voids can be identified,

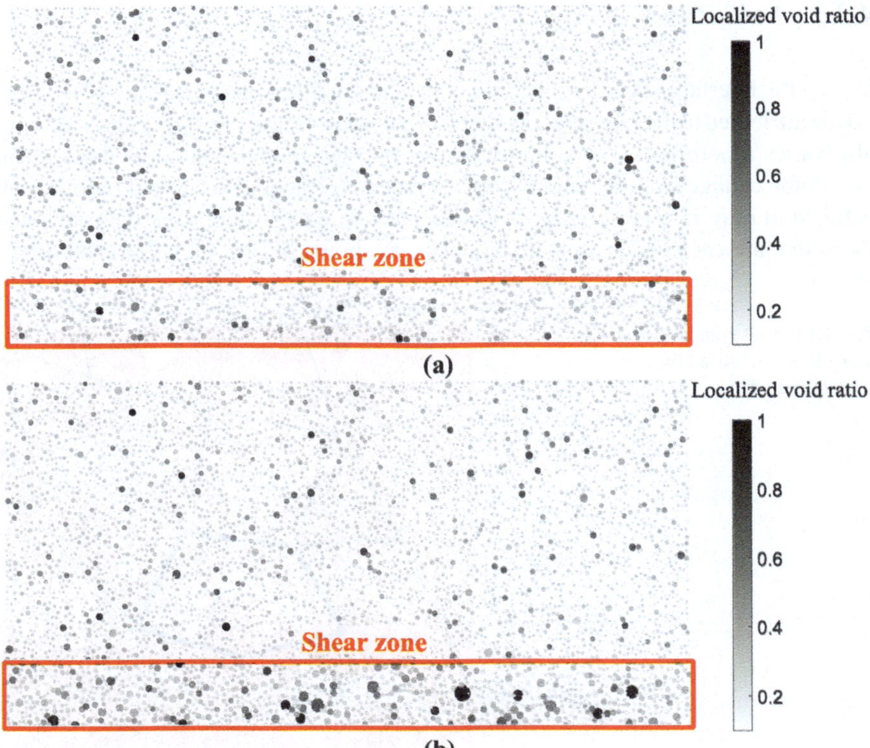

Fig. 4.14 Localized void ratios of the meso-loops in dense specimen with $k = 1000$ kPa/mm before **a** and after **b** shear test

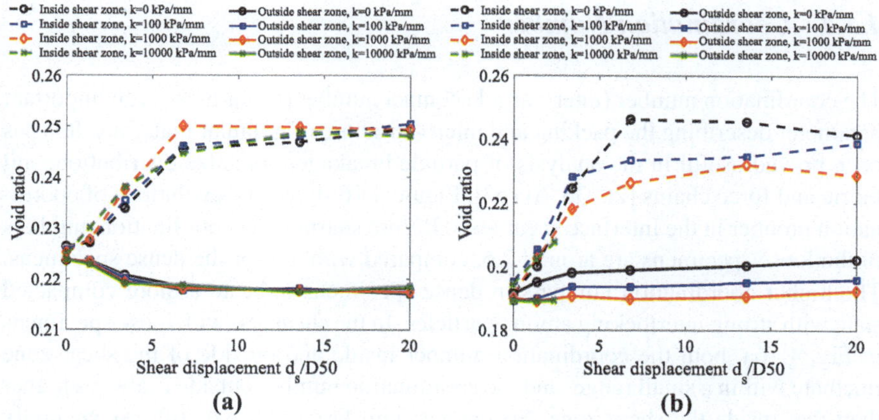

Fig. 4.15 Evolution of void ratio inside and outside of shear zone in interface shear test with **a** loose specimens and **b** dense specimens

indicated by large spheres in Fig. 4.14b. Besides the increased void size, the localized void ratios are also higher than those before the test. The above observation confirms that interface shear induces significant changes in terms of void size and localized void ratio for particles within the shear zone, but only has a slight effect outside of the shear zone. Similar results are also observed in other tests and will not be shown here.

The void ratios inside and outside of the shear zone in different tests with various normal stiffness are presented in Fig. 4.15. For the loose specimens, the void ratios inside the shear zones increase by around 10% at the beginning of the shear tests and then become stable at a shear displacement of 6.8 D_{50}. While the void ratios outside of the shear zone slightly decrease at the beginning and also stabilize at the shear displacement of 6.8 D_{50}. It is interesting to see that the macroscopic shear behaviors, including the shear force, normal force and volumetric strain, also plateau at the shear displacement of 6.8 D_{50} according to Figures from 5 to 7, which exhibits a strong dependence on the microscopic behaviors. The influence of normal stiffness is very slight in tests with loose specimens. For tests with dense specimens, both the void ratios inside and outside of the shear zones increase at first and then become stable. The increase of void ratio inside the shear zone (23% on average) is much higher than that outside of the shear zone (less than 5%). In addition, the value of void ratio decreases with the increase of normal stiffness, which can be attributed to the large normal stress generated during the shear test (Fig. 4.6).

4.3.4 Coordination Number

The coordination number (inter-particle contact number per particle) is an important parameter describing the packing and micro-structure of granular materials. It is has been proved useful in the analysis of particle breakage, pore size distribution, soil fabric and force chains [26, 38, 61–63]. Figure 4.16 shows the evolutions of coordination number in the interface shear tests. Before sharing, the coordination numbers of the loose specimens are around 2.6, compared with 3.5 for the dense specimens. The higher coordination numbers in dense specimens indicate a more compacted state with strong interlocking among particles. In the shear test with loose specimens in Fig. 4.16a, both the coordination number inside and outside of the shear zone fluctuate within a small range, and the coordination number outside is always higher than that inside the shear zone. Specimens with larger normal stiffness generally have lower coordination numbers, which can be attributed to the decrease of the normal stress during the shear test. However, contrary to the effect in the loose specimens, a significant increase of final coordination number is induced by increasing the normal stiffness, as shown in Fig. 4.16b. For example, the coordination number inside the shear zone increases from 2.55 to 2.81 when the constant normal stiffness is increased from 0 to 1000 kPa/mm. The increased coordination number in dense specimens can be explained by the high normal stress in the test. It is also noticeable that the coordination number decreases beyond the shear zone at the beginning of shearing, which can be explained by the uniform shear deformation before localization [49]. In summary, despite the huge difference in the normal stiffness, it is actually the normal stress generated during the shear test that determines the coordination number in the soil specimen.

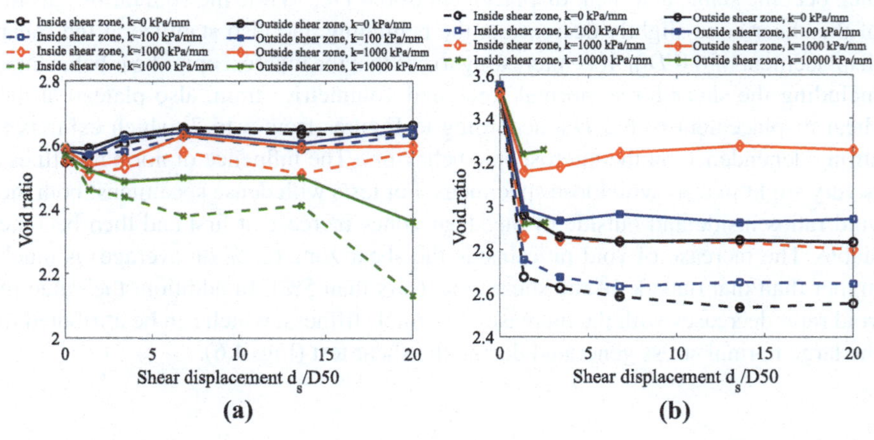

Fig. 4.16 Evolution of coordination number inside and outside of shear zone during interface shear test from **a** loose specimens and **b** dense specimens

4.3.5 Force Chain and Soil Fabric

One of the important advantages of DEM analysis is the ability to monitor and track the orientation and force magnitude of particle contacts. The force chains and soil fabric are useful micro-descriptors to understand the macroscopic behavior and to develop multi-scale constitutive models [64–68].

The force chain distribution before and after the shear test in the dense specimen with $k = 1000$ kPa/mm is shown in Fig. 4.17. Before the test, contacts have a preferential orientation in the vertical direction, which can be explained by the vertical compaction in the sample preparation process. The contact network is dense with a large number of weak contacts, indicating that the applied external force is shared by a large number of particles, i.e. low force concentration. In addition, the particle interlocking is weak and elastic energy among particles is low at this state [36, 69]. After the shear test, the initially vertically oriented weak force chains gradually change their preferential orientation to align to the shear direction, as shown in Fig. 4.17b. The contact network is dominated by thick force chains. Strong anisotropy and force concentration is generated. Compared with the initial state, the magnitude of the normal contact force also increases significantly.

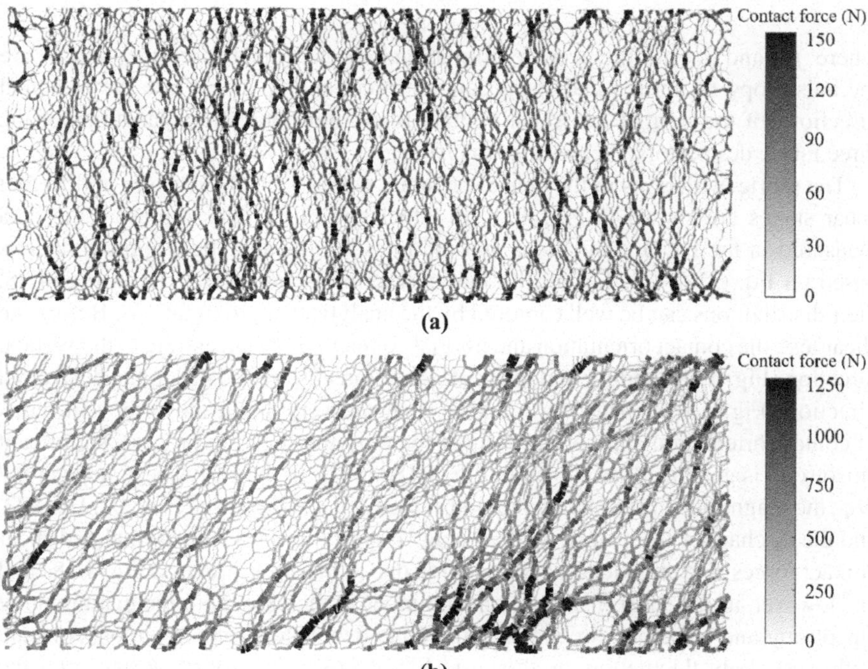

Fig. 4.17 Force chain distribution in dense specimen with $k = 1000$ kPa/mm before (**a**) and after (**b**) interface shear test

In order to quantify the evolution of anisotropy and internal structure during the interface shear test, three fabric descriptors are analyzed, i.e. particle contact orientation, normal contact force and shear contact force. For a given contact between two soil particles, the contact orientation is defined as the normalized vector connecting the centroids of the two particles. In a soil specimen, the angular contact orientation distribution, $E(\theta)$, is given by [70]

$$E(\theta) = \frac{1}{2\pi}[1 + acos2(\theta - \theta_a)] \tag{4.3}$$

where a defines the magnitude of anisotropy and θ_a defines the principal direction of anisotropy. In the same way, the normal and shear contact force distributions, $f_n(\theta)$ and $f_s(\theta)$, are respectively expressed as

$$f_n(\theta) = f_{n0}[1 + a_n cos2(\theta - \theta_n)] \tag{4.4}$$

and.

$$f_s(\theta) = -f_{s0} a_s sin2(\theta - \theta_s) \tag{4.5}$$

where f_{n0} and f_{s0} are the average normal and shear contact forces; a_n and a_s are the anisotropy of normal and shear contact force; and θ_n and θ_s are the principal directions of the normal and shear contact force. For detailed explanations of the three fabric descriptors, please refer to [70].

The fabric evolutions in the dense specimen with $k = 1000$ kPa/mm at different shear stages are shown in Fig. 4.18, in which the black lines represent the data measured in the DEM simulations and red lines are the analytical approximations based on Eq. (3) to (5). For all the fabric descriptors from (a) to (l) in Fig. 4.18, their distributions can be well captured by the analytical approximations. Before the shear test, the contact orientation and normal contact force concentrate in the vertical direction (Fig. 4.18a and b), and the peak shear force is around 45° to the horizontal direction (Fig. 4.18c). The low a and a_n indicate very slight anisotropy in terms of contact orientation and normal contact force. In the shearing process, significant anisotropy is generated, as shown in Fig. 4.18d to (l). In the early shear stage at 0.7 D_{50}, the magnitude of contact orientation anisotropy a is increased from 0.09 to 0.23, and the θ_a changes from 91.0° to 49.5° (Fig. 4.18d). Similarly, the normal and shear contact forces also experience some rotation due to the shear process (Fig. 4.18e and f). However, as the shear displacement continues to increase, both the magnitude of anisotropy and principal directions of the three descriptors remain almost constant with very slight fluctuation, as shown in Fig. 4.18g to l. In other words, after the initial anisotropy induced at the beginning of the test, the soil fabric keeps stable.

The evolutions of anisotropy in terms of contact orientation, normal contact force and shear contact force are summarized in Figs. 4.19 to 4.21. Despite the difference

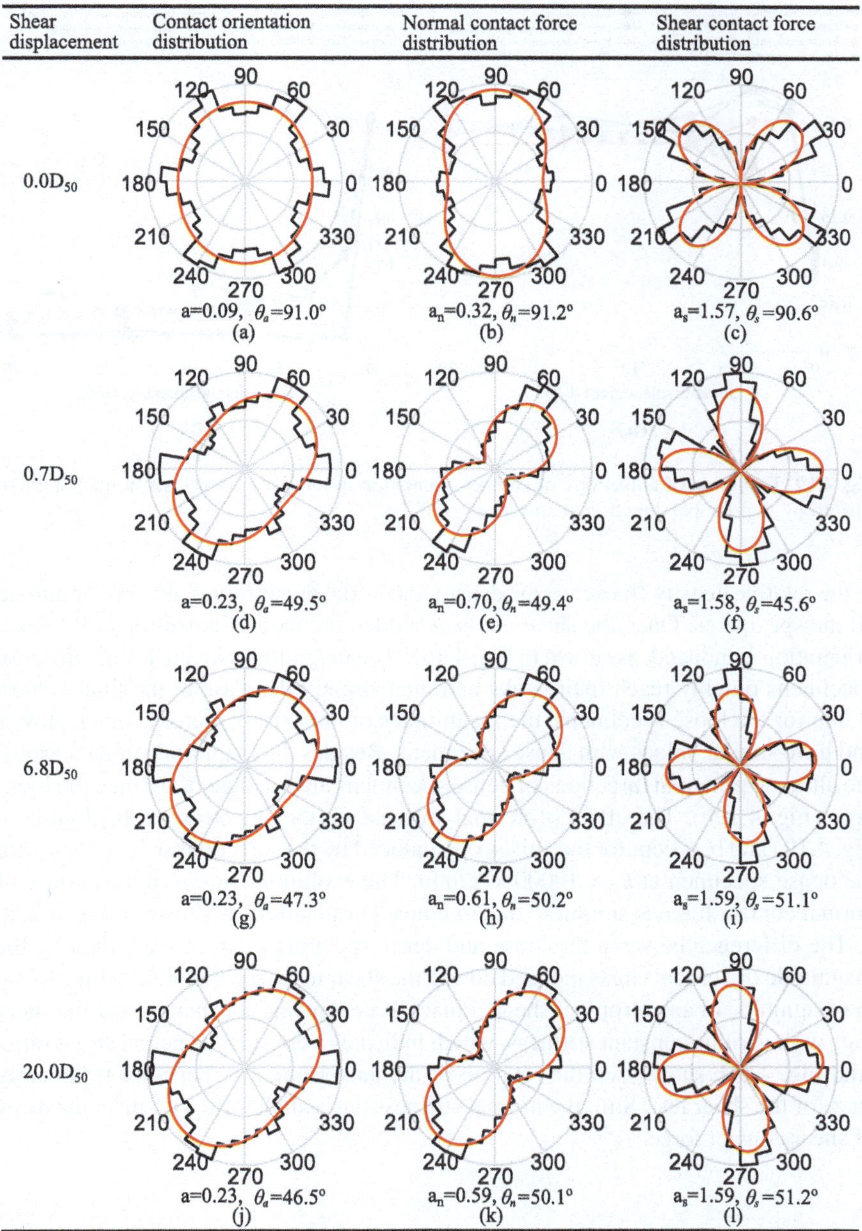

Fig. 4.18 Fabric evolution of particles in dense specimen with $k = 1000$ kPa/mm

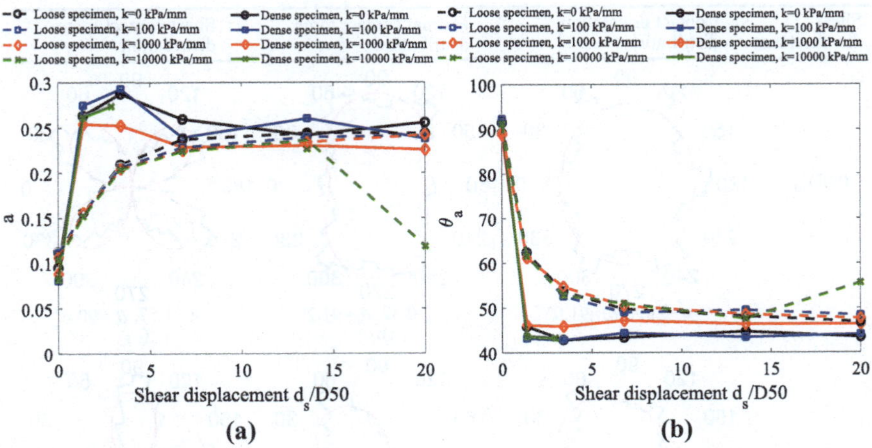

Fig. 4.19 Evolution of anisotropy of contact orientation in interface shear tests: **a** magnitude of anisotropy; **b** principal direction of anisotropy

in the relative density (loose versus dense), the initial anisotropies are similar among all the specimens. Once the shear starts, a sudden increase of anisotropy of contact orientation is induced, as shown in Fig. 4.19a. The magnitudes of anisotropy in dense specimens quickly reach their peaks and then slightly decrease to the final values. While for the loose specimens, the magnitudes of anisotropy increase much slower and have similar values with dense specimens. Besides the magnitude of anisotropy, the ultimate principal directions of dense specimens also change faster than the loose ones (Fig. 4.19b). The effect of normal stiffness on the anisotropy is negligible in Fig. 4.19a and b, except for the anisotropy induced by the soil collapse in the test with the dense specimen at $k = 10000$ kPa/mm. The evolution pattern of anisotropy of normal contact force is similar to that of contact orientation, as shown in Fig. 4.20a-b. The difference between the loose and dense specimens can be explained by the magnitude of vertical stress induced during the shear tests, see Fig. 4.6. In Fig. 4.21a, the magnitude of anisotropy of shear contact force remains constant during the shear tests with various constant stiffness, which indicates a constant principal stress ratio. And Fig. 4.21b shows that the rotation of the anisotropy also happens at the early stage of the shear test. Still, the normal stiffness has a slight effect on the anisotropy of shear contact force.

4.4 Discussions

Besides the normal stiffness, it has been reported that the surface roughness and particle characteristics also have a strong influence on the behavior of soil-structure interface. Therefore, their effects are investigated in this section. In addition, one 3D DEM simulation is conducted and the results are compared with the 2D simulations.

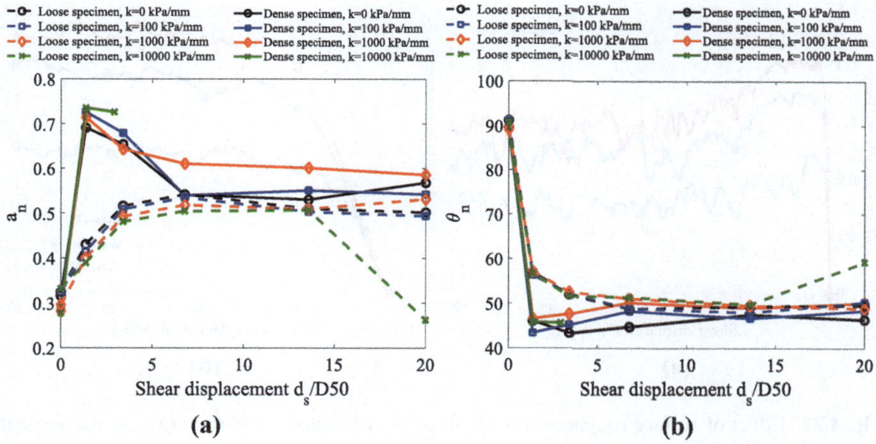

Fig. 4.20 Evolution of anisotropy of normal contact force in interface shear tests: **a** magnitude of anisotropy; **b** principal direction of anisotropy

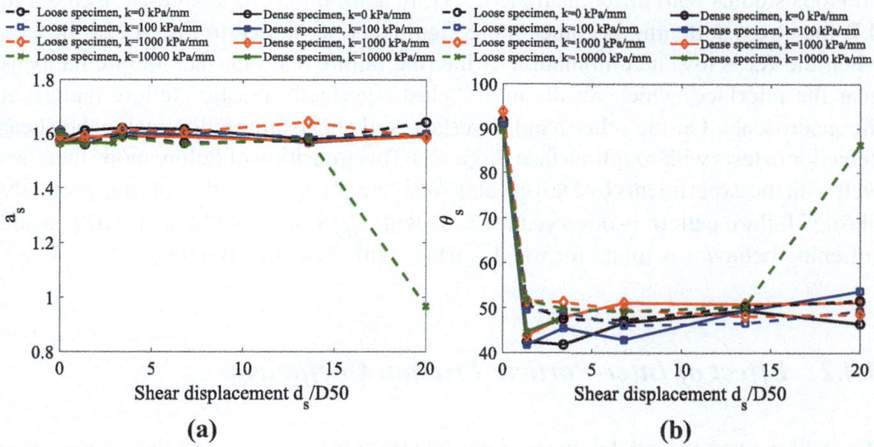

Fig. 4.21 Evolution of anisotropy of shear contact force in interface shear tests: **a** magnitude of anisotropy; **b** principal direction of anisotropy

4.4.1 Effect of Surface Roughness

Three interface shear tests with R_n respectively equal to 0.3, 0.5 and 0.7 are simulated. Except for the surface roughness, these tests are conducted with the same DEM parameters, sample preparation method, and boundary conditions, as discussed previously in Sect. 4.2. A constant normal stiffness of 0 kPa/mm is adopted for the tests. The effect of surface roughness on stress ratio and vertical displacement is shown in Fig. 4.22. The peak and residual stress ratios, as well as the dilation of the

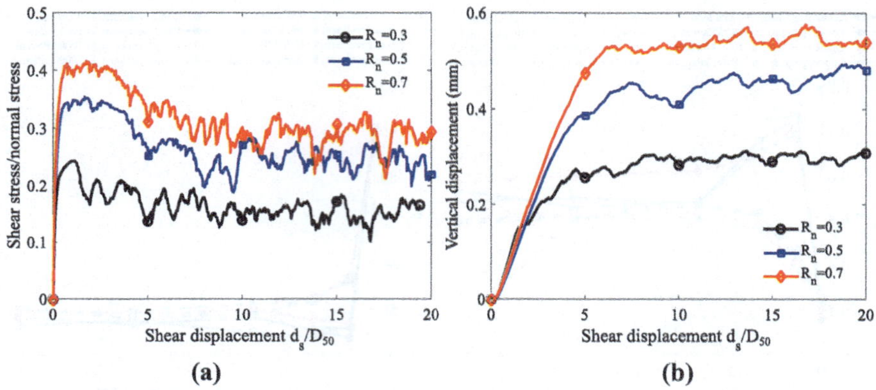

Fig. 4.22 Effect of surface roughness on the shear stress/normal stress ratio (**a**) and the vertical displacement (**b**)

specimens increase with the increase of R_n. These observations are consistent with previous studies with different R_n [51, 71]. In addition, as R_n increases from 0.3 to 0.7, the stress softening becomes more significant, as shown in Fig. 4.22a. In fact, when the R_n is low, a combination of internal failure and interface failure happens near the interface, which results in an "elastic-perfectly plastic" failure pattern at the macroscale. On the other hand, interface soil shear failure dominates the shear behavior in tests with rough surface (high R_n).This transition of failure mode matches well with the experiments by DeJong and Westgate [7], in which the "elastic-perfectly plastic" failure pattern is observed for tests with R_n equal to 0.008 and 0.074, while softening behavior is found for rough surface with R_n equal to 0.95.

4.4.2 Effect of Inter-Particle Friction Coefficient

It is well-acknowledged that the interparticle friction (μ) influences the macroscopic shear resistance and friction angle of granular materials. In the interface shear tests in this study, because the interface failure is induced by the internal failure of soil particles, the inter-particle friction coefficient plays a significant role in determining the peak and residual frictional angle. Therefore, the effect of inter-particle friction coefficient is investigated with three interface tests with $k = 100$ kPa/mm and μ ranging from 0.3 to 0.7, and the simulations results are presented in Fig. 4.23. As expected, both peak and residual stress ratios increase with the increase of μ. The peak/residual friction angles for the tests with μ equal to 0.3, 0.5 and 0.7 are 26.6°/19.3°, 24.0° /17.7°, and 18.8° /14.6°, respectively. Note that the low friction angles are due to 2D condition of the simulations, which will be discussed in the following section. In addition, as shown in Fig. 4.23b, the vertical displacement also increases with the increase of μ, which is consistent with previous observations in

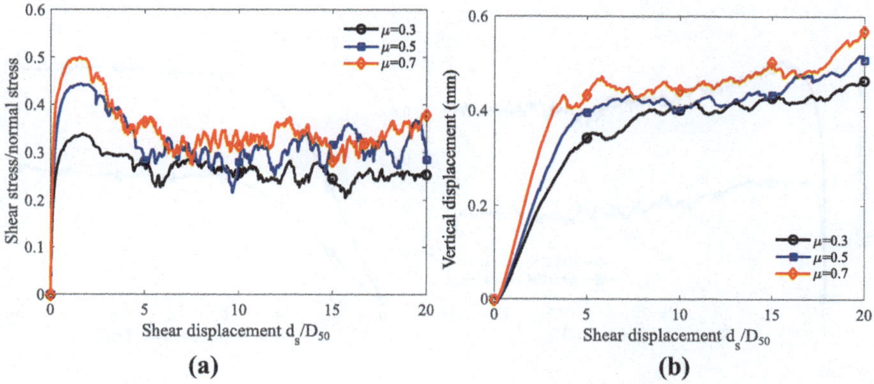

Fig. 4.23 Effect of inter-particle friction coefficient on the shear stress/normal stress ratio (**a**) and the vertical displacement (**b**)

[72]. At the microscale, the effect of μ on the vertical displacement can be explained by the fact that the self-stability of force chains in a specimen is increased with μ, thus resulting in a more dilative behavior [72].

4.4.3 2D Versus 3D Simulations

In order to investigate the difference between 2 and 3D DEM simulations, one interface test is conducted in 3D. The parameters in Table 4.2 are used in the 3D case, and the surface roughness and normal stiffness are respectively 0.7 and 0 kPa/mm. The sample preparation method in Sect. 4.2.1 is used to prepare a dense specimen. As stated in the Introduction, 3D simulations are much more computationally expensive than the 2D ones because of the increased number of particles and higher degrees of freedom. To reduce the computational cost, the parallel PSD method is adopted, in which the mean particle size is increased by 2 times to 1.156 mm while the shape of the new PSD curve remains the same. With this method, the total number of particles in the 3D DEM simulation is around 100,000.

Figure 4.24 compares the shear stress/normal stress ratio and vertical displacement in 2D and 3D simulations. As shown in Fig. 4.24a, the evolution trends of 2D and 3D simulations are generally similar, and post-peak softening is observed in both tests. The peak and residual stress ratios in 3D are respectively 0.83 and 0.65, which correspond to peak and residual friction angles of 39.7° and 33.0°. Compared with the angles of 23.8° and 17.2° in 2D, the 3D specimen has much larger shear resistance, even though the same DEM parameters are used. The different shear resistance is also confirmed by the numerical study in [73], in which the peak stress ratios in 3D and 2D are 0.76 and 0.32, respectively. Similarly, in some other studies with DEM, low shear strengths and friction angles are observed in 2D than 3D [51, 61, 74]. This

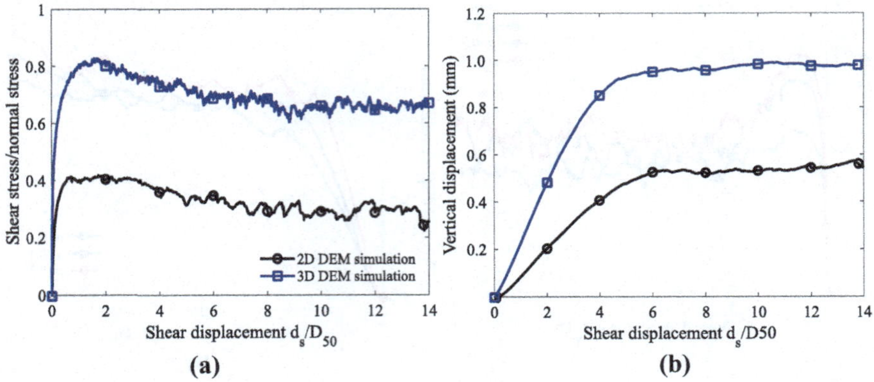

Fig. 4.24 Comparison between 2 and 3D DEM simulations: **a** ratio of shear stress to normal stress versus shear displacement; and **b** vertical displacement versus shear displacement

difference can be explained by a lower coordination number in 2D, which imposes less resistance to particle movement at the microscale. As a result, the macroscopic frictional angle is decreased. The dilation in 3D is also more significant than that in 2D (Fig. 4.24b), which is consistent with the observations in [75]. In addition, the 3D simulation significantly decreases the fluctuation of stress ratio compared with the 2D counterparts, as shown in Fig. 4.24a. This difference may be attributed to the high self-stability and the low possibility of force chain buckling in 3D.

4.5 Conclusions

The condition of constant normal stiffness is likely to prevail in many soil-structure interfaces. Therefore, in this study, the soil-structure interface behavior at the condition of constant normal stiffness has been investigated with DEM. Soil-structure interface shear tests with both loose and dense soil specimens under constant normal stiffness condition are simulated. A wide range of normal stiffness from 0 to 10,000 kPa/mm has been adopted to account for soils with various stiffness. The micro- and macroscopic mechanical responses of the interface tests have been discussed in detail, and the effect of the magnitude of normal stiffness has been clarified. Conclusions of this study are made as follows:

(1) An effective method has been developed to achieve the constant normal stiffness condition in an interface shear test. In the simulation of an interface shear test, for a given stress increment, the determination of displacement increment is adjusted iteratively based on the normal stress and soil stiffness. With the proposed method, the condition of constant normal stiffness is achieved with a wide range of normal stiffness.

(2) At the macroscale, both interface normal and shear stresses decrease for loose specimens and increase for dense specimens with the increase of normal stiffness. These observations are generally consistent with the experimental results and demonstrate the combined effects of normal stiffness and volumetric strain. Particularly, the normal and shear stress of a loose specimen at high normal stiffness may decrease to zero due to the strong volumetric contraction and static liquefaction. While normal and shear stresses of a dense specimen significantly increase (more than 10 times) at high normal stiffness. In addition, despite the difference in relative density and normal stiffness, the residual stress ratios are always the same.

(3) The thickness of shear zone is around 18 to 20 D_{50} for loose specimens and 14 to 15 D_{50} for dense specimens. Shear zone evolution is observed through the progressive development of displacement localization of soil particles, which includes stages of uniform shear deformation, initiation and growth of shear localization, and stabilized shear zone. The shear zone thickness of loose specimens is not affected by the magnitude of normal stiffness in our case, while the thickness for dense specimens slightly decreases with the increases of normal stiffness.

(4) The thickness of the dilation zone is sensitive to the void ratio and normal stiffness. The thickness of the dilation zone is 4 to 9 D_{50} for loose specimens, and 17 to 22 D_{50} for dense specimens, highly depending on the initial void ratio. Besides, the thickness of the dilation zone is also positively related to the magnitude of normal stress. As a result, the thickness of the dilation zone increases/decreases with normal stiffness respectively for loose and dense specimens.

(5) Within the shear zone, the void ratio increases by around 10 and 23% for the loose and dense specimens, respectively. A strong concentration of large voids can be identified from the meso-loops near the interface, which explains the macroscopic dilation. At the same time, the coordination numbers of loose specimens fluctuate within a small range due to the combination effect of shear zone dilation and force chain collapse. The coordination numbers of dense specimens significantly decrease as a result of volumetric dilation. After the formation of a stable shear zone, the void ratio and coordination remain almost constant.

(6) In the interface shear test, strong anisotropy and force concentration are generated. For dense specimens, the magnitude of anisotropy quickly reaches a peak and then decreases to a stable value, corresponding to the macroscopic softening behavior. While for the loose specimens, the magnitude of anisotropy increases much slower and has a similar final value with the dense specimen under the same normal stiffness. The effect of normal stiffness on the anisotropy is very slight.

References

1. Su LJ, Zhou WH, Chen WB, Jie XX (2018) Effects of relative roughness and mean particle size on the shear strength of sand-steel interface. Measurement 122:339–346
2. Liu SH, Sun DA, Matsuoka H (2005) On the interface friction in direct shear test. Comput Geotech 32:317–325
3. Ji K, Arson C (2020) Tensile strength of calcite/HMWM and silica/HMWM interfaces: a molecular dynamics analysis. Const Build Mater 251:118925
4. Wang HL, Zhou WH, Yin ZY, Jie XX (2019) Effect of grain size distribution of sandy soil on shearing behaviors at soil-structure interface. J Mater Civil Eng 31
5. Boukpeti N, White DJ (2017) Interface shear box tests for assessing axial pipe-soil resistance. Geotechnique 67:18–30
6. Tabucanon JT, Airey DW, Poulos HG (1995) Pile skin friction in sands from constant normal stiffness tests. Geotech Test J 18:350–364
7. DeJong JT, Westgate ZJ (2009) Role of initial state, material properties, and confinement condition on local and global soil-structure interface behavior. J Geotech Geoenviron 135:1646–1660
8. Ooi LH, Carter JP (1987) A constant normal stiffness direct shear device for static and cyclic loading. Geotech Test J 10:3–12
9. Porcino D, Fioravante V, Ghionna VN, Pedroni S (2003) Interface behavior of sands from constant normal stiffness direct shear tests. Geotech Test J 26:289–301
10. DeJong JT, Randolph MF, White DJ (2003) Interface load transfer degradation during cyclic loading: a microscale investigation. Soils Found 43:81–93
11. DeJong JT, White DJ, Randolph MF (2006) Microscale observation and modeling of soil-structure interface behavior using particle image velocimetry. Soils Found 46:15–28
12. Owen PJ, Cleary PW, Mériaux C (2009) Quasi-static fall of planar granular columns: comparison of 2D and 3D discrete element modelling with laboratory experiments. Geomech Geoeng 4:55–77
13. Wang JF, Gutierrez MS, Dove JE (2007) Numerical studies of shear banding in interface shear tests using a new strain calculation method. Int J Numer Anal Meth Geomech 31:1349–1366
14. Zhu HX, Zhou WH, Yin ZY (2018) Deformation mechanism of strain localization in 2D numerical interface tests. Acta Geotech 13:557–573
15. Khabazian M, Mirghasemi AA, Bayesteh H (2018) Compressibility of montmorillonite/kaolinite mixtures in consolidation testing using discrete element method. Comput Geotech 104:271–280
16. Suchorzewski J, Tejchman J, Nitka M, Bobinski J (2019) Meso-scale analyses of size effect in brittle materials using DEM. Granular Matter 21
17. Cundall PA (1971) A computer model for simulating progressive, large-scale movement in blocky rock system. In: Proceedings of the International Symposium on Rock Mechanics
18. Zhou Y, Shi Z, Zhang Q, Liu W, Peng M, Wu C (2019) 3D DEM investigation on the morphology and structure of landslide dams formed by dry granular flows. Eng Geol 258:105151
19. Zhang J, Wang X, Yin Z-Y, Liang Z (2020) DEM modeling of large-scale triaxial test of rock clasts considering realistic particle shapes and flexible membrane boundary. Eng Geol 279:105871
20. Frost JD, DeJong JT, Recalde M (2002) Shear failure behavior of granular-continuum interfaces. Eng Fract Mech 69:2029–2048
21. Wang J, Gutierrez M (2010) Discrete element simulations of direct shear specimen scale effects. Geotechnique 60:395–409
22. Liu X, Zhou A, Shen S-l, Li J, Sheng D (2020) A micro-mechanical model for unsaturated soils based on DEM. Comput Methods Appl Mech Eng 368:113183
23. Wood DM, Maeda K (2008) Changing grading of soil: effect on critical states. Acta Geotech 3:3–14
24. Jiang MJ, Konrad JM, Leroueil S (2003) An efficient technique for generating homogeneous specimens for DEM studies. Comput Geotech 30:579–597

25. Wu MM, Wang JF (2020) A DEM investigation on crushing of sand particles containing intrinsic flaws. Soils Found 60:562–572
26. Zhao SW, Zhang N, Zhou XW, Zhang L (2017) Particle shape effects on fabric of granular random packing. Powder Technol 310:175–186
27. Goldenberg C, Goldhirsch I (2005) Friction enhances elasticity in granular solids. Nature 435:188–191
28. Potyondy DO, Cundall PA (2004) A bonded-particle model for rock. Int J Rock Mech Min Sci 41:1329–1364
29. Wang P, Arson C (2016) Discrete element modeling of shielding and size effects during single particle crushing. Comput Geotech 78:227–236
30. Shi JS, Guo PJ (2018) Induced fabric anisotropy of granular materials in biaxial tests along imposed strain paths. Soils Found 58:249–263
31. Ai J, Chen JF, Rotter JM, Ooi JY (2011) Assessment of rolling resistance models in discrete element simulations. Powder Technol 206:269–282
32. Jiang MJ, Yu HS, Harris D (2005) A novel discrete model for granular material incorporating rolling resistance. Comput Geotech 32:340–357
33. Wang X, Gong J, An A, Zhang K, Nie Z (2019) Random generation of convex granule packing based on weighted Voronoi tessellation and cubic-polynomial-curve fitting. Comput Geotech 113:103088
34. Zhao SW, Evans TM, Zhou XW (2018) Shear-induced anisotropy of granular materials with rolling resistance and particle shape effects. Int J Solids Struct 150:268–281
35. Liu YM, Liu HB, Mao HJ (2018) The influence of rolling resistance on the stress-dilatancy and fabric anisotropy of granular materials. Granular Matter 20
36. Barnett N, Rahman MM, Karim MR, Nguyen HBK (2020) Evaluating the particle rolling effect on the characteristic features of granular material under the critical state soil mechanics framework. In: Granular Matter. Springer, pp 1–24
37. Yin Z-Y, Wang P (2021) Micro-mechanical analysis of caisson foundation in sand using DEM: Particle shape effect. Appl Ocean Res 111:102630
38. Wang P, Yin Z-Y (2020a) Micro-mechanical analysis of caisson foundation in sand using DEM. Ocean Eng 203:107240
39. Wang P, Yin Z-Y (2020b) Micro-mechanical analysis of caisson foundation in sand using DEM: particle breakage effect. Ocean Eng 215:107921
40. Huang X, O'Sullivan C, Hanley KJ, Kwok CY (2014) Discrete-element method analysis of the state parameter. Geotechnique 64:954–965
41. Md. Mizanur R, Lo SR (2012) Predicting the onset of static liquefaction of loose sand with fines. J Geotech Geoenviron Eng 138:1037–1041
42. Yamamuro JA, Lade PV (1997) Static liquefaction of very loose sands. Can Geotech J 34:905–917
43. Chang CS, Yin ZY, Hicher PY (2011) Micromechanical analysis for interparticle and assembly instability of sand. J Eng Mech 137:155–168
44. Daouadji A, Darve F, Al Gali H, Hicher PY, Laouafa F, Lignon S, Nicot F, Nova R, Pinheiro M, Prunier F, Sibille L, Wan R (2011) Diffuse failure in geomaterials: experiments, theory and modelling. Int J Numer Anal Meth Geomech 35:1731–1773
45. Evgin E, Fakharian K (1996) Effect of stress paths on the behaviour of sand-steel interfaces. Can Geotech J 33:853–865
46. Afzali-Nejad A, Lashkari A, Shourijeh PT (2017) Influence of particle shape on the shear strength and dilation of sand-woven geotextile interfaces. Geotext Geomemb 45:54–66
47. Alshibli KA, Sture S (2000) Shear band formation in plane strain experiments of sand. J Geotech Geoenviron Eng 126:495–503
48. Wiebicke M, Ando E, Viggiani G, Herle I (2020) Measuring the evolution of contact fabric in shear bands with X-ray tomography. Acta Geotech 15:79–93
49. Sadrekarimi A, Olson SM (2010) Shear band formation observed in ring shear tests on sandy soils. J Geotech Geoenviron Eng 136:366–375

50. Gu XQ, Chen YW, Huang MS (2017) Critical state shear behavior of the soil-structure interface determined by discrete element modeling. Particuology 35:68–77
51. Jing XY, Zhou WH, Zhu HX, Yin ZY, Li YM (2018) Analysis of soil-structural interface behavior using three-dimensional DEM simulations. Int J Numer Anal Meth Geomech 42:339–357
52. Grabowski A, Nitka M, Tejchman J (2021) 3D DEM simulations of monotonic interface behaviour between cohesionless sand and rigid wall of different roughness. Acta Geotech 16:1001–1026
53. Sadrekarimi A, Olson Scott M (2010) Shear band formation observed in ring shear tests on sandy soils. J Geotech Geoenviron Eng 136:366–375
54. Iwashita K, Oda M (1998) Rolling resistance at contacts in simulation of shear band development by DEM. J Eng Mech 124:285–292
55. Uesugi M, Kishida H, Tsubakihara Y (1988) Behavior of sand particles in sand-steel friction. Soils Found 28:107–118
56. Rui S, Wang L, Guo Z, Cheng X, Wu B (2021) Monotonic behavior of interface shear between carbonate sands and steel. Acta Geotech 16:167–187
57. Chen W-B, Zhou W-H, dos Santos JA (2020) Analysis of consistent soil–structure interface response in multi–directional shear tests by discrete element modeling. Transport Geotech 24:100379
58. Ho TYK, Jardine RJ, Anh-Minh N (2011) Large-displacement interface shear between steel and granular media. Géotechnique 61:221–234
59. Zhao C, Zhang R, Zhao C, Wang W, Wang Y (2019) A three-dimensional evaluation of interface shear behavior between granular material and rough surface. J Test Eval 49:713–727
60. Vangla P, Latha GM (2015) Influence of particle size on the friction and interfacial shear strength of sands of similar morphology. Int J Geosynth Groun 1:6
61. Yin Z-Y, Wang P, Zhang F (2020) Effect of particle shape on the progressive failure of shield tunnel face in granular soils by coupled FDM-DEM method. Tunnell Undergr Space Technol 100:103394
62. Gao ZW, Zhao JD (2013) Strain localization and fabric evolution in sand. Int J Solids Struct 50:3634–3648
63. Wang G, Wei JT (2016) Microstructure evolution of granular soils in cyclic mobility and post-liquefaction process. Granular Matter 18
64. Yin ZY, Chang CS, Hicher PY (2010) Micromechanical modelling for effect of inherent anisotropy on cyclic behaviour of sand. Int J Solids Struct 47:1933–1951
65. Yin ZY, Xu Q, Chang CS (2013) Modeling cyclic behavior of clay by micromechanical approach. J Eng Mech 139:1305–1309
66. Yin Z-Y, Xu Q, Hicher P-Y (2013) A simple critical-state-based double-yield-surface model for clay behavior under complex loading. Acta Geotech 8:509–523
67. Xiong H, Nicot F, Yin ZY (2019) From micro scale to boundary value problem: using a micromechanically based model. Acta Geotech 14:1307–1323
68. He X, Wu W, Cai G, Qi J, Kim JR, Zhang D, Jiang M (2020) Work–energy analysis of granular assemblies validates and calibrates a constitutive model. Granular Matter 22:28
69. Wang P, Arson C (2018) Energy distribution during the quasi-static confined comminution of granular materials. Acta Geotech 13:1075–1083
70. Rothenburg L, Bathurst RJ (1989) Analytical study of induced anisotropy in idealized granular-materials. Geotechnique 39:601–614
71. Hu LM, Pu JL (2004) Testing and modeling of soil-structure interface. J Geotech Geoenviron Eng 130:851–860
72. Huang X, Hanley KJ, O'Sullivan C, Kwok CY (2014) Exploring the influence of interparticle friction on critical state behaviour using DEM. Int J Numer Anal Met 38:1276–1297

73. Cheung G, O'Sullivan C (2008) Effective simulation of flexible lateral boundaries in two- and three-dimensional DEM simulations. Particuology 6:483–500
74. Zhu HX, Zhou WH, Jing XY, Yin ZY (2019) Observations on fabric evolution to a common micromechanical state at the soil-structure interface. Int J Numer Anal Meth Geomech 43:2449–2470
75. Fleischmann JA, Plesha ME, Drugan WJ (2013) Quantitative comparison of two-dimensional and three-dimensional discrete-element simulations of nominally two-dimensional shear flow. Int J Geomech 13:205–212

Chapter 5
Three-Dimensional Analysis of Soil-Structural Interface

5.1 Introduction

From a microscopic viewpoint, the macro-behavior of the granular material is associated with the evolution of the material fabric due to the movement, rotation, and breakage of soil particles [1–6]. The microphysics underlying the macroscopic phenomena of the granular material must be evaluated under interface shearing. Moreover, fabric anisotropy has been observed in granular material when subjected to shearing [7–9], resulting in the changes of macroscopic mechanical behavior of the granular material. In this sense, the fabric anisotropy of granular material plays a significant role in its macro-response. Several types of fabric tensors are proposed to mathematically describe the distribution of contact orientation [10–12]. In this chapter, the fabric anisotropy of soil caused by interface shearing is also discussed to determine the relationship with the rotation of the major principal stress.

The discrete element method [13] has been used to explore the mechanical behaviors of two-dimensional (2D) granular assemblies under interface shearing [14–17]. However, the macro- and micro-mechanical responses of the three-dimensional granular soil may vary when an additional dimension is added to the system. For instance, Rothenburg and Bathurst [18] revealed that in a drained biaxial loading path, a 2D granular assembly dilated at a greater rate than did real sand. This result was observed because in a 2D scenario, the specimen can only expand in one direction. Bathurst & Rothenburg [19] found that Poisson's ratio of a 2D assembly was different from that of a 3D assembly. In this context, it is essential to explore the mechanical behaviors of 3D granular systems. In the literature, some numerical simulations have investigated the interactions between 3D granular soils and geogrids or geotextiles using pullout tests [20–22]. However, 3D numerical simulation of soil-structural interface shear test (IST) has not been sufficiently discussed, and the relationship between the localized band and interface roughness is not fully known. These problems deserve further investigation and discussion.

As shearing progresses, the shear strain tends to be localized in a narrow zone. This zone is named the localized band, which develops and deforms during the shearing

W.-H. Zhou and Z.-Y. Yin, *Practice of Discrete Element Method in Soil-Structure Interface Modelling*, https://doi.org/10.1007/978-981-19-0047-1_5

process [23–27]. After the localized band appears, the granular assembly shows a discontinuous feature. The homogeneity of the material has been broken. To this extent, it is necessary to distinguish between the soil inside the localized band and all other soil regions. DeJong et al. [26] proved that the thickness of the localized band depends on the roughness of the interface using particle image velocimetry (PIV). The thickness of the localized band should be identified under various loading and modelling conditions.

In this chapter, a series of 3D interface shear tests (ISTs) with different roughnesses were modeled using PFC3D (Itasca Consultants, Inc.). This chapter analyzes three aspects of the macro- and micro-mechanical features of the soil-structural interface: (1) the effect of the artificial interface roughness on the shear strength and volumetric change of soil at the interface, (2) the formation of a localized band and the correlation between the thickness of the localized band and the roughness of the interface, and (3) the evolution of the material fabric in the localized band in terms of both its anisotropy and rotation. The numerical modeling is introduced in Sect. 5.2. The scale effect of the interface shear box is discussed in Sect. 5.3, and an appropriate dimension is proposed. The results are presented and discussed in Sect. 5.4 in terms of the macroscopic response, the evolution of the micro-structure and the material fabric.

5.2 DEM Simulation

5.2.1 Model of the Interface Shear Test

A numerical IST was conducted using the discrete element program PFC3D (Itasca Consultants, Inc.). This program uses an explicit scheme for analysis, and each calculation cycle includes two steps: (1) using Newton's second law to determine the particle movement resulting from the contact force and (2) updating the contact network by applying the contact model between particles to calculate the contact force and relative displacement. A local damping scheme was applied on each ball to dissipate additional kinetic energy. The magnitude of the damping force was proportional to the unbalanced force taking the damp coefficient as the slope.

As shown in Fig. 5.1a, this shear box is composed of six rigid walls. Four vertical walls are fixed, the top wall can only move in the vertical direction, and the bottom wall can only move horizontally. Although random interface geometry exists in nature, in this chapter, a standard sawtooth pattern is adopted to represent the rough surface. A front view of the rough surface is shown in Fig. 5.1b. The inclination of the saw teeth alternates between 45° and 135°, while the span of the sawtooth plates in the x-direction is λ. The normalized roughness index R_n [28] is used to quantify the roughness of the rough plates, which is equal to $\lambda/2d_{50}$, where d_{50} is the mean particle diameter. As listed in Table 5.1, six standard sawtooth plates with various R_n values are used in the tests.

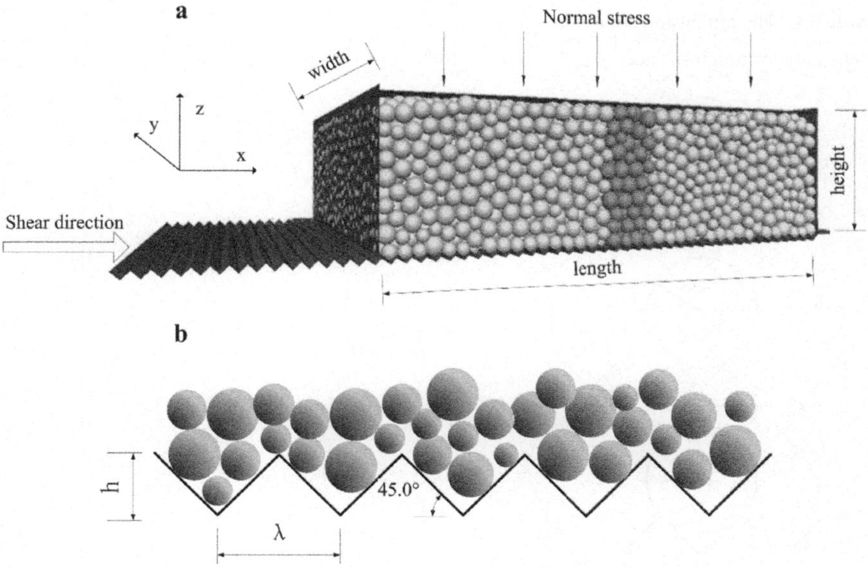

Fig. 5.1 The PFC3D model: **a** schematic diagram of an interface shear test and **b** geometry of the rough surface

5.2.2 Interaction Law

In conventional DEM analyses, free rolling is allowed without any resistance at the contact. However, a rolling resistance moment exits when a particle rolls over another due to the irregular particle shape or particle surface roughness of the real grains. Several types of rolling resistance contact models are proposed and applied in both 2D and 3D simulations [29–32]. The interaction law used in the chapter is briefly introduced here.

Two relative movements occur at a contact, sliding and rolling. The force–displacement relationship between inter-particles follows the elastic contact law. The contact forces increase linearly with the relative displacement between contacting particles according to the normal stiffness k_n and tangential stiffness k_t until sliding occurs when the tangential contact force reaches the Coulomb limit. The k_n between two contacting elements with radii r_1 and r_2 is calculated through Young's modulus E by Eq. (5.1):

$$k_n = \pi r^2 E/L \tag{5.1}$$

with $L = \begin{cases} r_1 + r_2, \textit{ball} - \textit{ballcontact} \\ r_1, \textit{ball} - \textit{facetcontact} \end{cases}$ and $r = \begin{cases} \min(r_1, r_2), \textit{ball} - \textit{ballcontact} \\ r_1, \textit{ball} - \textit{facetcontact} \end{cases}$

Table 5.1 The dimensions of distinct rough plates

Schematic of rough surface	Number of rough plates	λ(mm)	R_n
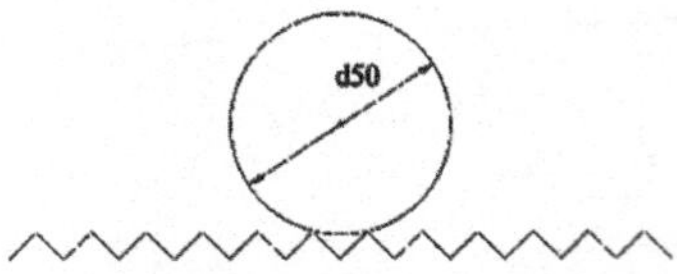	No. 1	0.87	0.125
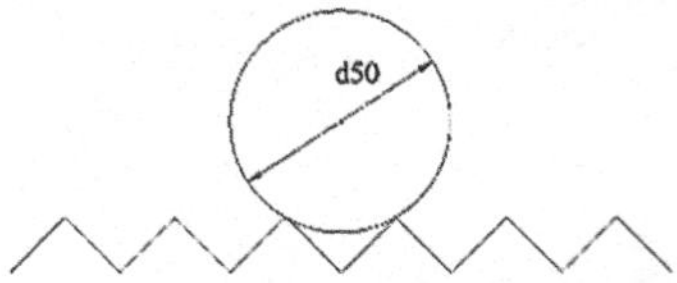	No. 2	1.74	0.25
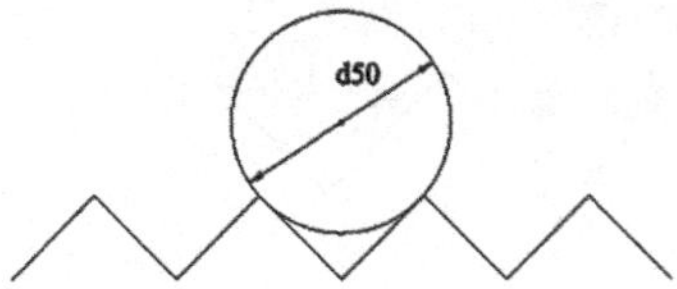	No. 3	2.61	0.375
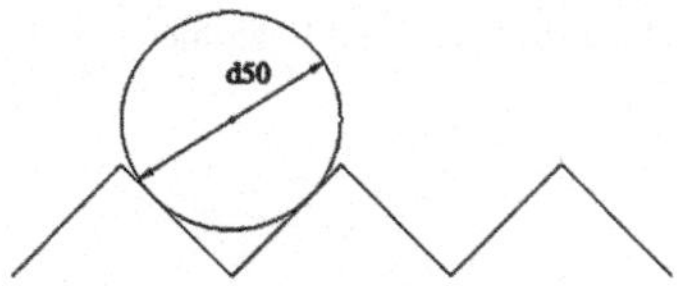	No. 4	3.48	0.5
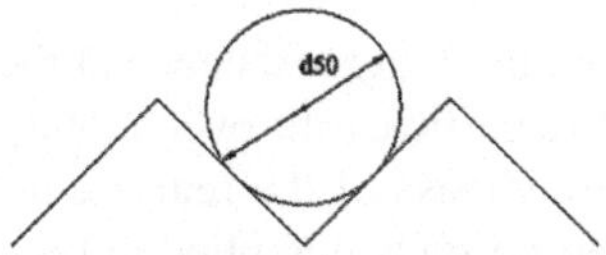	No. 5	5.22	0.75
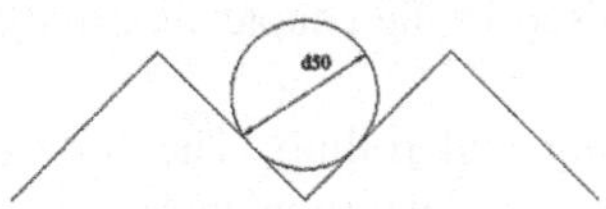	No. 6	6.96	1.0

The rolling leads to a relative rotation between two contacting particles. The rolling resistance moment is denoted by M_r. The incremental rolling resistance moment ΔM_r is calculated from the rolling stiffness k_r [29] and the incremental relative rotation $\Delta\theta_r$ between the two particles:

$$\Delta M_r = -k_r \Delta\theta_r \tag{5.2}$$

$$k_r = k_t \bar{r}^2 \tag{5.3}$$

where k_t is the tangential stiffness and $\bar{r}$ is the effective contact radius equaled to $r_1r_2/(r_1+r_2)$. A threshold value of M_r is denoted by M^*, which is defined as follows:

$$M^* = \mu_r \bar{r} F_n \tag{5.4}$$

where F_n is the current normal contact force and μ_r is the rolling resistance coefficient. As discussed in [31], the rolling resistance coefficient μ_r corresponds to the tangent of the maximum angle of a slope on which a rolling resistance moment counterbalances the torque produced by gravity acting on the body.

5.2.3 *Simulation Process and Measurement*

The input parameters are summarized in Table 5.2. Experimental comparisons for the input parameters have been presented in the study of [33]. The spherical particles were generated inside the shear box, with diameters that uniformly disperse in a given range. The radius of each particle increased gradually inside a fixed space, and the inter-particle friction coefficient μ_p was zero in this phase to create a dense specimen [34]. After the granular assembly was prepared, a constant vertical force was applied to the top wall and maintained as a constant using a servo system. Then, the bottom plate started to move horizontally in the x-direction at a steady rate v_x to induce a quasistatic interface shearing. The macroscopic behaviors of specimens were measured from the boundaries of the shear box. These behaviors included the shear stress τ on the interface, the normalized shear displacement d_s/d_{50} of the bottom plate, the normal stress σ_n on the top plate, and the relative vertical displacement Δh, which was defined as $\Delta h = h_{instant} - h_{initial}$, where $h_{instant}$ was the instant height of the top wall and $h_{initial}$ was the initial height of the sample before shearing.

Table 5.2 Summary of input parameters used in DEM simulations

Parameters	Value
Range of particle's diameter (mm)	2.464 ~ 4.032
d_{50}(mm)	3.48
$C_u(d_{60}/d_{10})$	1.2
Density (kg/m^3)	2880
Young's modulus (MPa)	50
Normal stiffness of the wall k_{nw} (N/m)	9.0*10^{10}
k_n/k_t	2
μ_p	0.35
Particle–wall friction coefficient μ_{pw}	0.2
Damp coefficient	0.7
μ_r	0.2
v_x(mm/s)	0.015

Table 5.3 Summary of tests of different shear boxes

No of shear box	Length (mm)	Length/d_{50}	Width (mm)	Width/d_{50}	Height (mm)	Height/d_{50}	Number of particles	Normal stress (kPa)
IST 1	100	30	100	30	25	8	13,359	40/60/80/100
IST 2	120	35	100	30	30	9	19,267	40/60/80/100
IST 3	140	41	100	30	35	10	26,212	40/60/80/100
IST 4	200	58	100	30	50	14	53,504	40/60/80/100

5.3 Determining the Dimension of the Shear Box

The dimension of the shear box may affect the soil's mechanical properties in terms of shear strength, volumetric change, and peak friction angle [35–37]. The original purpose of this section is to choose an appropriate dimension of the shear box for the numerical modeling to ensure a valid result and an affordable computational cost. A limited study of scale effect has been done here. Four interface shear boxes with different lengths (L) and heights (H) were designed and named IST 1, IST 2, IST 3, and IST 4. The same ratio of H/L was kept. The previous research works proved that the behavior of soil-structural interface is mainly attributed to the localized band and the common thickness of the localized band is 5 times d_{50} [23, 38], so we chose 8–10 times d_{50} which exceeded the thickness of the localized band. Sixteen ISTs with $R_n = 0.5$ are listed in Table 5.3, which gives the dimensions of the shear boxes.

The similar initial porosities n_0 were controlled for all specimens (Fig. 5.2a). The peak shear stresses τ_p measured under the same confining stress approach each other (Fig. 5.2b). Moreover, the difference in the vertical displacement of the specimen is not significant (Fig. 5.2c). To maintain a reasonable computational efficiency, IST 1 (100 mm × 100 mm × 25 mm) is chosen for the following simulations.

5.4 DEM Modeling Results and Discussion

5.4.1 *Macro-Mechanical Behavior of the Interface*

The evolutions of τ during the loading process are plotted in Fig. 5.3 for the ISTs with various values of R_n under σ_n of 80 kPa. The peak shear stress measured in the IST with a smooth interface is 20 kPa, which approaches the value of $\mu_{pw} \times \sigma_n$. It suggests that for smooth case, the soil particles at the bottom tend to slide along the interface. Thus, the shear resistance is mainly attributed to the friction between the bottom particles and the interface. Periodic oscillation is observed in the curve of τ during the residual stress phase when R_n is equal to 0.25. In this case, the soil particles in the bottom layer cannot fit in such small volumes between sawteeth (Table 5.1). Thus,

Fig. 5.2 Effect of the shear box dimension on: **a** the initial porosity n_0 of the specimen; **b** the peak shear stress τ_p measured at the interface; and **c** the vertical displacement of the top wall under $\sigma_n = 80$ kPa

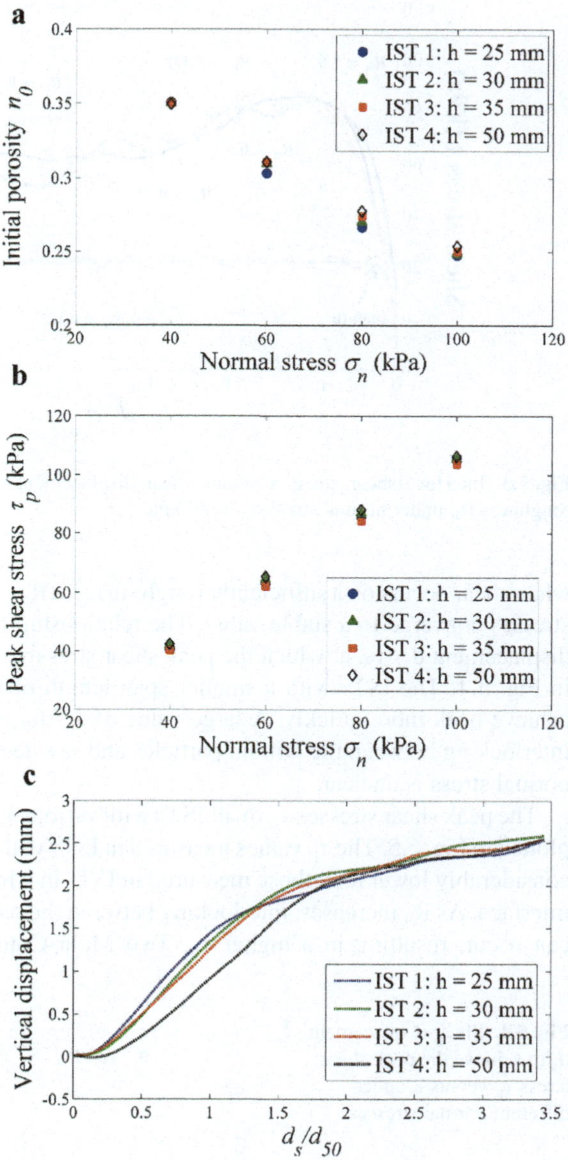

the bottom layer of particles moves to the tops of teeth and the areas between teeth alternately, which results in periodic oscillation in the total contact number between the particles and interface, and induce this kind of evolution of τ. By contrast, when $R_n > 0.25$, the bottom particles can be trapped in the areas between sawteeth and move with the surface. Two deformation patterns are observed in ISTs with different R_n values: (1) an "elastic-perfectly plastic" pattern in tests with a relatively smooth surface ($R_n < 0.375$) and a "stress softening" pattern detected in granular material

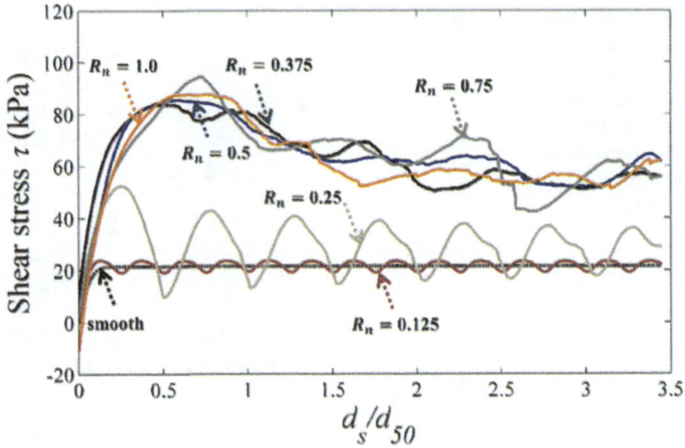

Fig. 5.3 Interface shear stress τ versus shear displacement of ISTs with different normalized roughness R_n under normal stress $\sigma_n = 80$ kPa

when shear occurs on a sufficiently rough surface ($R_n \geq 0.375$) and the residual shear stresses converge to a stable value. The relationship between the normalized shear displacement d_s/d_{50} at which the peak shear stress τ_p is attained and λ is illustrated in Fig. 5.4. The ISTs with a smaller span length of the saw-teeth, i.e., smaller λ, achieve peak more quickly. A large value of d_s/d_{50} is required to fully disturb the interlocking between the bottom particles and saw-teeth with larger λ. The effect of normal stress is unclear.

The peak shear stresses τ_p of all ISTs with various R_n values under different σ_n are plotted in Fig. 5.5. The τ_p values measured in ISTs with relative smooth interfaces are considerably lower than those measured in ISTs in which shearing occurs at a rough interface. As R_n increases, interlocking between the particles and the rough interface can occur, resulting in a higher τ_p. Two Mohr-Coulomb parameters, cohesion c

Fig. 5.4 Shear displacement d_s to achieve the peak shear stress τ_p versus λ under different normal stresses

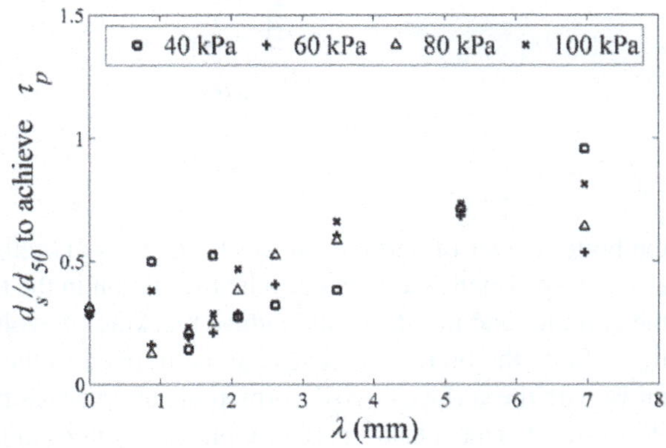

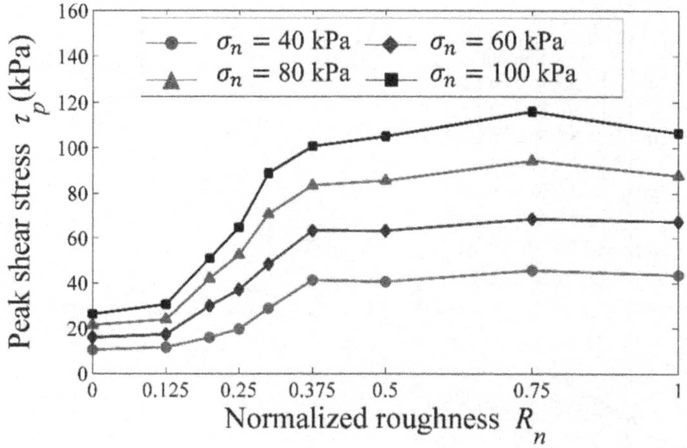

Fig. 5.5 Peak interface shear stress τ_p versus normalized roughness R_n under different normal stresses σ_n

and peak internal friction angle ϕ_p are obtained, as shown in Fig. 5.6a. For the cohesionless soil in this chapter, c is set to zero. As a reference, a direct shear test (DST) with the same input parameters was modeled. The height of the interface shear box was twice of the specimen in IST. The ϕ_p measured in DST is 52.97 °. Figure 5.6b illustrates that ϕ_p values in the ISTs vary with R_n. Notably, when R_n increases from 0.125 to 0.375, the increase in ϕ_p is almost 30 °. The shear strength measured on interface is reinforced by the rough surface when $R_n \geq 0.375$ and gradually approaches (still less than) the shear strength of the granular material. The above results illustrate the correlation between the shear strength of the soil-structural interface and R_n.

The curves of vertical displacement of the top wall versus d_s/d_{50} of ISTs with various R_n under σ_n of 80 kPa are illustrated in Fig. 5.7. It shows that the volumetric change of the granular material subjected to interface shear is associated with R_n. Notably, the curves of ISTs with sufficiently small R_n values ($R_n = 0.125$) approach zero, as do the curves of smooth interfaces. In these cases, the soil particles tend to slide along the surface during shearing, which results in a negligible volumetric change. The granular assembly intensely dilates when shear occurs at a rough interface ($R_n \geq 0.375$), especially at high values of σ_n. Moreover, the rate of dilatancy decreases at a d_s/d_{50} of approximately 1.0 and gradually converges to zero. This dilatancy in the bulk material suggests that a dense, stable specimen deforms into a loose, unstable specimen that cannot sustain such a higher shear stress, resulting in stress softening. The above results demonstrate that a critical value of R_n exists. This value is 0.375 in this chapter. The macro-behaviors of ISTs with $R_n \geq 0.375$ are quite different from those of ISTs with relative smooth surfaces in terms of the shear strength and volumetric change.

As shown in Fig. 5.8, the macroscopic behaviors of numerical ISTs with different R_n are similar to the phenomena observed from sand-steel interface experiments

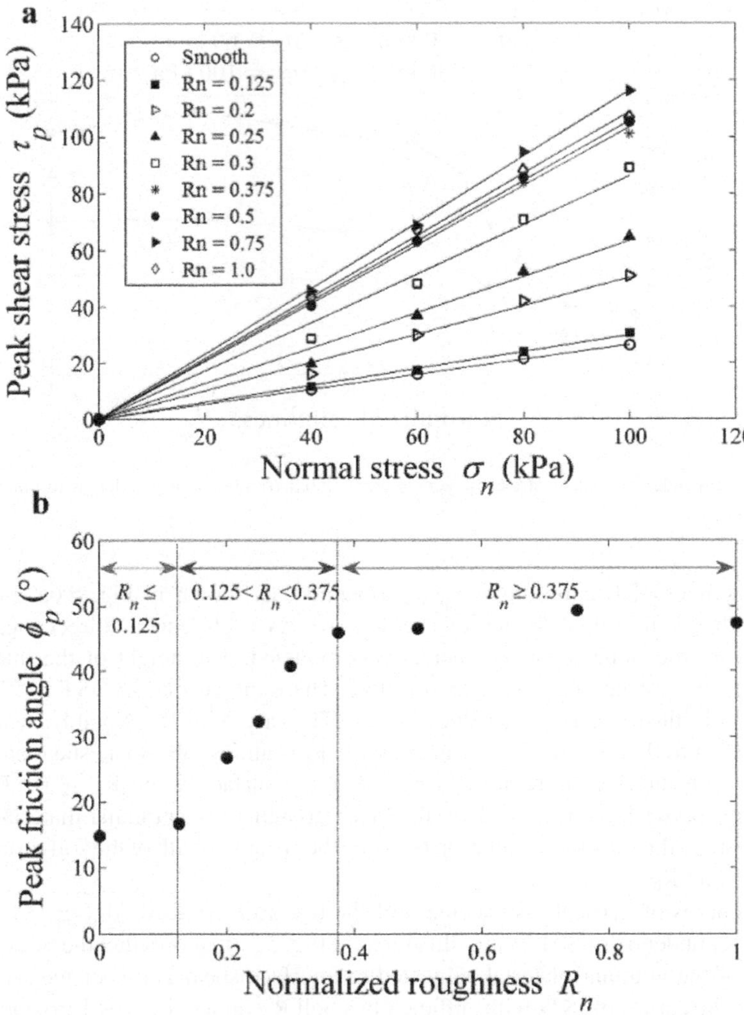

Fig. 5.6 Analysis of interface shear strength parameters: **a** fitting data according to the Mohr–Coulomb criterion, and **b** the variation in the peak interface friction angle ϕ_p with R_n

[38]. The peak shear stress ratio defined as τ_p/σ_n is compared with the one in DEM simulation. A similar increasing tendency of τ_p/σ_n as a function of R_n is observed. The different growth rate is probably induced by the dissimilar geometry of rough interface and soil properties used in laboratory experiments and DEM simulations. The numerical results also correspond to the classification of interfaces with different geometries and roughnesses proposed by Paikowsky et al. [39]. Based on the results of their laboratory experiments, interfaces are generally classified into three categories according to the shear strength measured at the interface: (1) smooth ($R_n < 0.02$), intermediate ($0.02 \leq R_n \leq 0.5$), and (3) rough ($R_n > 0.5$). However, to quantitatively

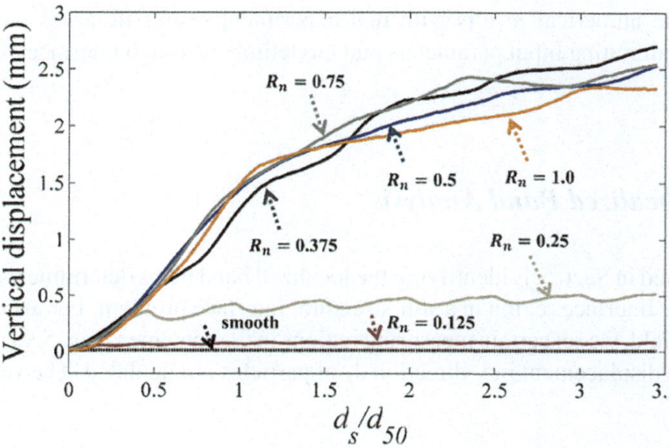

Fig. 5.7 **a** Evolution of shear stress of interface shear test with different R_n [38], and **b** comparison of peak shear stress ratio τ_p/σ_n with DEM simulation

Fig. 5.8 Vertical displacement of the top wall versus normalized shear displacement d_s/d_{50} of ISTs with different normalized roughness R_n under normal stress $\sigma_n = 80$ kPa

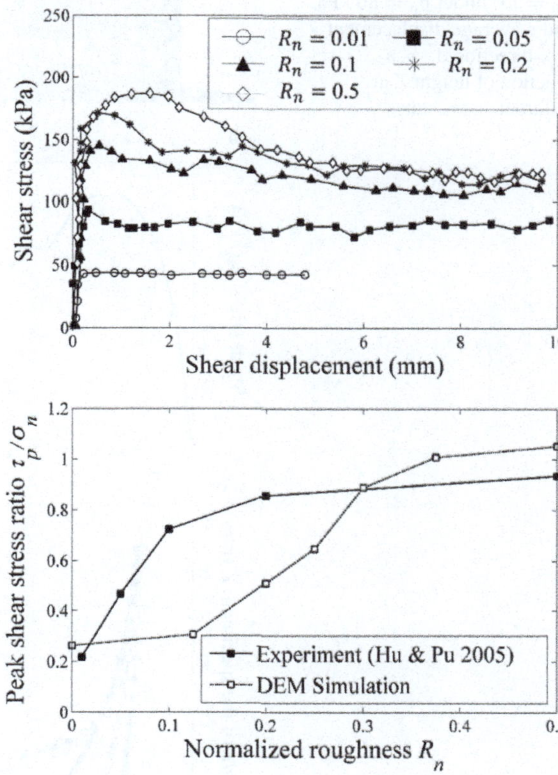

compare the numerical results with real experiment results deserves more study, including calibrating input parameters and modelling the rough interface of the same geometry.

5.4.2 Localized Band Analysis

As mentioned in Sect. 5.1, identifying the localized band helps determine the primary scope of the interface region in a soil-structural interface problem. For averaging the kinematic field, we set certain measuring regions in the specimen (Fig. 5.9a), in which the average displacement in x-direction $\overline{d_x}$ of particles is calculated. The values of $\overline{d_x}$

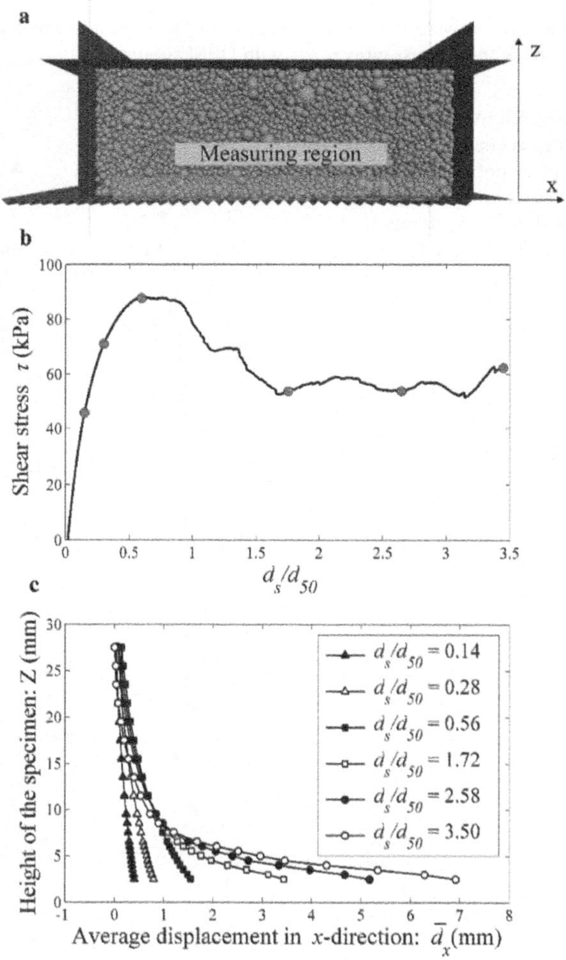

Fig. 5.9 **a** Measuring region set-up in the specimen; **b** shear stress versus shear displacement of IST with $R_n = 1.0$ under $\sigma_n = 80$ kPa, and **c** average displacement in x-direction $\overline{d_x}$ as a function of height Z at different stress states

as a function of Z at different shearing states (Fig. 5.9b) are plotted in Fig. 5.9c. Each dot in the figure represents one measurement at a given height Z. With the increasing of the shear stress, $\overline{d_x}(Z)$ gradually enlarges its non-linearity and becomes stable at the steady stress state. The bottom particles close to the interface show a larger displacement than the upper ones. The evolution of $\overline{d_x}(Z)$ illustrates the formation of a localized band. This is consistent with the result reported by Wang et al. [27] regarding the formation of the localized band in 2D numerical ISTs. In their study, a narrow zone with an intense shear strain emerged above the rough surface around the peak stress state.

The inflection point of the curve of $\overline{d_x}(Z)$ at the steady stress state (when $d_s/d_{50} = 3.5$) is used to define the thickness of the localized band. Spline interpolation is applied to get a smooth $\overline{d_x}$ - Z curve f(Z). The first derivative $f'(Z)$ and the second derivative $f''(Z)$ are calculated using the finite difference method. The curvature κ of f(Z) is calculated by

$$\kappa = \frac{|f''(Z)|}{(1 + f'(Z)^2)^{3/2}} \tag{5.5}$$

The value of κ approaches zero at the higher position of the specimen. It means that the $\overline{d_x}$ changes approximately linearly with the value of Z in that region. With the decreasing of Z, the value of κ sharply increases at a certain value of Z. The inflection point of the κ is considered as a sign of the top boundary of the localized band. The authors suggest that where the κ equals 0.02 is the inflection point.

The normalized average shear displacement $\overline{d_x}/d_{50}$ and normalized average vertical displacement $\overline{d_v}/d_{50}$ measured in ISTs have been compared with the experimental results performed by [38] in Fig. 5.10. They are performed under the same confining stress (200 kPa) with similar normalized roughness R_n. The numerical data are obtained at a steady shear stress state ($d_s/d_{50} = 10$) as well as those obtained

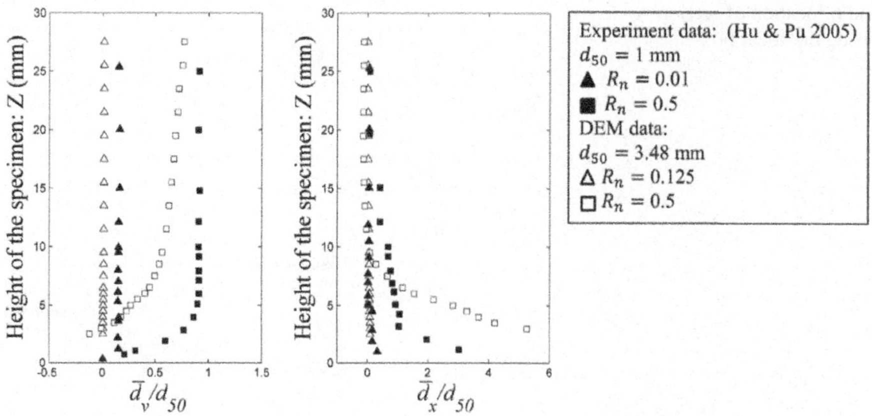

Fig. 5.10 Comparison of particle displacements from DEM simulations with laboratory data

in experiments. It shows that the numerical results match with the laboratory data. It is seen that both numerical and laboratory curves show an inflection point which can be used to identify the localized band. The thicknesses of the localized band measured in the DEM simulation and experiment test are 3.4 times d_{50} and 5 times d_{50} respectively by applying the curvature analysis. It is found that the thicknesses of the localized band δ_h are close in the two cases. Therefore, it can be concluded that the comparison, to a certain degree, validates the DEM model and demonstrates that the DEM model can reproduce a similar realistic interface shearing behavior. It shall be noted that the dissimilar features for the DEM and laboratory experiment data are due to many factors, such as the different d_{50}, soil properties, friction behavior, and geometry of interface. On the other hand, the thicknesses of the localized band measured in our 3D simulations are around 4~5 times d_{50}, which are thinner than the ones measured in 2D simulations (8~10 times d_{50}) performed by Wang et al. [27]. This might be because the particles restricted in a plane tend to exhibit a larger dilation.

To better visualize the localized band inside the specimen, the spatial discretization approach was used to compute the field of displacement. At a given strain state, the movement of the centroid of each particle $\vec{u}(x,\ y,\ z)$ was measured. The domain of the material was discretized by constructing a grid. Then, the displacement of nodes in the grid was interpolated according to the $\vec{u}(x,\ y,\ z)$ values of nearby particle centroids using a linear interpolation function (Fig. 5.11a). Accordingly, the 3D fields of $\vec{u}(x,\ y,\ z)$ can be obtained. The cross-section A in the 3D field, located in the center of the shear box, is chosen to visualize the displacement field (Fig. 5.11b).

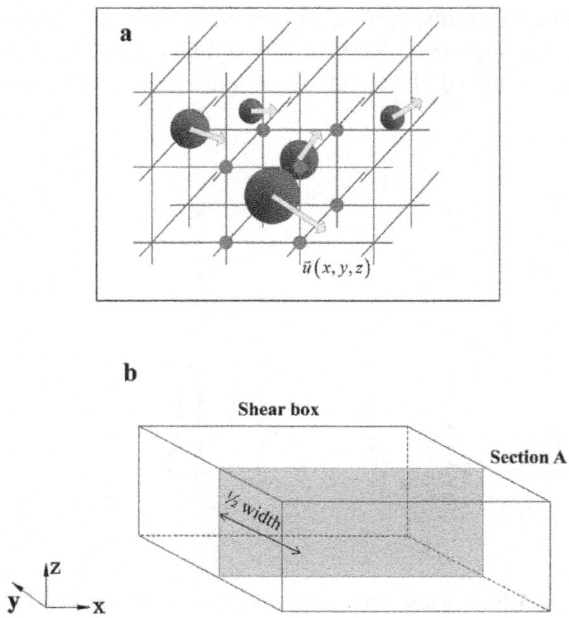

Fig. 5.11 **a** Schematic diagram of interpolation on a three-dimensional grid according to the particle displacement $\vec{u}\,(x,\ y,\ z)$, and **b** location of section A inside the shear box

The effects of R_n and σ_n on the thickness of the localized band will be discussed. The $\overline{d_x}$ – Z curves and corresponding fields of u_x measured at a steady stress state ($d_s/d_{50} = 3.5$) for ISTs ($\sigma_n = 80$ kPa, $R_n = 0.125/0.3/0.75/1.0$) are compared in Fig. 5.12. Displacement fields in the x-direction are colored according to the values of u_x. Note that the boundary effect results in an arc-shaped localized band as the u_x values of the bilateral particles are restricted by the vertical boundaries. A few bottom particles are trapped in sawtooth valleys and move almost the same distance as the interface. The u_x values of all the particles are approximately zero when R_n is equal to 0.125. This supports the argument in Sect. 5.4.1 that only a few particles in the bottom layer slide at the interface, resulting in a negligible volumetric change. A large-scale localized band is only observed when $R_n > 0.3$. A thicker localized band develops if the granular material shears on a rougher surface.

The thickness of the localized band δ_h is normalized by the mean particle diameter (δ_h/d_{50}). As illustrated in Fig. 5.13, the δ_h/d_{50} increases with increasing R_n and tends to be stable after R_n exceeds 0.375. The specimen under small normal stress tends to form a thicker localized band, because the material dilates more under a lower confining stress. Although this tendency does not strictly occur in every R_n, it is generally valid. The maximum difference of δ_h is around 1.5 times d_{50} and occurs when $R_n = 0.75$. Figure 5.13 reminds us the evolution of ϕ_p as a function of R_n (Fig. 5.6b). It suggests that δ_h/d_{50} is associated with the shear strength at the interface. There is a transition process of the failure location inside the material subjected to an interface shearing. For the smooth case, the failure occurs at the interface due to the sliding between the bottom particles and the interface. With the increasing of R_n, a few layers of particles adjacent to the interface would gradually form a localized band. Thus, the failure would shift from the interface into the soil layer.

5.4.3 Local Average Stress Analysis

As discussed in Sect. 5.4.2, the granular material becomes heterogeneous after the localized band is formed during the shearing process. The granular material is stratified into two regions: the upper zone and the interface zone (Fig. 5.14a). The depth of the boundary between the two regions in this chapter is δ_h. The average stress tensor $\overline{\sigma}_{ij}$ of the granular assembly is proposed based on the microscopic quantities given by Christoffersen et al. [40]:

$$\overline{\sigma}_{ij} = -\frac{1}{V}\sum_{N_c}\frac{f_i^{\alpha}l_j^{\alpha} + f_j^{\alpha}l_i^{\alpha}}{2}(i, j = x, y, z) \tag{5.6}$$

where V is the total volume of the measured space, N_c is the total contact number in the measure spaced, f_i^{α} and f_j^{α} are the contact forces at the contact α, and l_i^{α} and l_j^{α} are the branch vectors joining the centroids of the two particles at the contact α. The average principal stresses, ordered by their magnitudes, are $\overline{\sigma}_1, \overline{\sigma}_2$ and $\overline{\sigma}_3$. The average

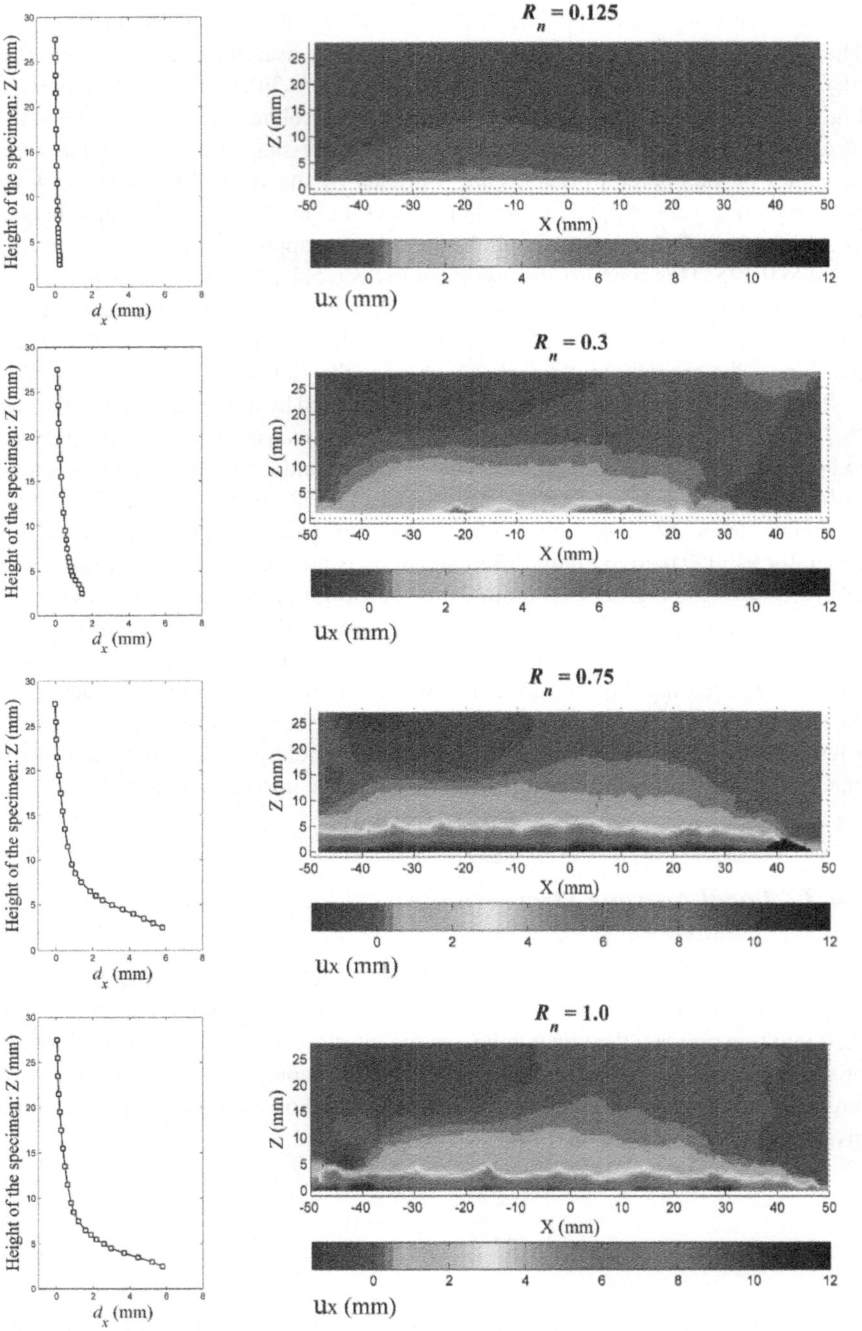

Fig. 5.12 The $\overline{d_x}$–Z curves and the shear displacement fields for the ISTs (σ_n = 80 kPa, R_n = 0.125/0.3/0.75/1.0)

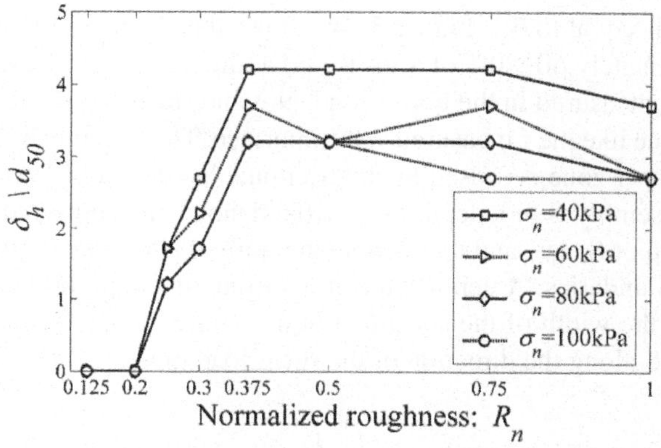

Fig. 5.13 Normalized thickness of the interface zone δ_h/d_{50} as a function of normalized roughness R_n under various normal stress σ_n

a

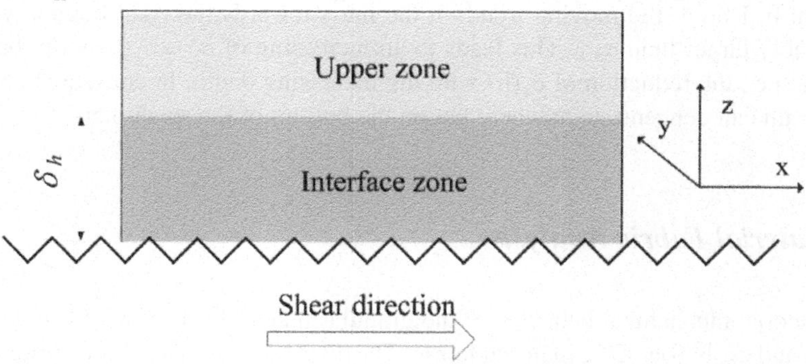

b

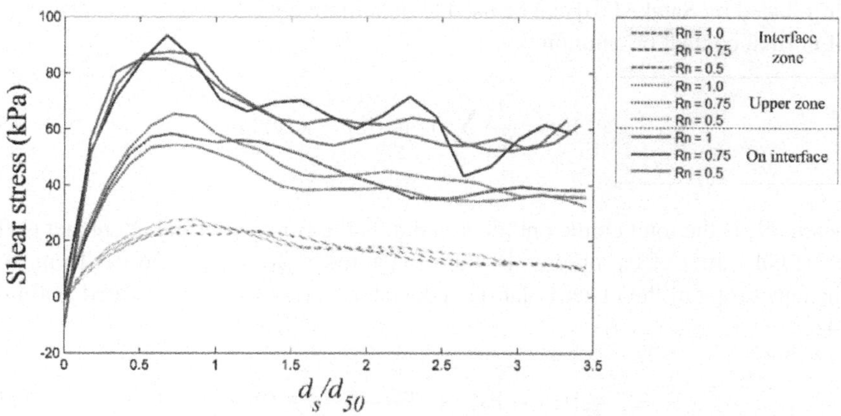

Fig. 5.14 Average local shear stress analysis: **a** the measured rectangular regions, and **b** average shear stress values measured in the interface zone and upper zone

shear stress $\overline{\sigma}_s$ is equal to $\overline{\sigma}_{xz}$. Figure 5.14b shows that $\overline{\sigma}_s$ measured in the interface zone is approximately 60~80% of τ measured at the interface, which is remarkably larger than that measured in the upper zone. Notable stress softening is observed in the interface zone like the τ measured at the interface. The different $\overline{\sigma}_s$ in the interface zone and the upper zone is caused by the frictional lateral walls. For a lower part of the granular assembly S in a depth of h, $\overline{\sigma}_s(h)$ denotes the subjected average shear stress on the top of S, $\sigma_{n,1}$ and $\sigma_{n,2}$ denote the normal stress respectively on the left and the right boundaries, A denotes the area of the upper and bottom section of S, and w denotes the width of the specimen in the y-direction. Considering the force equilibrium of S along the direction of the shear load gives:

$$(\overline{\sigma}_s - \overline{\sigma}_s(h))A = \int_0^h \int_0^w (\sigma_{n,2} - \sigma_{n,1})dydz \quad (5.7)$$

At a given state, the average shear stress at the bottom of S ($\overline{\sigma}_s$) is a constant to the height h. Due to the moving trend on the mass towards the right boundary, $\sigma_{n,2}$ is arguably larger than $\sigma_{n,1}$. This leads to an increasing of $\overline{\sigma}_s - \overline{\sigma}_s(h)$ with the increasing h, i.e., the reduction of $\overline{\sigma}_s(h)$ with the increasing depth. In consequence, the shear strain can concentrate nowhere but on the bottom of the specimen.

5.4.4 *Material Fabric Analysis*

The macroscopic mechanical behavior of the granular material originates from the distribution and evolution of the material fabric. The distribution of the contact orientation is commonly used to describe the fabric of a material. A second-order tensor F_{ij} introduced by Satake [10] can be used to quantitatively characterize the distribution of normal contact orientation:

$$F_{ij} = \frac{1}{N_c}\sum_{\alpha}^{N_c} n_i^{\alpha} n_j^{\alpha} (i, j = x, y, z) \quad (5.8)$$

where N_c is the total contact number and n_i is the contact normal at contact α. The principal values of F_{ij}, ordered by their magnitudes, are F_1, F_2 and F_3. To measure the anisotropy of the material fabric, a deviator fabric of F_{ij} is calculated as follows [41].

$$\delta_D = \frac{1}{\sqrt{2}}[(F_1 - F_2)^2 + (F_2 - F_3)^2 + (F_1 - F_3)^2]^{0.5} \quad (5.9)$$

The coordination number C_n is defined as the average contact per particle:

$$C_n = (\sum_{N_p} n_c^p)/N_p \tag{5.10}$$

where N_p is the number of particles in the measured region and n_c^p is the contact number of particle p in the measured region.

The evolutions of porosity n and C_n of the IST ($R_n = 0.75$) measured in the interface zone and upper zone are compared. The soil at the interface gradually deforms from dense to loose (Fig. 5.15a). At the particle scale, a continuous loss of contact is detected in the interface zone (Fig. 5.15b). This leads to the dilatancy of the entire specimen. By contrast, a stable structure is preserved in the upper zone, in which the steady values of n and C_n are measured. A minor difference in the initial C_n and n in the two regions is observed, which implies the heterogeneity of the specimen caused by the normal pressure. In the conventional laboratory interface shear experiment, a normal stress is applied on the top wall of the shear box before shearing. Correspondingly in the DEM tests, a vertical movement is applied on the top wall to produce such a normal pressure. This procedure induces the initial heterogeneity along the z-direction of the specimen, accordingly slight differences in the porosity at the different heights of the specimen.

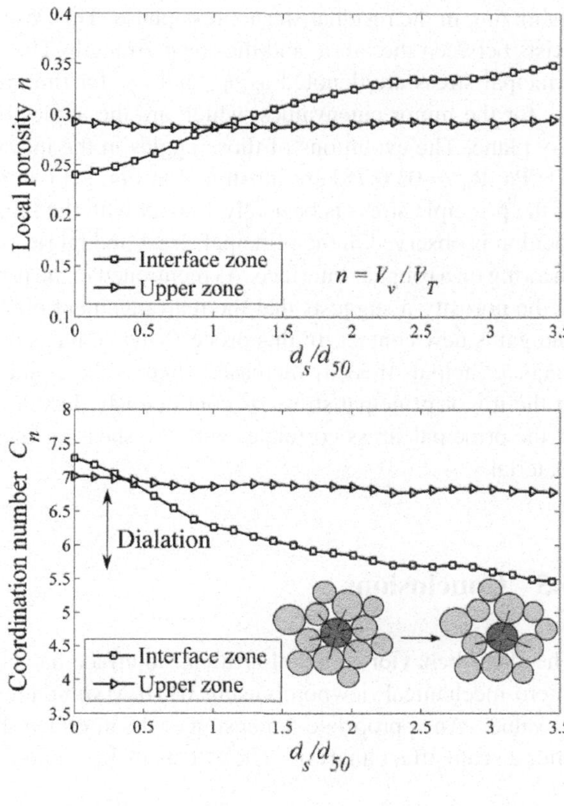

Fig. 5.15 Measurements in the interface zones and the upper zones of ISTs with $R_n = 0.75$: **a** porosity n versus shear displacement; **b** coordination number C_n versus shear displacement

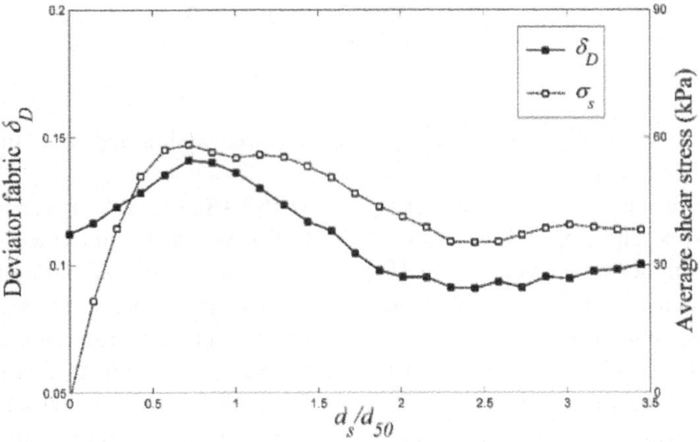

Fig. 5.16 The evolution of deviator fabric δ_D and average shear stress $\overline{\sigma_s}$ in the interface zone of the IST with $R_n = 0.75$ under $\sigma_n = 80$ kPa

The evolutions of deviator fabric δ_D in the interface zone of the IST with R_n equals 0.75 (Fig. 5.16) illustrated that the evolution trends of δ_D is consistent with the $\overline{\sigma}_s$, which increases with increasing d_s/d_{50} and decreases during stress softening before stabilizing in the residual shear stress phase. This result implies that a link possibly exists between the fabric and the shear strength. The orientations of the fabric and principal stress are denoted as θ_{F1} and θ_{S1} for the major eigenvalues and θ_{F3} and θ_{S3} for the minor eigenvalues, which are the angles between the eigenvectors and x–y plane. The evolutions of those angles in the interface zone for the ISTs (σ_n = 80 kPa, R_n = 03/0.75) are illustrated in Fig. 5.17. It illustrates that the orientation of the principle stress is basically coaxial with the fabric tensor. A higher degree of rotation is observed in the principal stress and fabric eigenvalue when a specimen is shearing on a rougher interface. As mentioned in the previous paragraph, the increase in the porosity n suggests that the rearrangement process of particles loses contact and gains new contact. In this process, the contacts oriented in the direction of the major principal stress $\overline{\sigma}_1$ increase, whereas the contacts orientated in the direction of the minor principal stress $\overline{\sigma}_3$ continuously decrease. As a result, the orientation of the principal stress correlates with the shearing-induced fabric anisotropy of the material.

5.5 Conclusions

The shear behavior of a soil-artificial interface has been studied from macro- and micro-mechanical viewpoints using 3D DEM simulations of ISTs with various R_n and σ_n values. An appropriate dimension of the interface shear box has been determined after a scale effect analysis. The effects of R_n and σ_n on shear strength, volumetric

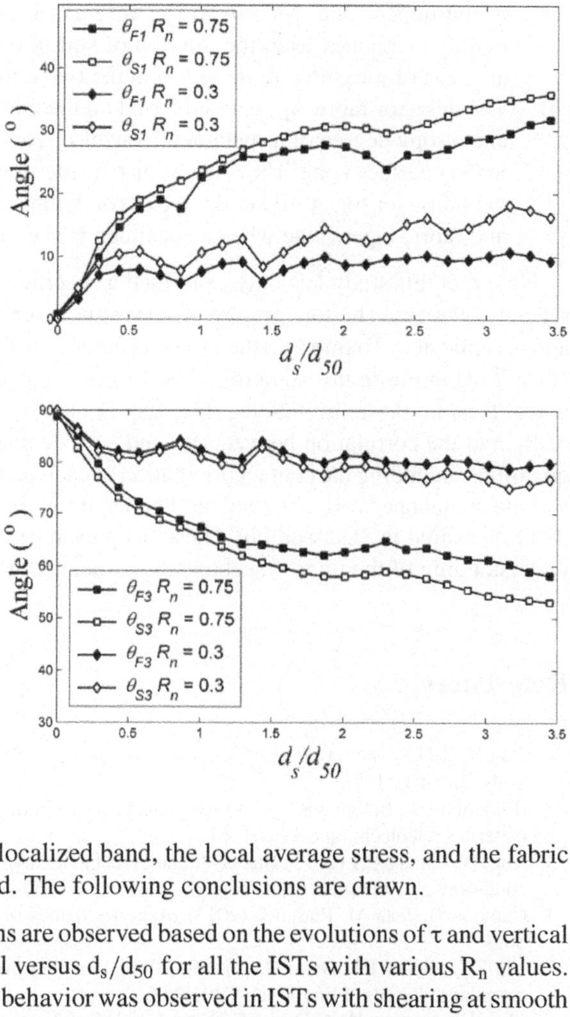

Fig. 5.17 The orientation evolutions of fabric and principal stress in the interface zone for the IST ($\sigma_n = 80$ kPa, $R_n = 0.3/0.75$)

changes, the thickness of the localized band, the local average stress, and the fabric anisotropy have been analyzed. The following conclusions are drawn.

(1) Two deformation patterns are observed based on the evolutions of τ and vertical displacement of top wall versus d_s/d_{50} for all the ISTs with various R_n values. Elastic-perfectly plastic behavior was observed in ISTs with shearing at smooth interfaces ($R_n < 0.375$), stress softening, and volumetric dilation was observed in ISTs with rough interfaces ($R_n \geq 0.375$).

(2) The shear strength of the interface, i.e., τ_p and ϕ_p, increases with increasing R_n and a plateau is reached when R_n is equal to 0.75. These findings suggest that the R_n plays a significant role in the macroscopic mechanical properties of the soil-structural interface.

(3) The $\overline{d_x}(Z)$ curves illustrate the formation of a localized band. The thicknesses of the localized bands δ_h is affected by R_n and σ_n and varies between 4 and 5 times the median grain size d_{50}. A thicker localized band is formed in ISTs with higher values of R_n.

(4) After the localized band appears, the granular material structuralizes into two regions: the interface zone and the upper zone. The local average shear stress $\overline{\sigma}_s$ measured in the interface zone is considerably higher than the value measured

in the upper zone. Moreover, the volumetric change of the entire specimen mainly originates from the dilation of soil in the interface zone based on the analysis of measured porosities n in the two regions.

(5) The deviator fabric δ_D, defined based on the material fabric tensor, reflects the anisotropy degree of a specimen. Anisotropy induced by shearing is observed in the interface zone. The orientation of principle stress is basically coaxial with the fabric tensor. A higher degree of rotation is observed in the principal stress and fabric eigenvalue when a specimen is shearing on a rougher interface.

Note that this study has only examined a specific type of artificial interface with regular sawteeth. The topography of a structural surface in nature is more random and complicated. To improve the model accuracy, different rough surfaces should be considered in future investigations. The effect of the grain shape could also be evaluated. Despite these limitations, this chapter clearly indicates the significant effect of R_n and the correlation between R_n and δ_h. The analysis according to the micro-quantities, including the contact orientation of each particle, the motion of the particle and the contact network obtained based on a 3D assembly, are more convincing than those measured in 2D assembly. These insights at the particle scale can improve our understanding of the micro-mechanisms associated with soil-interface shearing.

References

1. Oda M (1972) The mechanism of fabric changes during compressional deformation of sand. Soils Found 12:1–18
2. Rothenburg L, Bathurst RJ (1989) Analytical study of induced anisotropy in idealized granular materials. Géotechnique 39:601–614
3. Yin ZY, Chang CS (2013) Stress–dilatancy behavior for sand under loading and unloading conditions. Int J Numer Anal Methods Geomech 37:855–870
4. Cambou B, Jean M, Radjaï F (2013) Micromechanics of granular materials. John Wiley & Sons
5. Yin ZY, Hicher PY, Dano C, Jin YF (2016) Modeling mechanical behavior of very coarse granular materials. J Eng Mech C4016006
6. Zhu HX, Nguyen HNG, Nicot F, Darve F (2016) On a common critical state in localized and diffuse failure modes. J Mech Phys Solids 95:112–131
7. Oda M, Konishi J (1974) Microscopic deformation mechanism of granular material in simple shear. Soils Found 14:25–38
8. Calvetti F, Combe G, Lanier J (1997) Experimental micromechanical analysis of a 2D granular material: relation between structure evolution and loading path. Mech Cohesive-Frictional Mater 2:121–163
9. Chang CS, Yin ZY (2009) Micromechanical modeling for inherent anisotropy in granular materials. J Eng Mech 136:830–839
10. Satake M (1982) Fabric tensor in granular materials. In: Proc., IUTAM Symp. on Deformation and Failure of Granular materials. Delft, The Netherlands, pp 1525–1533
11. Kanatani K (1984) Distribution of directional data and fabric tensors. Int J Eng Sci 22:149–164
12. Yin ZY, Chang CS, Hicher PY (2010) Micromechanical modelling for effect of inherent anisotropy on cyclic behaviour of sand. Int J Solids Struct 47:1933–1951
13. Cundall PA, Strack ODL (1979) A discrete numerical model for granular assemblies. Geotechnique 29:47–65

14. Jensen RP, Besscher PJ, Plesha ME, Edil TB (1999) DEM simulation of granular media—structure interface: effects of surface roughness and particle shape. Int J Numer Anal Methods Geomech 23:531–547
15. Frost JD, DeJong JT, Recalde M (2002) Shear failure behavior of granular-continuum interfaces. Eng Fract Mech 69:2029–2048. https://doi.org/10.1016/S0013-7944(02)00075-9
16. Wang JF, Marte SG, Joseph ED (2007) Numerical studies of shear banding in interface shear tests using a new strain calculation method. Int J Numer Anal Methods Geomech 31:1349–1366
17. Wang JF, Jiang MJ (2011) Unified soil behavior of interface shear test and direct shear test under the influence of lower moving boundaries. Granul Matter 13:631–641
18. Rothenburg L, Bathurst RJ (1992) Micromechanical features of granular assemblies with planar elliptical particles. Géotechnique 42:79–95
19. Bathurst RJ, Rothenburg L (1988) Micromechanical aspects of isotropic granular assemblies with linear contact interactions. J Appl Mech 55:17–23
20. McDowell GR, Harireche O, Konietzky H, et al (2006) Discrete element modelling of geogrid-reinforced aggregates. In: Proceedings of the Institution of Civil Engineers-Geotechnical Engineering. pp 35–48
21. Bhandari A, Han J (2010) Investigation of geotextile-soil interaction under a cyclic vertical load using the discrete element method. Geotext Geomembranes 28:33–43
22. Tran VDH, Meguid MA, Chouinard LE (2013) A finite-discrete element framework for the 3D modeling of geogrid-soil interaction under pullout loading conditions. Geotext Geomembranes 37:1–9. https://doi.org/10.1016/j.geotexmem.2013.01.003
23. Uesugi M, Kishida H, Tsubakihara Y (1988) Behavior of sand particles in sand-steel friction. Soils Found 28:107–118
24. Hu L, Pu J (2004) Testing and modeling of soil-structure interface. J Geotech Geoenviron Eng 130:851–860. https://doi.org/10.1061/(ASCE)1090-0241(2004)130
25. Zhang G, Liang DF, Zhang JM (2006) Image analysis measurement of soil particle movement during a soil-structure interface test. Comput Geotech 33:248–259
26. Dejong JT, White D, Randolph MF (2006) Microscale observation and modeling of soil-structure interface behavior using particle image. Soils Found 46:15–28. https://doi.org/10.3208/sandf.46.15
27. Wang J, Gutierrez MS, Dove JE (2007) Numerical studies of shear banding in interface shear tests using a new strain calculation method. Int J Numer Anal Methods Geomech 31:1349–1366. https://doi.org/10.1002/nag.589
28. Uesugi M, Kishida H (1986) Frictional resistance at yield between dry sand and mild steel. Soils Found 26:139–149
29. Iwashita K, Oda M (1998) Rolling resistance at contacts in simulation of shear band development by DEM. J Eng Mech 124:285–292
30. Jiang MJ, Yu HS, Harris D (2006) Bond rolling resistance and its effect on yielding of bonded granulates by DEM analyses. Int J Numer Anal Methods Geomech 30:723–761. https://doi.org/10.1002/nag.498
31. Ai J, Chen JF, Rotter JM, Ooi JY (2011) Assessment of rolling resistance models in discrete element simulations. Powder Technol 206:269–282
32. Wensrich CM, Katterfeld A (2012) Rolling friction as a technique for modelling particle shape in DEM. Powder Technol 217:409–417
33. Jing XY, Zhou WH, Li YM (2017) Interface direct shearing behavior between soil and saw-tooth surfaces by DEM simulation. In: Procedia Engineering. Delft, The Netherlands, pp 36–42
34. Cui L, O'Sullivan C (2006) Exploring the macro- and micro-scale response of an idealised granular material in the direct shear apparatus. Géotechnique 56:455–468. https://doi.org/10.1680/geot.56.7.455
35. Barton N, Bandis S (1980) Some effects of scale on the shear strength of joints. Int J Rock Mech Min Sci Geomech Abstr 17:69–73
36. Lutenegger AJ, Cerato AB (2006) Specimen size and scale effects of direct shear box tests of sands. Geotech Test J 29:1–10

37. Wang J, Gutierrez MS (2010) Discrete element simulations of direct shear specimen scale effects. Géotechnique 60:395–409
38. Hu LM, Pu JL (2005) Testing and modeling of soil-structure interface. J Geotech Geoenviron Eng 130:851–860
39. Paikowsky SG, Player CM, Connors PJ (1995) A dual interface apparatus for testing unrestricted friction of soil along solid surfaces. ASTM Geotech Test J 18:168–193. https://doi.org/10.1520/GTJ10320J
40. Christoffersen J, Mehrabadi MM, Nemat-Nasser S (1981) A micromechanical description of granular material behavior. J Appl Mech 48:339–344
41. Barreto D, O'Sullivan C, Zdravkovic L (2009) Quantifying the evolution of soil fabric under different stress paths. In: AIP Conference Proceedings. pp 181–185

Chapter 6
Influences of Particle Sphericity and Fabric on the Soil-Structure Interface

6.1 Introduction

A rich body of investigations has proved that the grain shape emerges as an essential soil property that affects the various mechanical behaviors of bulk soil. The relationship between the compactness of the soil and the shape parameter has been exploited in terms of the maximum and minimum void ratio [1, 2]. The motions of the particles, including movement and rotation, result in the macroscopic deformation of a granular system. The rotation of a particle with an irregular shape is restricted and accordingly increases the interlocking inside the soil, leading to a higher shear strength and a larger dilation [3]. Moreover, the shear-induced anisotropy of a granular material composed of non-spherical particles is emphasized due to the particle eccentricity [4, 5]. In this context, the particle shape emerges as an essential soil property that needs to be properly considered in the SSI issue. The particle shape is generally characterized using three scales: roundness, sphericity, and smoothness [6]. The sphericity S is correlated to the rotation of the particle and the arrangement of the granular material, which are crucial to the macroscopic behaviors of the granular material. Thus, this chapter will focus on the effects of particle sphericity. Furthermore, the orientations of irregular particles will lead to an initial anisotropy of the specimen [7, 8]. Thus, the effect of initial fabric on SSI shearing behavior should be discussed as well.

The discrete element method (DEM) as a numerical tool has been widely used in the geotechnical field due to the fact that soil is discontinuous in nature. Two-dimensional (2D) and three-dimensional DEM simulations have been successfully applied in the SSI issue [9–12]. The particle used in the early DEM models was a disc in the 2D case and a spherical particle in the 3D case. Certain methods have been proposed to mimic the behavior of a non-spherical particle in DEM simulation. For example, the rolling resistant contact law between spherical particles has been proposed to manually prevent the rotation of particles by introducing a rolling friction coefficient [13–15]. However, real soil particles are generally with various shapes, different from idealized granular system with discs and spherical particles, which

W.-H. Zhou and Z.-Y. Yin, *Practice of Discrete Element Method in Soil-Structure Interface Modelling*, https://doi.org/10.1007/978-981-19-0047-1_6

significantly affects the mechanical behavior of soils. For this reason, non-spherical elements have been successfully applied in DEM simulation, such as clustered particles, polygons, and ellipsoids [9, 16–20]. Jensen et al. [9] employed a clustered element in 2D simulation of IST. However, how the shear resistance, material fabric, and particle motions are affected by the particle sphericity and initial fabric during interface shearing has not been fully studied. Furthermore, the thickness of localized band should be measured under various loading and modeling conditions.

In this chapter, the effect of particle shape was thoroughly investigated by 3D DEM. Different types of clustered particles were used to represent the irregular particles with various sphericities. Specimens were randomly generated and sheared on interfaces with different roughnesses. Based on the DEM interface shear test results, the following 4 aspects were explored: (1) the effect of particle sphericity on extreme porosities of granular material, the effect of particle sphericity on both macro- and micro- shearing behaviors of SSI, (3) the effect of interface roughness on the behaviors of SSI and the effect of initial fabric on the shearing behaviors of SSI.

6.2 DEM Simulation

6.2.1 Input Parameters

PFC 3D 5.0 software based on the discrete element method proposed by [21] was employed in this chapter. The Hertz-Mindlin contact law was used to describe the non-linear force–displacement relationship between two contacting particles [22]. The shear modulus G and Poisson's ratio ν were used to describe the deformability of the granular material. The values of the input parameters used in this chapter refer to the 3D simulation performed by [16] using arrays of ellipsoids, in which the shear modulus G was 28.957 GPa, the Poisson's ratio ν was 0.15 and the inter-particle friction coefficient μ_p was 0.5. A damping coefficient with a value of 0.7 was applied to dissipate the energy together with the sliding and guarantee a quasistatic analysis.

6.2.2 Geometries of the Clumps

A clustered particle, named clump, can be formed by adding certain particles together with or without overlapping. Efforts were made to bring the geometry of the clump close to that of real sand grain by composing more particles with the help of a 3D scanning technique or specific algorithms. Those sophisticated approaches validated the significance of the particle shape in the DEM simulation but created another problem. It was time-consuming, because of the remarkably increased particle numbers, to

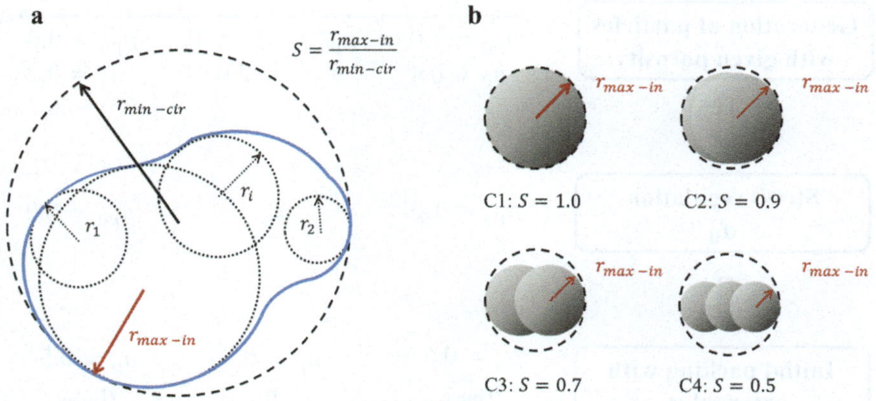

Fig. 6.1 **a** Definition of the sphericity S [6]; **b** clumps used in this chapter

form a clustered element that would be closer to the real one. It has been asserted that a clump having asymmetry geometry is sufficiently close to the mechanical behavior of real soil material. Thus, clumps composed of two or three single particles were enough for the simulation, which took into account the effect of particle shape [19, 23].

The sphericity S is characterized as shown in Fig. 6.1a [6]. The r_{max_in} is the radius of the maximum inscribed sphere, and the r_{min_cir} is the radius of the minimum circumscribed sphere of the irregular particle. The clumps, composed of different numbers of spherical particles representing various sphericity S used in the model, are named C1, C2, C3, and C4, respectively (Fig. 6.1b).

6.2.3 Specimen Generation Process

The specimen preparation method refers to the procedure proposed by Muir and Kenichi [24] to obtain a granular material with varied porosity n (Fig. 6.2). The specimen followed a given particle-size distribution, and a specific initial porosity n_0 was randomly generated inside a container with six frictionless walls. To obtain the densest granular material, the initial friction coefficient between particles μ_{p0} was set to zero and the initial porosity n_0 of the specimen was set to 0.2. Overlapping particles immediately spread out or separated to achieve an equilibrium state. Then the walls of the container were controlled by a servo system until the mean stress on the walls reached a given value σ_0 by moving slowly inward or outward. The friction coefficient of particle μ_{p0} changed to the eventual value μ_p and was maintained as a constant in the shearing stage. Then the final porosity of the specimen regained

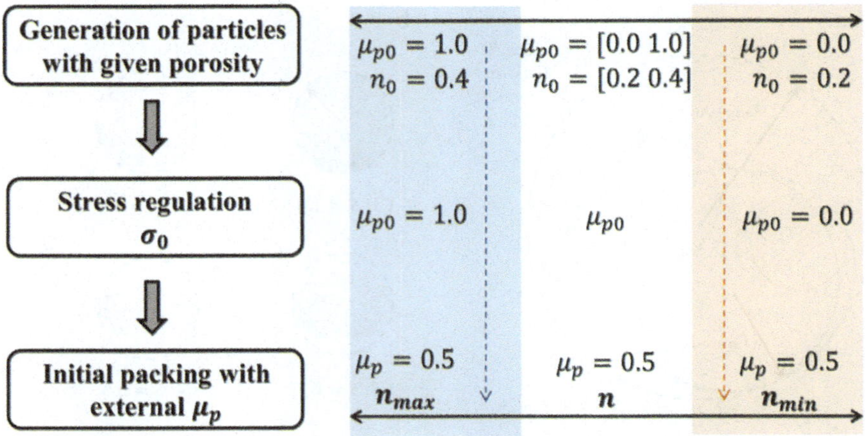

Fig. 6.2 Specimen generation procedure after Muir and Kenichi [24]

the equilibrium state, which was defined as the minimum porosity n_{min}. In contrast, to obtain a "loosest" specimen, the initial friction coefficient of particle μ_{p0} was set to 1.0 to generate a specimen with a high n_0 equals 0.4. Then the same procedure was performed to obtain the loosest sample. The eventual porosity n of the granular material can be altered by inputting a different value of μ_{p0} and n_0.

6.2.4 The Simulation of Soil-Rough Interface Shearing

The numerical model of soil-rough interface shear test is illustrated in Fig. 6.3. The dimension of the shear box is described using length (L), width (W), and height (H). A regular saw-tooth wall is used in the model with an inclined angle θ equals 45° and

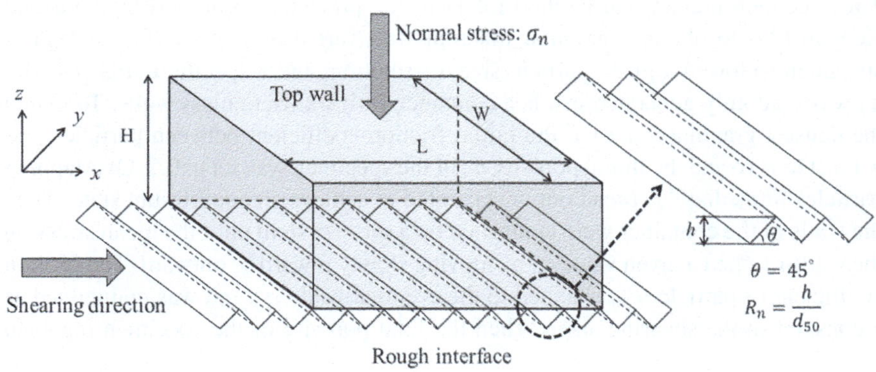

Fig. 6.3 Schematic diagram of interface shear test in the DEM simulation

a depth of each valley h. The normalized roughness R_n of this continuum interface is defined as h/d_{50} referring to the definition proposed by Uesugi and Kishida [25], where d_{50} is the mean particle diameter. The value of R_n is 0.5 in the following simulations. Four specimens consisting of clumps C1, C2, C3, and C4, respectively, have been generated with a desired porosity n. Equivalent diameter d_{eq} is denoted for the clumps with a non-spherical shape, which is defined as the diameter of a spherical ball with the same volume as the clump. All specimens follow a same linear grain size distribution. The value of d_{eq} ranges between 1.8 mm and 3.6 mm, and the $d_{50(eq)}$ equals 2.7 mm.

Once the granular material reached an equilibrium state, a constant normal stress σ_n was applied on the top wall. The bottom rough interface wall began to move horizontally in the x-direction at a low speed once the granular system was stabilized. The four lateral walls were fixed, and the top wall was vertically moveable during the shearing loading process. The top wall was controlled by a servo system to maintain a constant normal stress.

The macroscopic mechanical behaviors were measured according to the displacements and forces of the walls. The shear stress τ was the shear force measured on the interface wall divided by the area of horizontal section of the shear box. The shear displacement d_s was the displacement of the bottom wall in the direction of shearing. The normal stress σ_n was measured on the top wall. The vertical displacement d_v of the top wall was recorded to reflect the volumetric change of the specimen.

6.3 Compactness of the Specimen

In this chapter, the maximum porosity n_{max} and minimal porosity n_{min} of a specimen composed of different clumps were obtained using the procedure introduced in Sect. 6.2.3. The values of n_{max} and n_{min} of various specimens are illustrated in Fig. 6.4, which shows that the specimen composed of spherical particles ($S = 1.0$) tends to form a loose configuration. Non-spherical particles allow a better filling of the void space compared to spherical particles, and as a result, a dense packing is achieved for the specimen with a smaller value of S. On the other hand, rolling easily occurs with spherical particles ($S = 1.0$) and leads to a similar configuration of the granular assembly at the loosest and densest configurations. Accordingly, the difference between the n_{max} and n_{min} for the specimen with spherical particles ($S = 1.0$) is smaller than the others with irregular particles. It should be noted that the most elongated clump ($S = 0.5$) can form a structure with more void space and correspondingly results in a higher value of n_{max}. As mentioned by Salot [19], the extreme porosities obtained in the numerical simulation cannot compare directly with those obtained in the experimental tests because of the difference in preparation procedure. However, it is necessary to control the relative density of the granular material when taking into account the particle shape effect in the DEM tests.

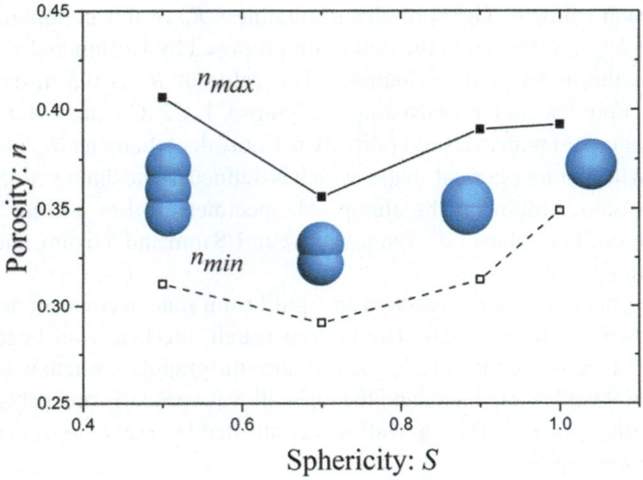

Fig. 6.4 The extreme porosities n_{max} and n_{min} of the specimen featuring various sphericity S

6.4 Effect of Particle Shape and Interface Roughness

The relative density D_r of the granular material is Eq. (6.1) calculated by below,

$$D_r = \frac{(n_{max} - n)(1 - n_{min})}{(n_{max} - n_{min})(1 - n)} \tag{6.1}$$

Four specimens consisting of spheres and three types of clumps were generated, named S1, S2, S3, and S4, respectively. Each specimen comprised around 30,000 spheres or clumps. The dense configuration was guaranteed by controlling the D_r = 90% for all specimens. The desired initial porosities n_0 of each specimen were derived according to Eq. (6.1) as listed in Table 6.1. To demonstrate the effect of particle sphericity on the macroscopic mechanical behavior of the SSI, sixty ISTs of specimen S1/2/3/4 shear on a rough interface featuring R_n = 0.1/0.25/0.5/0.75/1.0 under a normal stress σ_n equals 25 MPa/50 MPa/100 MPa respectively were modeled. The generation procedure is presented in Sect. 6.2.3.

6.4.1 Macroscopic Response

The macroscopic mechanical behaviors of the ISTs comprising particles of various S are illustrated in Fig. 6.5 in terms of the stress ratio τ/σ_n and the vertical displacement d_v as a function of shear displacement d_s. As shown in Fig. 6.5a, the evolutions of τ/σ_n of the four tests display a similar tendency. Stress softening occurs once the τ/σ_n peaks. Note that the peak shear stress at the interface is affected by particle

Table 6.1 Summary of the numerical tests with various elements

Specimen	Sphericity	Normalized roughness of interface: R_n	Initial porosity: n_0
S1	$S = 1.0$	0.1	0.359
		0.25	0.353
		0.5	0.355
		0.75	0.352
		1.0	0.357
S2	$S = 0.9$	0.1	0.325
		0.25	0.329
		0.5	0.326
		0.75	0.322
		1.0	0.321
S3	$S = 0.7$	0.1	0.306
		0.25	0.309
		0.5	0.301
		0.75	0.297
		1.0	0.297
S4	$S = 0.5$	0.1	0.323
		0.25	0.326
		0.5	0.323
		0.75	0.325
		1.0	0.325

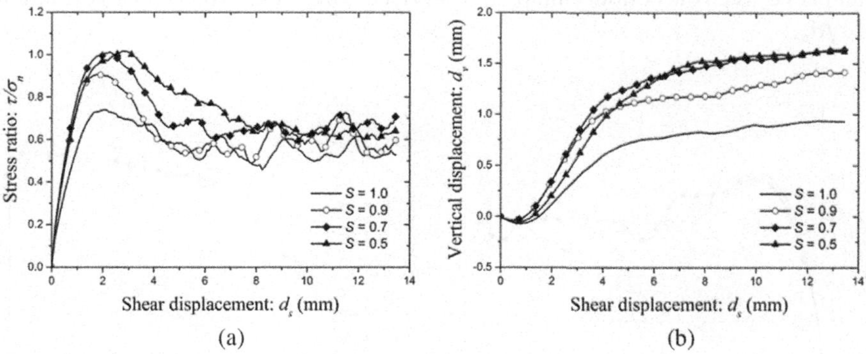

Fig. 6.5 Macroscopic responses of the ISTs comprising particles of various S ($D_r = 90\%$, $R_n = 0.5$, $\sigma_n = 50$ MPa): **a** stress ratio τ/σ_n versus shear displacement d_s; **b** vertical displacement d_v versus shear displacement d_s

sphericity S. The specimens composed of non-spherical particles show a higher peak shear stress than those composed of spherical balls ($S = 1.0$). The difference in shear resistance is attributed to the interlocking phenomenon between the particles. Unlike the way a spherical particle easily rotates when making contact with another one, an irregular particle tends to interlock with other particles or the rough interface. The evolution of vertical displacement of the top wall d_v reflects the volumetric change in the specimen, showing that all specimens contract at the beginning of shearing and then gradually dilate. The growing rate of dilation slows down at shear displacement d_s where shear stress softening appears. This suggests that the volumetric change in the specimen is also affected by the particle sphericity. A larger dilatancy can be observed in the specimen with non-spherical particles. From the perspective of micro-mechanics, the volumetric change of a granular material is the result of the micro-physics of individual particles, i.e., movement and rotation. To help explain the macroscopic responses we obtained in the simulations, the micro-physics of the particles will be analyzed in the following sections.

The macroscopic mechanical behaviors of the ISTs ($S = 0.5$) featuring varying R_n under $\sigma_n = 50$ MPa are illustrated in Fig. 6.6. As shown in the figure, the peak shear stress ratio and volumetric change are affected by the R_n. A higher peak shear stress and larger dilation are observed when the specimen is shearing on a rougher interface. This result is consistent with the existing experimental findings [26, 27]; the shear strength of interface generally increases as the increasing of R_n. Note that periodic oscillation is observed in the curve of τ/σ_n when $R_n = 0.25$. In this case, the clumps in the bottom layer cannot fit into such small volumes between sawteeth. Thus, the bottom layer of clumps moves alternately between the tops of the teeth and the areas between teeth, which results in periodic oscillation in the total contact number between the bottom clumps and interface. This induces this kind of evolution of τ/σ_n.

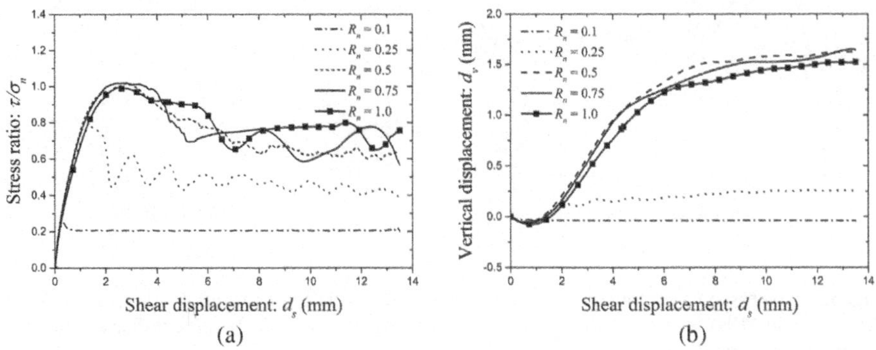

Fig. 6.6 Macroscopic responses of the ISTs ($S = 0.5$, $\sigma_n = 50$ MPa) featuring varying normalized roughness R_n: **a** stress ratio τ/σ_n versus shear displacement d_s; **b** vertical displacement d_v versus shear displacement d_s

6.4.2 *Interface Friction Angle Analysis*

The peak shear stress τ_p and steady shear stress τ_s (at $d_s = 13.5$ mm) were obtained for the ISTs under various normal stress 25 MPa/50 MPa/100 MPa. According to the Mohr-Coulomb criterion, the peak friction angle ϕ_p and steady friction angle ϕ_s can be obtained by linearly fitting the τ_p and τ_s under various normal stress conditions (Fig. 6.7). The cohesive force was assumed to be zero since a non-cohesive soil was considered in this chapter.

The friction angles of all ISTs are obtained by this criterion to discuss the effects of S and R_n on the shear resistance of SSI. As a reference, the direct shear tests (DSTs) with the same input parameters under σ_n equals 25 MPa/50 MPa/100 MPa are modeled. The height of the interface shear box is twice of the specimen in IST. The peak friction angles of ISTs (ϕ_p) and DSTs (ϕ_p^d) are summarized in Table 6.2.

6.4.2.1 Effect of Sphericity

The peak friction angles ϕ_p and steady friction angles ϕ_s measured in all ISTs are plotted in Fig. 6.8. Existing research studies reveal that the interface shear strength is profoundly correlated with the shear strength of pure soil. The friction angle

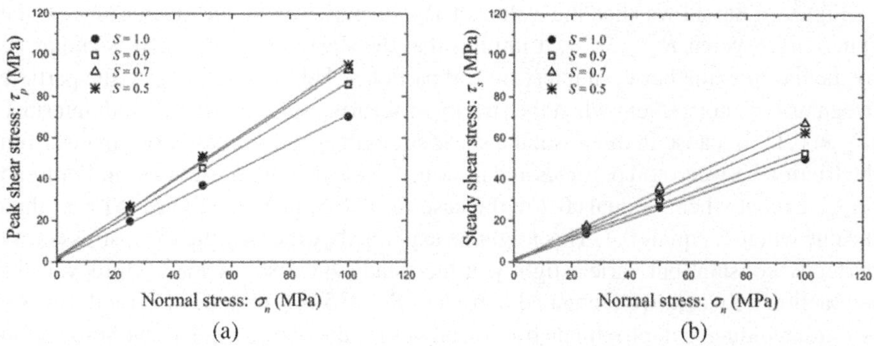

Fig. 6.7 **a** Fitting the peak shear stress τ_p as a function of normal stress σ_n; **b** fitting the steady shear stress τ_s as a function of normal stress σ_n ($R_n = 0.5$)

Table 6.2 Summary of the peak friction angles of ISTs and DSTs

Sphericity	Peak friction angle (°)					
	$R_n = 0.1$	$R_n = 0.25$	$R_n = 0.5$	$R_n = 0.75$	$R_n = 1.0$	DST
$S = 1.0$	11.89	21.70	35.51	36.07	34.85	35.74
$S = 0.9$	12.54	26.10	41.10	40.89	38.02	43.84
$S = 0.7$	13.14	30.15	43.53	45.05	41.48	49.42
$S = 0.5$	13.18	34.53	44.05	45.06	43.19	47.30

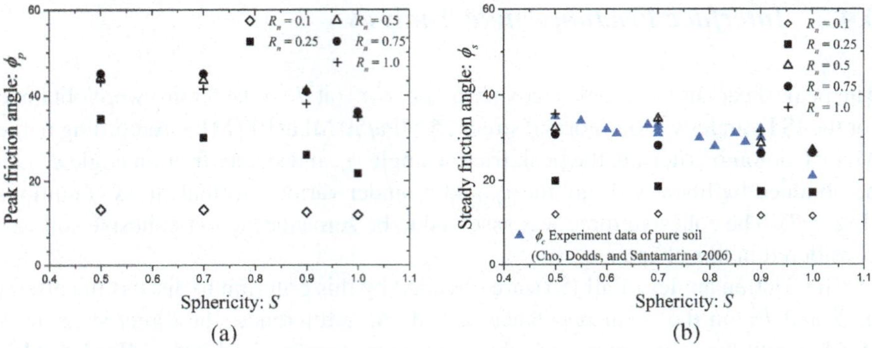

Fig. 6.8 **a** Peak friction angle ϕ_p obtained in the DEM ISTs; **b** comparison of the steady friction angle ϕ_s obtained in the DEM ISTs to the critical friction angle ϕ_c of pure soil obtained in the laboratory experiments [32] at varying sphericity S

measured on a rough IST is close to the friction angle of pure soil [12, 28–31]. For this reason, the steady friction angles ϕ_s obtained in the numerical ISTs are compared to the critical friction angles ϕ_c of pure soil obtained in the laboratory experiments in Fig. 6.8b. The experimental databases are derived from the study of [32]. The tested soils include crushed sands and natural sands from various places and some other materials such as glass beads and granite powder.

Figure 6.8a shows that the value of ϕ_p increases with the decreasing of the sphericity S when $R_n \geq 0.25$. It implies that the shear strength of SSI is enhanced by the interlocking between interface and particles. This augment due to the particle irregularity is not evident when the specimen shearing on a relative smooth interface ($R_n = 0.1$). Because in this case, the shear strength at SSI primarily originates from the friction between soil particles and interface. Note that the ϕ_s shows a similar trend for ϕ_p except when S equals 0.7 in the case $R_n = 0.5$, in which the ϕ_s is lower than the one where S equals 0.7. This might be explained by the way the shear stress is not perfectly constant but varies slightly at the steady shear stress state. Moreover, the interaction between two elongated particles ($S = 0.5/0.7$) and the saw-tooth surface is similar, inducing approximate friction angles for the two cases. The evolution trend of ϕ_s at various S is similar to that of the ϕ_c obtained in the laboratory experiment. This result verifies the accuracy of the numerical simulation to a certain degree. It suggests a correlation between the particle sphericity and the friction angle of SSI in the case of a relative rough interface.

6.4.2.2 Effect of Interface Roughness R_n

The peak friction angle ϕ_p measured on SSI is affected by R_n as well as S as illustrated in Fig. 6.9. In general, the value of ϕ_p increases as the increasing of R_n. This tendency is valid for the specimens featuring varying sphericity S. To compare the numerical results to the laboratory experiment results, the friction angles ϕ_p measured in IST is

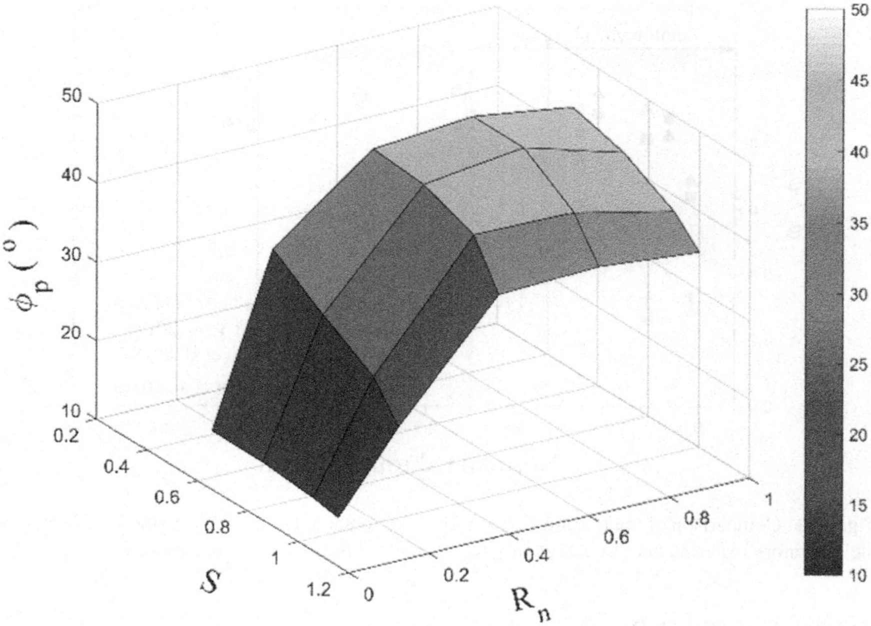

Fig. 6.9 The peak friction angle ϕ_p at varying normalized roughness of interface R_n and sphericity S

normalized by the ϕ_p^d obtained in DST. The ratios of ϕ_p/ϕ_p^d at varying R_n are plotted in Fig. 6.10. The experimental data are derived from the ISTs between natural soil and steel plate [33, 34]. Figure 6.10 illustrates that the value of ϕ_p/ϕ_p^d increases significantly in the range of R_n between 0 and 0.5. The growing rates of these tests are different, which depend on the properties of soil material, e.g. friction, grading, water content, particle size, particle shape, etc. When the value of R_n is greater than 0.5, the ratios of ϕ_p/ϕ_p^d achieve to a plateau value. It implies that the interaction between particles and interface similar to the interaction among pure particles when ϕ_p/ϕ_p^d is close to 1.0. Note that the ϕ_p/ϕ_p^d of the IST of $R_n = 1.0$ are slightly less than those of $R_n = 0.5$ and 0.75 in the numerical tests. In this case, the bottom particles of an approximately uniform distributed sample ($C_n \approx 1.45$) will be trapped in the valley between sawteeth of the interface, which weakens the interlocking between particles and interface. In contrast, for the well graded soil sample ($C_n = 19.2$) used by Wu and Yang [33], the soil particles can properly fit in the space of rough interface, leading to a stronger interlocking.

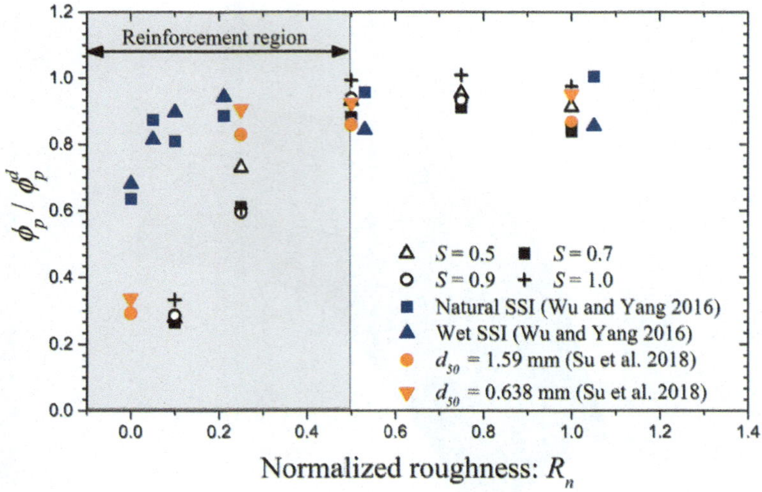

Fig. 6.10 Comparison of the friction angle ratio ϕ_p/ϕ_p^d obtained in the DEM to those measured in the laboratory experiments [33, 34] at varying normalized roughness R_n of interface

6.4.3 Localized Band Analysis

Shearing deformation is largely localized in a narrow zone during the shearing process, named the localized band. The localized band can be analyzed by tracing the movements of each particle at a specific stress state. To average the kinematic field, we set certain measuring windows at different heights for the specimen with a dimension of 100 mm × 100 mm × 5 mm (Fig. 6.11). The average shear displacement in x-direction $\overline{d_x}$ of the elements in each measuring window is calculated.

The values of $\overline{d_x}$ as a function of Z at different shear stress states ($R_n = 0.5, \sigma_n = 50$ MPa) are plotted in Fig. 6.12. Each dot in the figure represents one measurement at a specific height Z. As the shear stress increases, $\overline{d_x}(\mathrm{Z})$ shows a non-linearity, and an inflection point appears. The phenomenon of stratification becomes more evident at the steady stress state. The shear displacement induced by the interface shearing largely concentrates in the bottom layer of particles adjacent to the interface, named the localized band, rather than in the upper zone separate from the interface. It is consistent with the numerical result regarding the formation of the localized band in 2D/3D DEM simulations [12, 35] as well as the laboratory experiments using image analysis [26].

The inflection point of the curve of $\overline{d_x}(\mathrm{Z})$ at the steady stress state (when $d_s/d_{50} = 3.5$) is used to define the thickness of the localized band δ_h. Spline interpolation is applied to get a smooth $\overline{d_x}$ - Z curve $f(\mathrm{Z})$. The first derivative $f'(\mathrm{Z})$ $f^{'}(\mathrm{Z})$ and second derivative $f^{''}(\mathrm{Z})$ are calculated using the finite difference method. The curvature κ of $f(\mathrm{Z})$ is defined by Eq. (6.2),

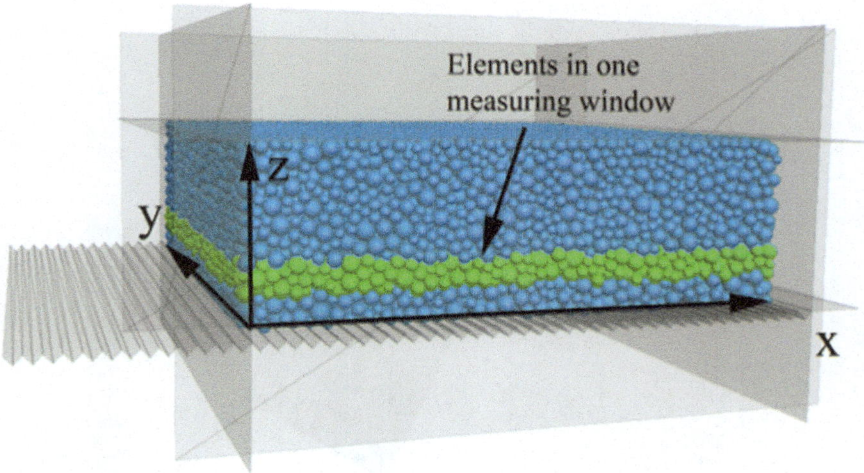

Fig. 6.11 Set-up of measuring window at different heights Z of the specimen

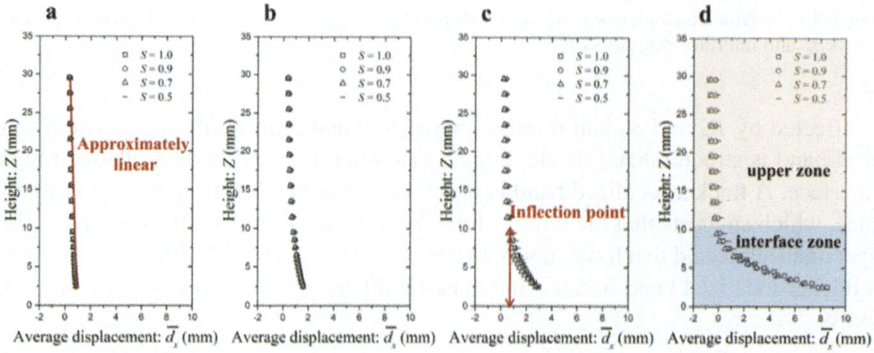

Fig. 6.12 Average shear displacement in x-direction $\overline{d_x}$ of four ISTs ($R_n = 0.5$, $\sigma_n = 50$ MPa) at different shear states: **a** $d_s = 1.0$ mm; b $d_s = 2.0$ mm; **c** $d_s = 4.0$ mm; and d $d_s = 13.5$ mm

$$\kappa = \frac{|f''(Z)|}{(1 + f'(Z)^2)^{3/2}} \tag{6.2}$$

The $\overline{d_x}$ changes approximately linearly with the height Z toward the higher position of the specimen where the value of κ approaches zero. As Z decreases, the κ sharply increases at a certain value of Z because of the localization of shear deformation. Thus, the inflection point of the κ is considered as a sign of the top boundary of the localized band. Jing et al. [12] suggest that the inflection point is where the κ equals 0.02.

According to this criterion, the thicknesses of the localized band δ_h is rarely affected by the particle sphericity S (Fig. 6.12). However, Fig. 6.13 shows that δ_h

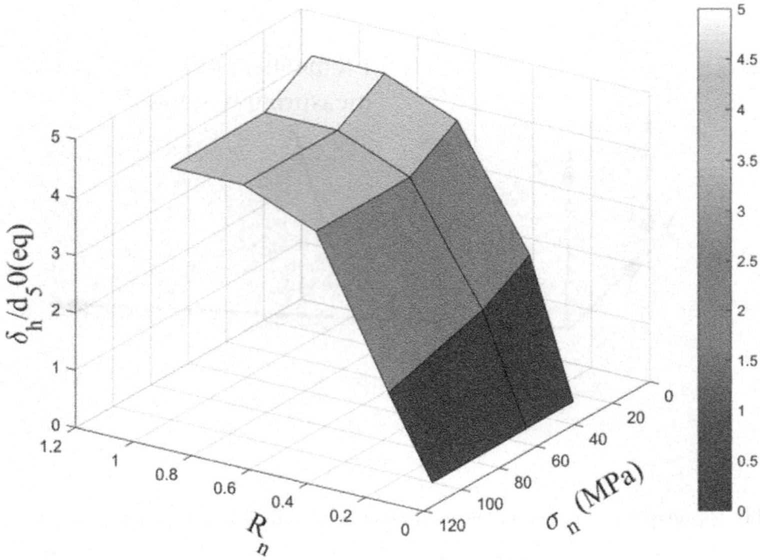

Fig. 6.13 Normalized thicknesses of localized band $\delta_h/d_{50(eq)}$ of ISTs ($S = 0.5$) at different normal stress σ_n and interface roughness R_n

is affected by R_n and σ_n and it ranges between 0 and 5 times of $d_{50(eq)}$. The localized band is structuralized inside the material when it shearing on a relative rough interface. A thicker localized band is observed in the IST featuring a rougher interface, which suggests that the failure shifts from the interface into the soil layer. The specimen subjected to a lower normal stress condition ($\sigma_n = 25$ MPa) tends to form a thicker localized band because the material dilates more under a lower confining stress.

6.4.4 *Local Porosity and Coordination Number*

To help visualize the local porosity distribution inside the specimen, a grid is constructed to compute the contour of local porosity. Certain measuring balls are set inside the shear box. All the centers of measuring balls are located in the central cross-section of the shear box, which represents the nodes of the grid. The porosity obtained in each measuring ball represents the local porosity at the position of the center of the ball, in other words, the node of grid. Accordingly, the contour of local porosity can be obtained. The contours of local porosity for the IST ($S = 0.7$) at different strain states are plotted in Fig. 6.14, showing that the initial distribution of porosity is almost homogenous. As shearing progresses, the particles gradually accumulate on the right side and accordingly lead to the dilation on the bottom left corner of the specimen. The dilation region enlarges from the bottom left corner to

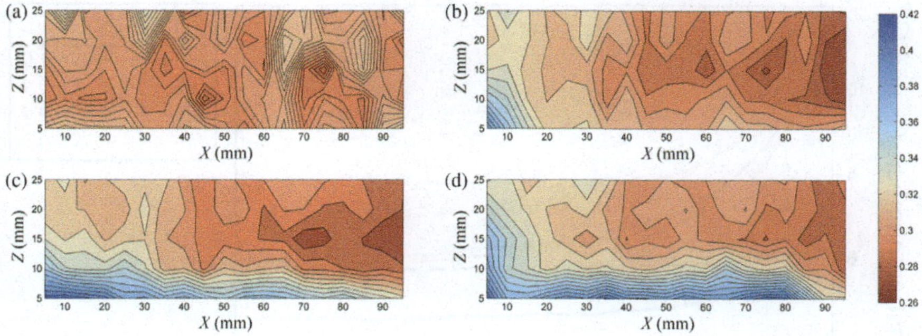

Fig. 6.14 Local porosity inside the central section of the specimen ($R_n = 0.5$, $S = 0.7$) at different strain states: a $d_s = 0.0$ mm; **b** $d_s = 2.0$ mm; **c** $d_s = 4.0$ mm; and **d** $d_s = 13.5$ mm

the bottom part of the entire specimen. The difference in the local porosity inside the specimen reaffirms that the granular material is structuralized into two regions when shearing on an interface (Sect. 6.4.3). The top line of the localized region is not strictly horizontally straight because of the fixed lateral walls that prevent the movement trend of particles.

The coordination number C_n is used to describe the local contact at particle scale, which is profoundly correlated to the porosity of the granular assembly. It is defined as the average contact number per particle Eq. (6.3),

$$C_n = (\sum_{N_p} n_c^p)/N_p \tag{6.3}$$

where N_p is the total number of particles in the measured region, and n_c^p is the contact number of particles p in the measured region. As discussed in Sect. 6.4.3, the specimen structuralizes into two regions after shearing, the interface zone and the upper zone (Fig. 6.12d). The evolutions of the coordination number inside the interface zone C_n^i and the upper zone C_n^u for the ISTs with various S are illustrated in Fig. 6.15a. The initial coordination number of the specimen composed of irregular clumps is much higher than the one consisting of spherical balls, which suggests that more contacts exist between the irregular particles. It explains why interlocking tends to occur inside such granular material. A sharp decrease for C_n^i is observed in all cases; in contrast, the change in C_n^u is minor. The dilation primarily occurs in the interface zone, as the contour of local porosity illustrates. The micro-structure of particles in the upper zone is almost preserved. Figure 6.15b shows the difference between the values measured in the interface zone and upper zone $C_n^u - C_n^i$. The values of $C_n^u - C_n^i$ increase gradually and approach a steady value. Note that the value in the case of spherical balls is much smaller than the others, in which the total volumetric change is the smallest.

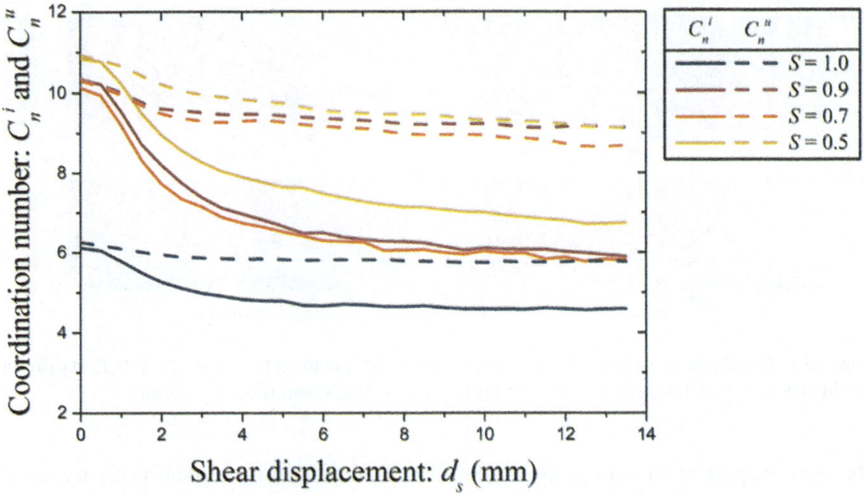

(a)

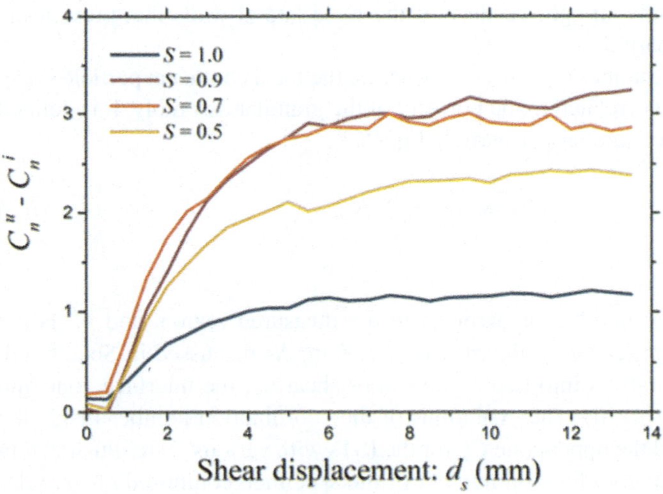

(b)

Fig. 6.15 **a** Coordination number inside the interface zone C_n^i and upper zone C_n^u of the ISTs ($R_n = 0.5$, $\sigma_n = 50$ MPa) with varying sphericity S; **b** the difference between the value measured in interface zone and upper zone $C_n^u - C_n^i$

6.4.5 Material Fabric Analysis

The macroscopic mechanical behavior of the granular material originates in the distribution and evolution of the material fabric. The distribution of the contact orientation is frequently used to describe the material fabric. A second-order tensor F_{ij} introduced by Satake [36] is used to quantitatively characterize the distribution in normal contact orientation:

$$F_{ij} = \frac{1}{N_c}\sum_{\alpha}^{N_c} n_i^{\alpha} n_j^{\alpha} (i, j = x, y, z) \tag{6.4}$$

where N_c is the total contact number, and n_i is the contact normal vector at contact α. The principal values of F_{ij}, ordered by decreasing magnitude, are F_1, F_2, and F_3. To measure the anisotropy of the material fabric, a deviator fabric δ_D of F_{ij} is calculated as follows [37]:

$$\delta_D = \frac{1}{\sqrt{2}}[(F_1 - F_2)^2 + (F_2 - F_3)^2 + (F_1 - F_3)^2]^{0.5} \tag{6.5}$$

The evolution of δ_D measured in the interface zone for the ISTs under $\sigma_n = 50$ MPa is plotted in Fig. 6.16. The initial values of δ_D are slightly higher than zero because anisotropy is induced by the one-dimensional normal pressure before shearing. The δ_D increases with the increase of shear displacement d_s and decreases once the stress softening appears. The peak value of δ_D depends on particle sphericity. The clumps with smaller sphericity S induce higher anisotropy during the interface sharing, in which a higher interface shear strength is measured. This implies that a correlation exists between δ_D and interface shear strength.

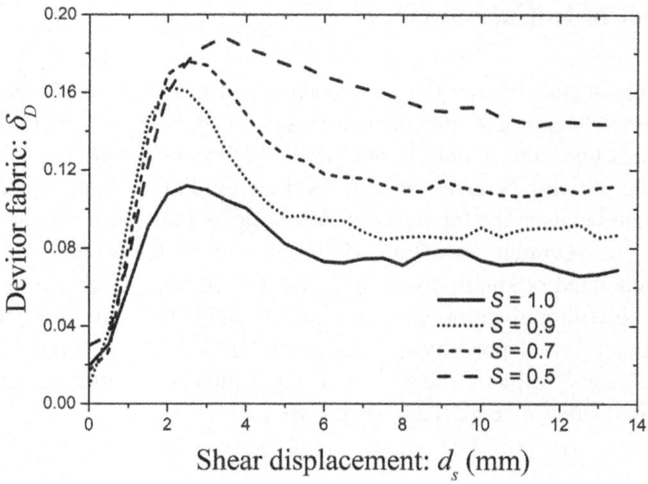

Fig. 6.16 The evolution of deviator fabric δ_D in the interface zone of the ISTs with various sphericities $S = 1.0/0.9/0.7/0.5$ ($\sigma_n = 50$ MPa, $R_n = 0.5$)

The probability density distribution $P(\vec{n})$ of a unit vector of contact normal $\vec{n}$ is characterized to better visualize the contact distribution inside a granular material. The unit vector $\vec{n}(\theta, \varphi)$ of contact normal between two contacting clumps is obtained based on the spherical coordinate system. The $P(\vec{n})$ can be obtained according to Eq. (6.6) below

$$P(\vec{n}) = \frac{N_c(d\Omega)}{N_c} \tag{6.6}$$

where N_c is the total contact number and $N_c(d\Omega)$ is the contact number of contact normal vectors pointing in the direction of a range of angle d_Ω.

The $P(\vec{n})$ measured in the interface zone of the four ISTs at initial state, peak shear stress state, and steady shear stress state are shown in Fig. 6.17. The shape of $P(\vec{n})$ is close to a spherical ball at the initial state because the specimen is approximately isotropic. As shearing stress increases, the contact orientation gradually accumulates in a certain direction. The concentration of contact orientation is a rearrangement process of particles, increasing the material's anisotropy. The anisotropy at the peak shear stress state is affected by the particle shape, and correspondingly, the shape of $P(\vec{n})$ is different. The anisotropy direction for all the tests featuring various S at peak shear stress state ranges between 40° and 60°. When the shear stress softening occurs and approaches a steady state, the decrease of anisotropy results in the reshaping of $P(\vec{n})$.

6.5 Effect of Initial Fabric

In the previous section, the particles were generated randomly inside the shear box, and approximately isotropic specimens were produced. However, the initial material fabric depends upon the initial orientation of the irregular particles, which has an impact on the shearing behavior of SSI. As shown in Fig. 6.18, θ_p is defined as the included angle between the long axis of the clump and the shear direction (positive x-direction). A specimen consisting of 29,058 clumps featuring $S = 0.7$ with a randomly generated orientation was prepared. In addition, another four specimens were prepared with a given orientation ($\theta_p = 0°/45°/90°/135°$) for each particle. An approximate initial porosity n_0 was controlled for all specimens as listed in Table 6.3. These specimens sheared on a rough interface featuring R_n equals 0.5 under a normal stress σ_n equals 25/50/100 MPa.

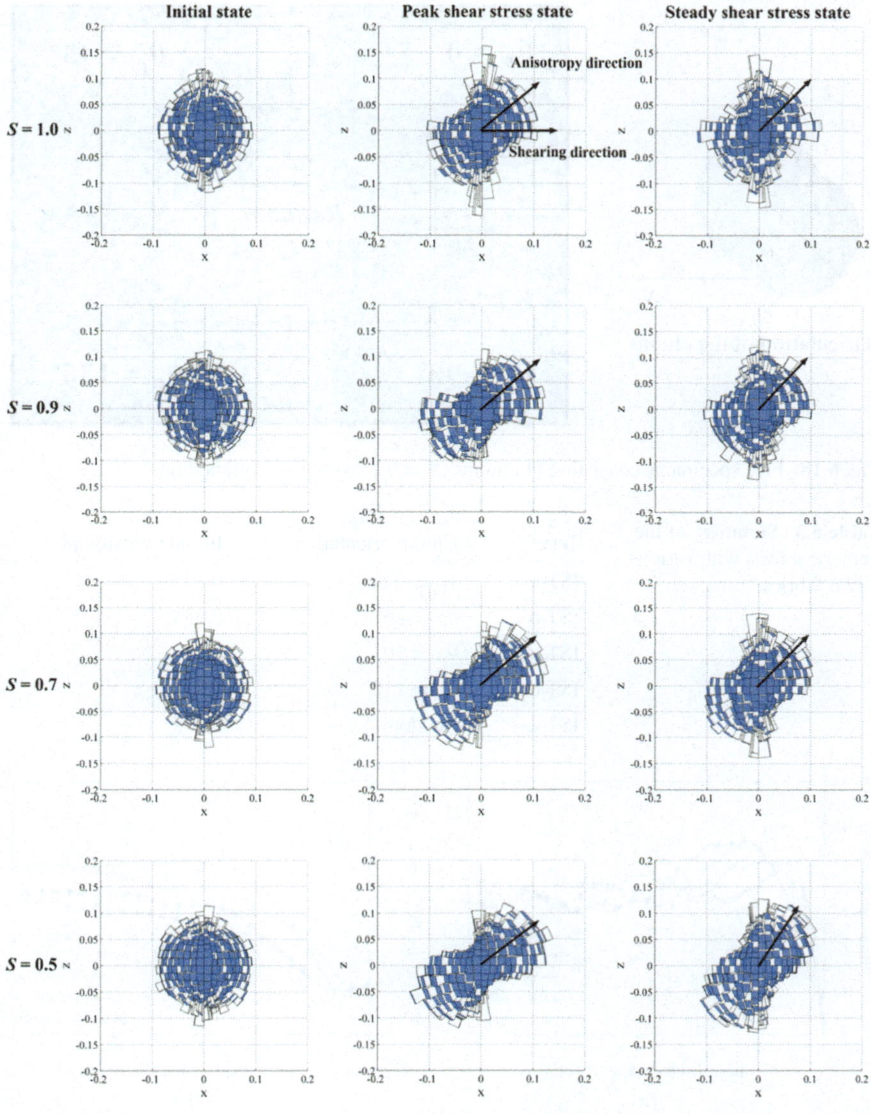

Fig. 6.17 The contact normal distribution in the interface zone of the four ISTs ($\sigma_n = 50$ MPa, $R_n = 0.5$) at initial state, peak shear stress state, and steady shear stress state

6.5.1 Macroscopic Response

The evolutions of stress ratio τ/σ_n and vertical displacement d_v are illustrated in Fig. 6.19. The peak shear stress τ_p is affected by the initial orientation of clumps. The specimen consisting of horizontally placed clumps ($\theta_p = 0°$) shows the lowest shearing resistance. As the θ_p increases, the shearing resistance increases. The peak

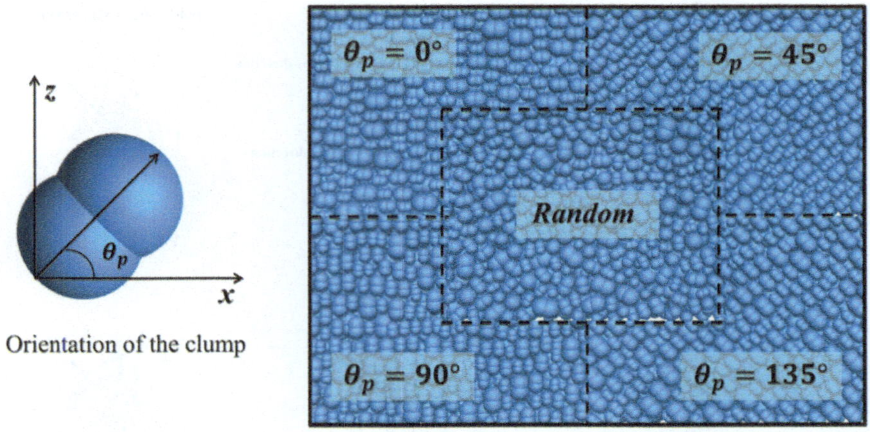

Fig. 6.18 Five specimens consisting of clumps (S = 0.7) with given orientations

Table 6.3 Summary of the numerical tests with various initial fabrics

Test	Clump orientation	Initial porosity:n_0
IST-a	$\theta_p = 0°$	0.337
IST-b	$\theta_p = 45°$	0.335
IST-c	$\theta_p = 90°$	0.339
IST-d	$\theta_p = 135°$	0.336
IST-e	Random	0.338

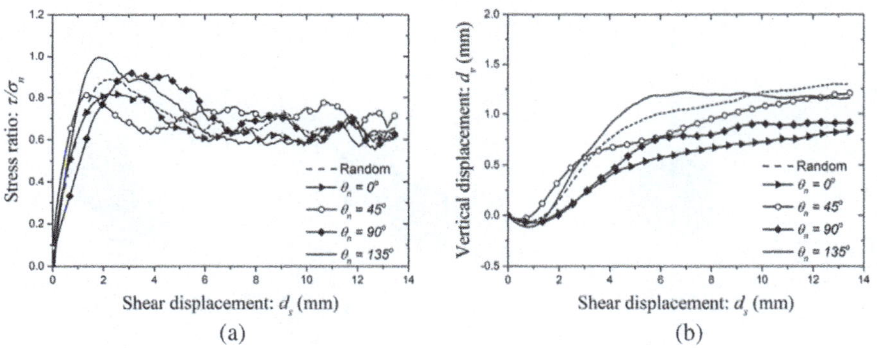

Fig. 6.19 Macro-responses of the ISTs featuring various included angle θ_p ($\sigma_n = 50$ MPa, $R_n = 0.5$): **a** stress ratio τ/σ_n versus shear displacement d_s; **b** vertical displacement d_v versus shear displacement d_s

shear stress for the case with randomly distributed clumps is between the extreme cases ($\theta_p = 0°$ and $\theta_p = 135°$). Stress softening is observed among all cases. Moreover, the values of d_s at which the peak shear stress ratio τ_p/σ_n is achieved are different for the five tests. This implies that a different value of d_s is required to fully trigger the interlocking inside the granular materials. Figure 6.19b illustrates a similar evolutionary trend of volumetric change for various specimens. Before the peak shear stress is achieved, the specimen with an included angle $\theta_p = 135°$ shows the largest dilation; in contrast, the one with horizontally placed clumps dilates less than the others. These results suggest that the vertical movement tends to be easily triggered when the clumps are randomly placed and $\theta_p = 135°$. In contrast, horizontally placing the clumps restricts the interaction between the bottom layer clumps and the rough interface. Accordingly, both the shear strength and dilatation for that case are the smallest.

6.5.2 Localized Band Analysis

The curves of $\overline{d_x}$-Z for the IST-a/b/c/d/e at different shear stress states are plotted in Fig. 6.20. The evolution pattern of $\overline{d_x}(Z)$ curves is similar to those of ISTs with varying S. According to the analysis of curvature κ, the thickness of the localized band can be obtained. Figure 6.21 illustrates the normalized thickness $\delta_h/d_{50(eq)}$ under varying normal stress σ_n, where $d_{50(eq)}$ is the equivalent mean particle diameter. Generally, the $\delta_h/d_{50(eq)}$ is larger when the specimen subjected to a smaller σ_n, because the material dilates more under a lower confining stress. Besides, it shows that the $\delta_h/d_{50(eq)}$ depends on the particle orientation rather than the particle sphericity at the steady stress state. A thicker localized band is formed in the specimen with inclined clumps (i.e. $\theta_p = 45°$ and 135°) and randomly distributed clumps. It is noted that the value of $\delta_h/d_{50(eq)}$ varies between 4 and 6, which is slightly higher than that ($\delta_h/d_{50(eq)} = 4$) measured from the previous tests presented in Sect. 6.4. This is because the n_0 for these tests are relatively higher than the previous ones. The loose specimen tends to form a thicker localized band.

6.5.3 Local Coordination Number

The evolutions of coordination numbers inside the interface zone C_n^i and the upper zone C_n^u for the five ISTs are illustrated in Fig. 6.22a. The dilation primarily occurs in the interface zone, which is consistent with the tests with varying S (Fig. 6.15). On the other hand, the C_n^u also decreases during the shearing test; especially in the case $\theta_p = 90°$, it almost decreased the same as the C_n^i. Figure 6.22b shows the difference between the values measured in the interface zone and upper zone $C_n^u - C_n^i$. The values of $C_n^u - C_n^i$ increase gradually and approach a steady value. It can be noted that

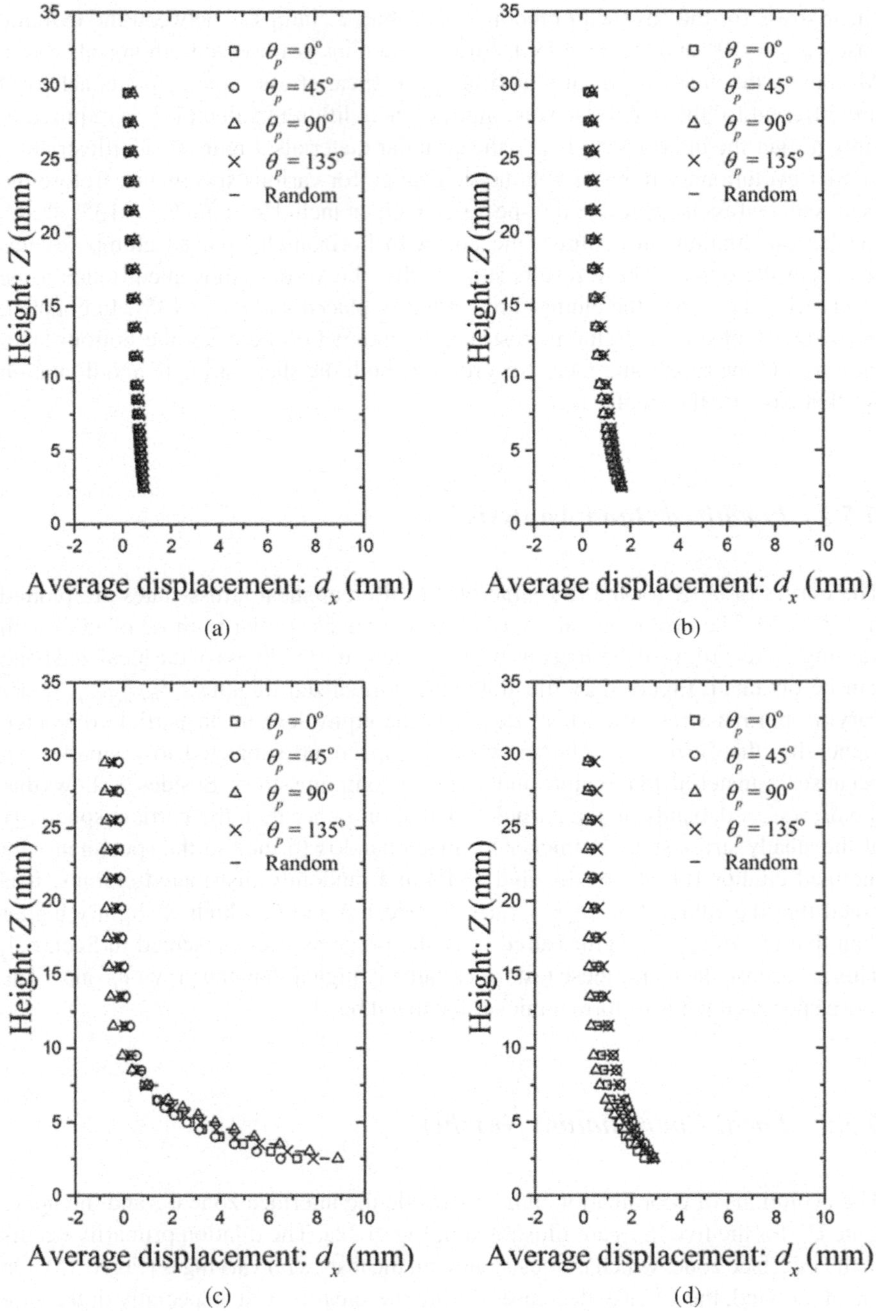

Fig. 6.20 Average shear displacement in x-direction $\overline{d_x}$ of five ISTs (random distribution, $\theta_p = 0°/45°/90°/135°$) at different strain states: **a** $d_s = 1.0$ mm; **b** $d_s = 2.0$ mm; **c** $d_s = 4.0$ mm; and **d** $d_s = 13.5$ mm

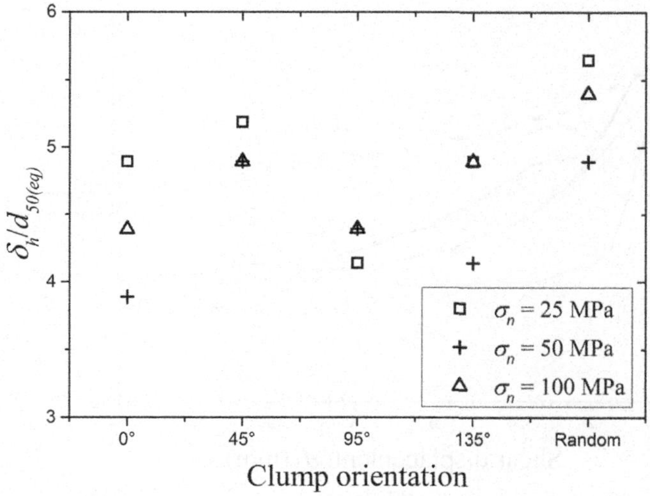

Fig. 6.21 The normalized thickness of localized band $\delta_h/d_{50(eq)}$ of the specimen comprising of different orientated particles under varying normal stress σ_n

the value for case $\theta_p = 90°$ is quite different from the others because the vertically placed clumps are easily disturbed by the shearing even in the upper zone.

6.5.4 Material Fabric Analysis

The evolutions of δ_D measured in the interface zone for the ISTs under $\sigma_n = 50$ MPa are plotted in Fig. 6.23, and the $P(\vec{n})$ at various states are shown in Fig. 6.24. The specimen consisted of randomly generated clumps, almost isotropic before shearing. Over the progress of shearing, the contacts accumulate in a specific direction, correlated with the shear direction. This anisotropy is purely induced by the shearing, which increases gradually and approaches a peak value at the peak shear stress state. The shear-induced "anisotropy direction" is shown in the figure of $P(\vec{n})$. On the other hand, the initial fabric of specimens consisting of variously oriented particles is anisotropic since the contacts are initially concentrated in various directions. The initial anisotropic δ_D for those specimens are about 0.17. As the shear displacement d_s increases, the δ_D increases and then decreases once the stress softening occurs for the cases with $\theta_p = 0°/90°/135°$, as well as the case with a randomly generated specimen. Especially when $\theta_p = 135°$, the contact normal has already been concentrated in the direction of pure shear-induced anisotropy. Thus, the highest level of anisotropy is observed, and accordingly, the largest shearing stress is measured. By contrast, the initial contacts ($\theta_p = 45°$) gather in a direction perpendicular to the pure shear-induced anisotropy direction, preventing the development of the shear-induced anisotropy. For this reason, the δ_D decreases continuously, and a minimum peak

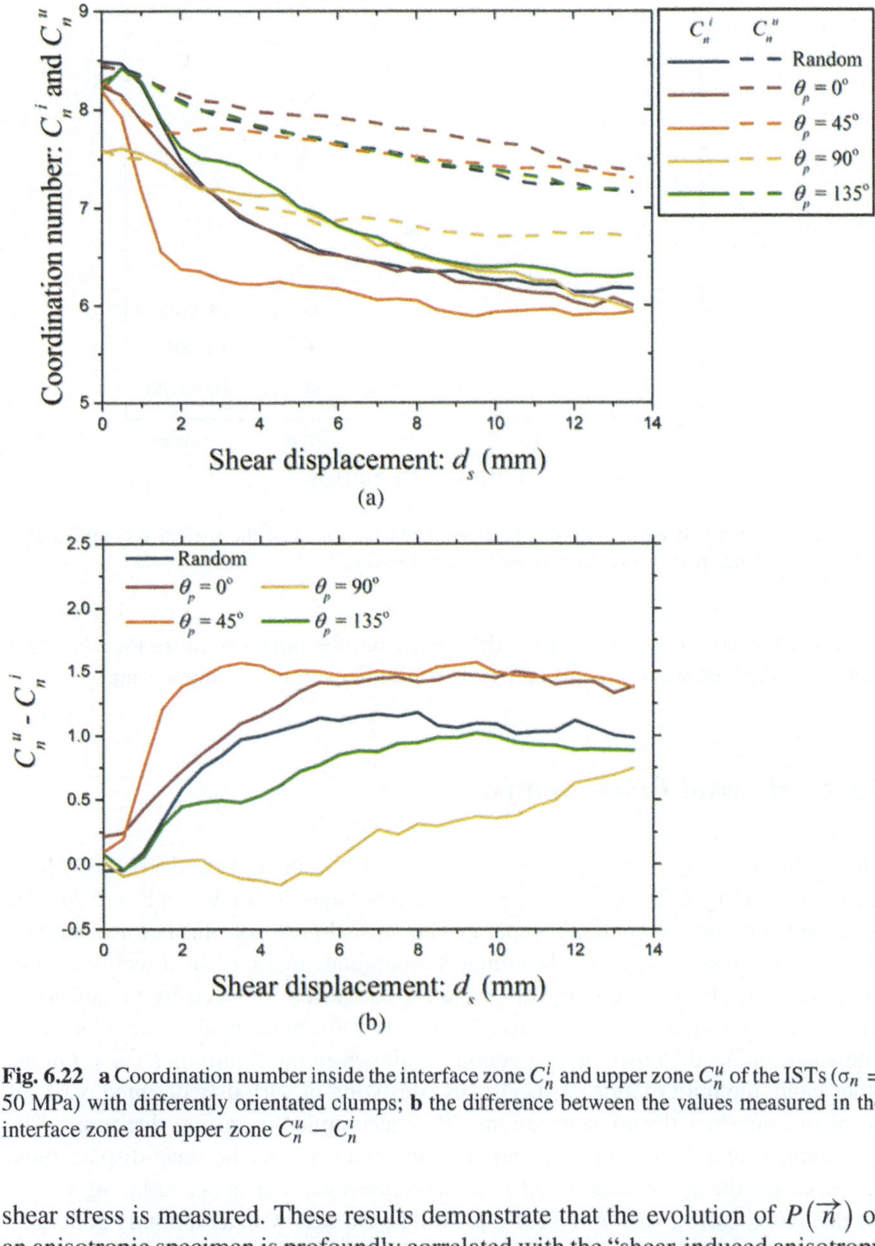

Fig. 6.22 **a** Coordination number inside the interface zone C_n^i and upper zone C_n^u of the ISTs (σ_n = 50 MPa) with differently orientated clumps; **b** the difference between the values measured in the interface zone and upper zone $C_n^u - C_n^i$

shear stress is measured. These results demonstrate that the evolution of $P(\vec{n})$ of an anisotropic specimen is profoundly correlated with the "shear-induced anisotropy direction" in the isotropy specimen.

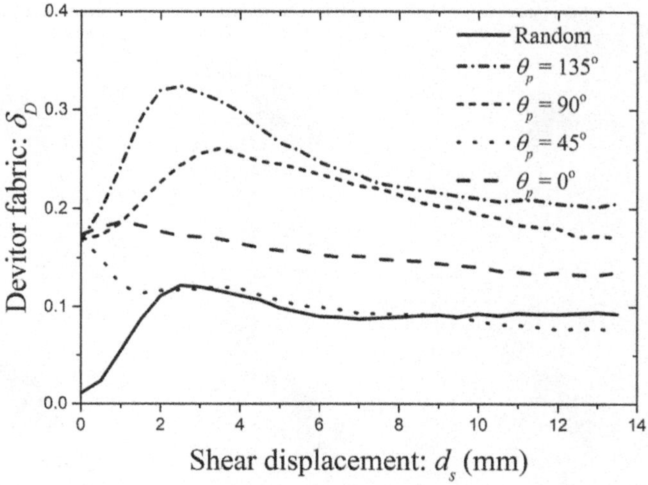

Fig. 6.23 The evolution of deviator fabric δ_D in the interface zone of five ISTs (random, $\theta_p = 0°/45°/90°/135°$) under $\sigma_n = 50$ MPa and $R_n = 0.5$

6.6 Conclusions

The macro- and micro- shearing behaviors of a soil-structural interface have been studied using 3D DEM simulations of ISTs that feature varying sphericity S, interface roughness R_n and initial fabric. The effects of S and σ_n on shear strength, volumetric changes, the thickness of the localized band, local porosity, contact normal distribution, and material fabric anisotropy have been analyzed. The following conclusions are drawn.

(1) Particle sphericity S plays a significant role in the mechanical properties of the SSI. The shear strength of the interface (i.e. τ_p/σ_n, ϕ_p and ϕ_s) increases as S decreases. The volumetric change in the specimen also depends on S. A larger dilation is observed for the specimen composed of non-spherical particles. The granular material structuralizes into two regions during interface shearing: the interface zone and the upper zone. Anisotropy in the interface zone is increased and a higher deviator fabric δ_D is induced by shearing when S is smaller.

(2) The interface roughness R_n affects the shearing behavior of the interface. The interface friction angle ϕ_p ascends with the increase of R_n and reaches to a plateau value. The growing rate is associated with particle sphericity S. A thicker localized band is observed in the IST featuring a rougher interface.

(3) The shear strength of the interface is affected by the initial fabric (particle orientation) of the specimen. The peak shear stress increases as the particle orientation increases. The initial fabric is associated with the interaction between the particles and rough interface, i.e., restricts or triggers the motions of particles. The specimen with an inclined angle $\theta_p = 135°$ shows the largest dilation; in contrast, the one with horizontally placed clumps dilates less than

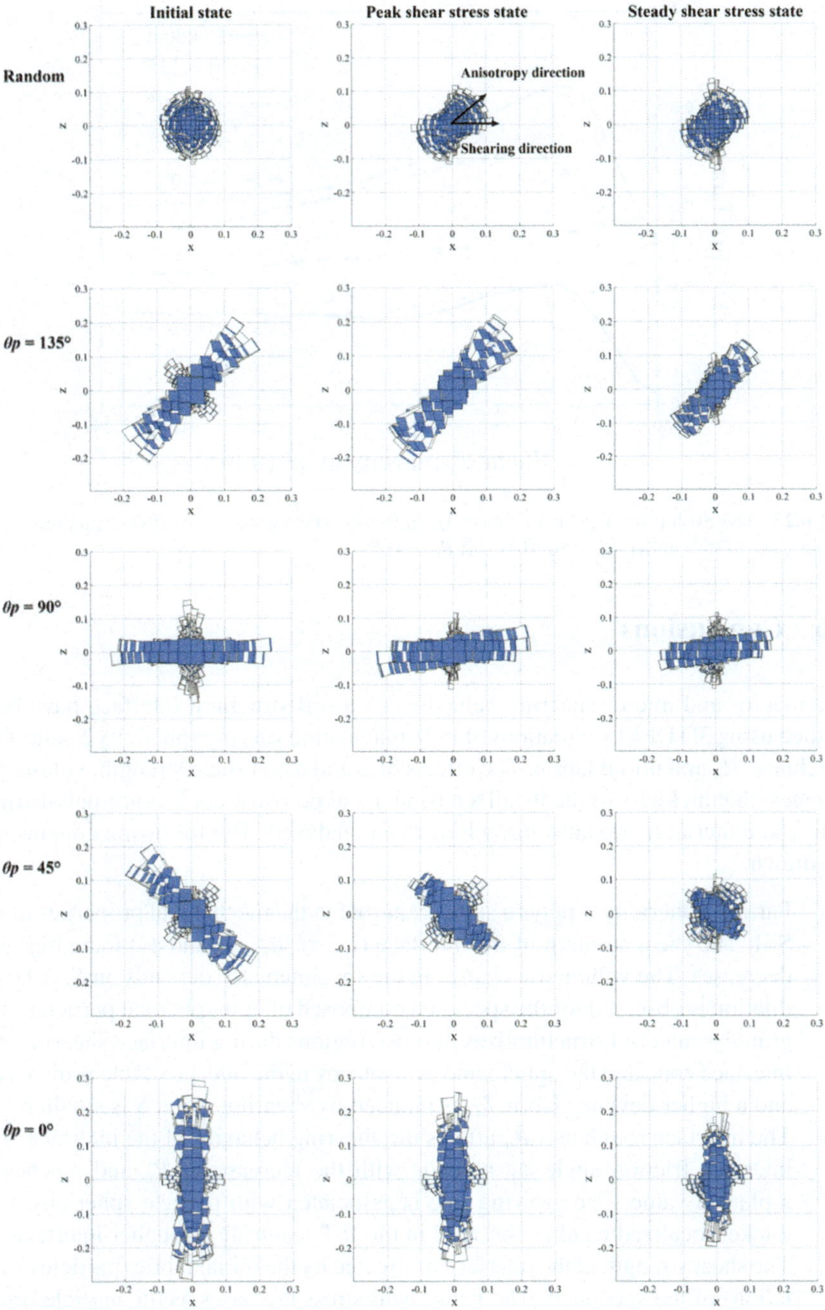

Fig. 6.24 The contact normal distribution in the interface zone of the five ISTs (random, $\theta_p = 0°/45°/90°/135°$) under $\sigma_n = 50$ MPa and $R_n = 0.5$ at initial state, peak shear stress state, and steady shear stress state

the others. The thickness of the localized band δ_h depends on the initial fabric. A thicker localized band is formed in a specimen with inclined clumps ($\theta_p = 45/135°$) and randomly distributed clumps. This tendency is generally valid under varying normal stress conditions. The given particle orientation leads to the different initial fabric of the specimen. The initial fabric affects the evolution of $P(\overrightarrow{n})$ and δ_D.

It is noted that this chapter has only examined the effect of sphericity S of irregular particles. Particle shape in nature is more random and complicated. To extend the study, other shape parameters should be considered in the future. Despite these limitations, this chapter clearly indicates the significant effect of S and its correlation with interface shear strength. The analysis of the micro-quantities, including the contact normal distribution, the motion of the particle, and the local porosity distribution, improve our understanding of the micro-mechanisms associated with soil-interface shearing.

References

1. Miura K, Maeda K, Furukawa M, Toki S (1998) Mechanical characteristics of sands with different primary properties. Soils Found 38:159–172
2. Nakata Y, Kato Y, Hyodo M et al (2001) One-dimensional compression behaviour of uniformly graded sand related to single particle crushing strength. Soils Found 41:39–51
3. Santamarina JC, Cho GC (2004) Soil behaviour: the role of particle shape. In: Advances in geotechnical engineering: The skempton conference. London, pp 604–617
4. Oda M, Nemat-Nasser S, Konishi J (1985) Stress-induced anisotropy in granular masses. Soils Found 25:85–97
5. Rothenburg L, Bathurst RJ (1992) Micromechanical features of granular assemblies with planar elliptical particles. Géotechnique 42:79–95
6. Krumbein WC, Sloss LL (1951) Stratigraphy and sedimentation. LWW
7. Chang CS, Yin ZY (2009) Micromechanical modeling for inherent anisotropy in granular materials. J Eng Mech 136:830–839
8. Yin ZY, Chang CS, Hicher PY (2010) Micromechanical modelling for effect of inherent anisotropy on cyclic behaviour of sand. Int J Solids Struct 47:1933–1951
9. Jensen RP, Bosscher PJ, Plesha ME, Edil TB (1999) DEM simulation of granular media - structure interface: effects of surface roughness and particle shape. Int J Numer Anal Methods Geomech 23:531–547
10. Frost JD, Dejong JT, Recalde M (2002) Shear failure behavior of granular—continuum interfaces. 69:2029–2048
11. Wang J, Jiang M (2011) Unified soil behavior of interface shear test and direct shear test under the influence of lower moving boundaries. Granul Matter 13:631–641. https://doi.org/10.1007/s10035-011-0275-2
12. Jing XY, Zhou WH, Zhu HX, et al. (2017) Analysis of soil-structural interface behavior using three-dimensional DEM simulations. Int J Numer Anal Methods Geomech. https://doi.org/10.1002/nag.2745
13. Iwashita K, Oda M (1998) Rolling resistance at contacts in simulation of shear band development by DEM. J Eng Mech 124:285–292
14. Ai J, Chen JF, Rotter JM, Ooi JY (2011) Assessment of rolling resistance models in discrete element simulations. Powder Technol 206:269–282

15. Wensrich CM, Katterfeld A (2012) Rolling friction as a technique for modelling particle shape in DEM. Powder Technol 217:409–417
16. Lin X, Ng TT (1997) A three-dimensional discrete element model using arrays of ellipsoids. Géotechnique 47:319–329
17. Ni Q, Powrie W, Zhang X, Harkness R (2000) Effect of particle properties on soil behavior: 3-D numerical modeling of shearbox tests. In: Numerical methods in geotechnical engineering. ASCE Geotechnical Special Publication, pp 58–70
18. Lu M, McDowell GR (2007) The importance of modelling ballast particle shape in the discrete element method. Granul Matter 9:69–80. https://doi.org/10.1007/s10035-006-0021-3
19. Salot C, Gotteland P, Villard P (2009) Influence of relative density on granular materials behavior: DEM simulations of triaxial tests. Granul matter 11:221–236
20. Bono JP, Mcdowell GR (2015) An insight into the yielding and normal compression of sand with irregularly-shaped particles using DEM. Powder Technol 271:270–277
21. Cundall PA, Strack ODL (1979) A discrete numerical model for granular assemblies. Géotechnique 29:47–65
22. Mindlin RD, Deresiewicz H (1953) Elastic spheres in contact under varying oblique forces. J Appl Mech 20:327–344
23. Coetzee CJ (2016) Calibration of the discrete element method and the effect of particle shape. Powder Technol 297:50–70
24. Wood DM, Maeda K (2008) Changing grading of soil: effect on critical states. Acta Geotech 3:3–14. https://doi.org/10.1007/s11440-007-0041-0
25. Uesugi M, Kishida H (1986) Influential factors of friction between steel and dry sands. Soils Found 26:33–46
26. Hu LM, Pu JL (2005) Testing and modeling of soil-structure interface. J Geotech Geoenviron Eng 130:851–860
27. Paikowsky SG, Player CM, Connors PJ (1995) A dual interface apparatus for testing unrestricted friction of soil along solid surfaces. ASTM Geotech Test J 18:168–193. https://doi.org/10.1520/GTJ10320J
28. Uesugi M, Kishida H, Tsubakihara Y (1988) Behavior of sand particles in sand-steel friction. Soils Found 28:107–118
29. Rao KSS, Allam MM, Robinson RG (1998) Interfacial friction between sands and solid surfaces. In: Proceedings of the ICE—Geotechnical Engineering. pp 75–82
30. Chen X, Zhang J, Xiao Y, Li J (2015) Effect of roughness on shear behavior of red clay—concrete interface in large-scale direct shear tests. Can Geotech J 52:1122–1135. https://doi.org/10.1139/cgj-2014-0399
31. Frost JD, DeJong JT, Recalde M (2002) Shear failure behavior of granular-continuum interfaces. Eng Fract Mech 69:2029–2048. https://doi.org/10.1016/S0013-7944(02)00075-9
32. Cho GC, Dodds J, Santamarina JC (2006) Particle shape effects on packing density, stiffness and strength: natural and crushed sands. J Geotech Geoenviron Eng 132:591–602
33. Wu XY, Yang J (2016) Direct shear tests of the interface between filling soil and bedrock of chongqing airport. In: Advances in Civil, Envrionmental, and Materials Research (ACEM16). Jeju Island, Korea
34. Su LJ, Zhou WH, Chen WB, Jie X (2018) Effects of relative roughness and mean particle size on the shear strength of sand-steel interface. Measurement 122:339–346
35. Wang JF, Marte SG, Joseph ED (2007) Numerical studies of shear banding in interface shear tests using a new strain calculation method. Int J Numer Anal Methods Geomech 31:1349–1366
36. Satake M (1982) Fabric tensor in granular materials. In: Proc., IUTAM Symp. on Deformation and Failure of Granular materials. Delft, The Netherlands, pp 1525–1533
37. Barreto D, O'Sullivan C, Zdravkovic L (2009) Quantifying the evolution of soil fabric under different stress paths. In: AIP Conference Proceedings. London, pp 181–184

Chapter 7
Multi-directional Soil-Structure Interface Shearing

7.1 Introduction

The surface's roughness is regarded as one of the key factors affecting the mechanical behavior of the SSI; its effect on the interactive mechanism between the granular assembly and the rigid surface under numerical and experimental interface shear tests (ISTs) has been highlighted repeatedly in previous studies [1–10]. In those studies, the surface roughness used to be quantified as one or more parameters. The most commonly used parameter is the relative roughness (R_n) as proposed by Uesugi and Kishida [11, 12]. The researchers have tended to assume, explicitly or implicitly, that roughness is an isotropic phenomenon, which ignores the role of the rough surface's directional properties. In fact, the surface normally possesses highly directional properties caused by the properties of the material and the manufacturing process. Samanta et al. [13] characterized the directional properties of the rough surface as surface texture and conducted a series of experimental interface shear tests under multi-shearing directions to investigate the influence of steel's surface texture orientation (relative to the shear direction) on the shear behavior of the sand-steel interface. The results demonstrated that the surface with ruled ribs, inclined at a right angle to the shear directions, showed the maximum interface shear strength. However, only the shear resistance along the shear direction was discussed from a macroscale perspective. The true macroscale shear behavior of an SSI and its relationship with microscale properties, when the granular assembly touching a directional rough surface is sheared by different directions, remain not well understood.

For a better understanding of the micro- and macro-mechanical interface behavior, laboratory tests have been undertaken to visualize and quantify the evolution of the soil microstructure during shearing, with the help of image analysis and the photo-elasticity technique [14, 15]. Such studies have yielded valuable and fundamental results. However, these methods only provide limited and qualitative information. Alternatively, the discrete element method (DEM) provides a powerful tool

W.-H. Zhou and Z.-Y. Yin, *Practice of Discrete Element Method in Soil-Structure Interface Modelling*, https://doi.org/10.1007/978-981-19-0047-1_7

for simulating and reproducing macroscale behavior and at the same time investigating the microstructure of a granular assembly. Two-dimensional (2D) and three-dimensional [16] DEM simulations have been successfully applied to a range of problems encountered with SSIs [8, 10, 17–30]. Compared with a 3D DEM simulation, the 2D approach is less time-consuming, but it fails to simulate multi-directional ISTs. Therefore, the 3D DEM is used to simulate the multi-directional ISTs in the present chapter.

The goal of this chapter is to provide a new perspective on the consistent soil-structure interface response in multi-directional shear tests using DEM. The consistent response of the macroscale shear behavior at the interface under multi-directional shearing is presented from the perspective of the resultant average shear stresses in x and y directions and volumetric strain. A determination of the thickness of the localized band in each specimen is followed by the discussion of particle displacement vector distribution, and some attempts are made to show the evolution of the coordination number and material fabric, in an effort to discuss their relationships with macroscale response and further confirm the consistent response of the macroscale shear behavior.

7.2 Numerical Modelling

7.2.1 Model of Interface Direct Shear Test

A shear box (length = 80 mm, width = 80 mm, and height = 34 mm), comprised of four rigid lateral walls and an upper boundary for applying normal stress, was placed on a bottom tooth-plate to simulate the interface direct shear apparatus. A regular arrangement of small teeth into a sequence formed the teeth-plate, and each tooth was shaped by two side segments, both of which were elevated 45° from the bottom, as schematically shown in Fig. 7.1a. The relative roughness R_n proposed by [11, 12] was used to characterize the roughness of the surface. This parameter can be expressed by $R_n = R_{max}/D_{50}$, where R_{max} is the perpendicular distance between the highest peak and the lowest valley on the surface profile and D_{50} is the medium particle size. To investigate the mechanical behavior of interface under multi-directional shear test, a shear orientation was introduced in this chapter. The shear orientation was defined as an angle θ between the shear direction and the principal rough direction (PRD), as shown in Fig. 7.1. The PRD on the rough surface was vertical to the texture line along the y' axis of local coordinate system. Different θ were achieved by changing the PRD, while the shear direction was kept along the x-direction of global coordinate system.

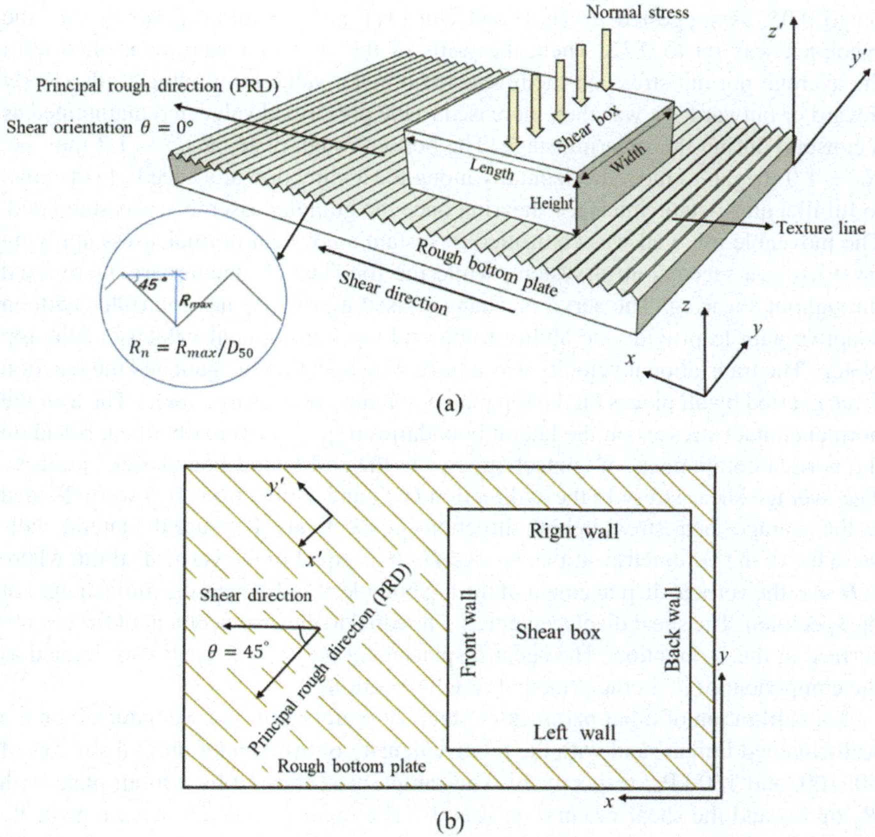

Fig. 7.1 Schematics of the numerical interface shear apparatus: **a** lateral view ($\theta = 0°$) and **b** vertical view ($\theta = 45°$)

7.2.2 *Simulation Process and Calibration of Input Parameters*

The numerical simulation in this chapter was carried out with a 3D-particle flow code (PFC^{3D}), based on DEM developed by Itasca. Particles with different shapes contact each other on surfaces in reality, and thus a moment of rolling resistance exists when one particle rolls over another. Therefore, to compensate the rolling resistance effect of grain shape on shear behavior in soil-structure interaction, a rolling resistance linear contact model was used in the present chapter. This model was also used to simulate the interaction between sand and structure in some previous studies [9, 21].

Based on a particle size distribution with medium particle size D_{50} of 1.7 mm and coefficient of uniformity C_u of 1.25, around 65,000 spherical particles in each test were generated inside the shear box shown in Fig. 7.1a. To achieve a relatively dense granular material, the initial frictional coefficient between particles f_p was set to 0.05, as suggested by Härtl and Ooi [31], and the initial porosity n of the specimen was set to 0.32. Then, the walls of the container were controlled until the average normal stress on them reached a given value by moving walls slowly inward or outward. f_p was then increased to the prescribed value and maintained as a constant during the shearing stage. The bottom rough plate ($R_{max} = 1.7$ mm, i.e. $R_n = 1.0$) began to move horizontally along the x-direction at a speed of 0.1 mm/s to fulfill a quasi-static interface shearing once the granular assembly was stabilized. The moveable top wall was controlled to sustain a constant normal stress applying on it using a servo control system, while the four lateral counterparts were fixed throughout shearing. The servo mechanism used a proportional controller with an adaptive gain to provide the ability to control the translational velocity of the top platen. The translational velocity was adjusted by updating the gain and the reaction force exerted by all pieces on the top platen to maintain a desired force. The average normal contact stresses on the lateral boundaries $\sigma^n_{ball-wall}$ were obtained, based on the normal components of contact stresses at the soil-lateral boundaries' contacts. The average shear stress in the x-direction (τ_{zx}) and y-direction (τ_{zy}) were defined as the average shear stress in both directions, respectively. Because the lateral walls were fixed, the volumetric strains $\varepsilon_v = \Delta H/H$ is equal to the vertical strain, where ΔH was the vertical displacement of the top boundary and H was the initial height of the specimen. The shear displacement d_x was equal to the displacement of the bottom surface in the x-direction. The shear displacement $d_p = d_x \times \cos\theta$ was defined as the component of d_x in the principal rough direction.

For calibration of input parameters, three laboratory tests were conducted on the well-rounded Fujian sand with the relative density of 90% under normal stresses of 50, 100, and 150 kPa, respectively. The sample was sheared by a tooth-plate with R_n of 1.0 and the shear orientation was 0°. The input parameters were repeatedly adjusted until the simulation results matched those of the laboratory tests. The calibration result is shown in Fig. 7.2. The curves of the shear stress or volumetric strain-displacement relationship, obtained from both laboratory and numerical ISTs,

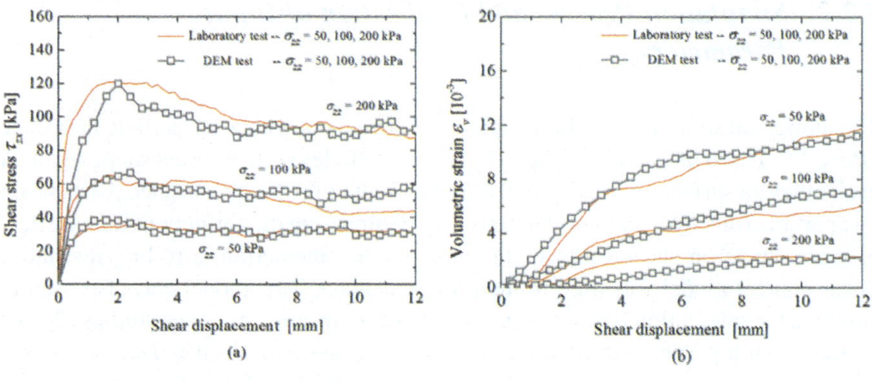

Fig. 7.2 Calibration results: **a** shear stress versus shear displacement and **b** volumetric strain versus shear displacement in laboratory tests and numerical simulations

Table 7.1 Summary of calibrated input parameters used in DEM simulations

Parameters	Value
Ball density (kg/m^3)	2650
Damping coefficient	0.7
Coefficient of uniformity C_u	1.25
Inter-particle normal stiffness k_n (N/m)	0.7×10^7
Inter-particle shear stiffness k_t (N/m)	0.35×10^7
Particle-wall normal stiffness k_{nw} (N/m)	1.0×10^8
Particle-wall shear stiffness k_{tw} (N/m)	0.5×10^8
Inter-particle frictional coefficient f_p	0.3
Particle-lateral boundaries frictional coefficient f_{pw}	0.2
Particle-bottom wall frictional coefficient f_{pbw}	0.05
Rolling resistance coefficient μ_r	0.05

demonstrate a similar tendency in general. The calibrated input parameters of the simulations are summarized in Table 7.1.

Using the said test process, in this chapter, a series of simulated ISTs placing on the directional rough plate ($R_n = 1.0$), under three normal stresses σ_{zz} (50, 100, and 150 kPa), were sheared under seven shear orientations ($\theta = 0°$, 10°, 30°, 45°, 60°, 75°, and 90°), respectively. The schematics of ISTs with different shear orientations is listed in Table 7.2. Additionally, the smooth IST ($R_n = 0.0$) and the soil–soil direct shear test were performed for comparison.

7.3 3D DEM Modelling Results

7.3.1 *Average Shear Stresses in Two Directions*

Figure 7.3 plots the relationship between average normal contact stresses on the four lateral boundaries $\sigma^n_{ball-wall}$ and shear displacement d_x for specimens, that were sheared along seven shear orientations ($\theta = 0$, 10, 30, 45, 60, 75, and 90°) under normal stresses of $\sigma_{zz} = 50$ kPa, $\sigma_{zz} = 100$ kPa, and $\sigma_{zz} = 150$ kPa. As shown in Fig. 7.3(1), as the shearing progresses, the average normal contact stress on the front boundary for all cases increases at the outset of shearing and remains a relatively stable state at the large shear displacement. In contrast, the reduction of average normal contact stress on the back boundary is followed by a stable state. Moreover, such increases and decreases of normal contact stresses on the front and back are more obvious as θ decreases. As for the average normal contact stresses on the right and left boundaries, they remain constant for $\theta = 0°$ and $\theta = 90°$ throughout the shearing. These two cases are similar to both the conventional numerical and laboratory interface shear tests [8, 11, 32], which only suffers the change of normal

Table 7.2 Schematics of multi-directional interface shear tests

Shear orientation	Schematic diagram of the numerical interface shear test (Overhead view)	Shear orientation	Schematic diagram of the numerical interface shear test (Overhead view)
$\theta = 0°$	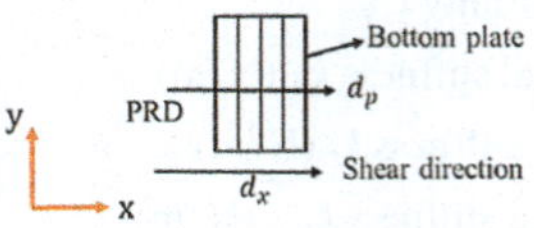	$\theta = 60°$	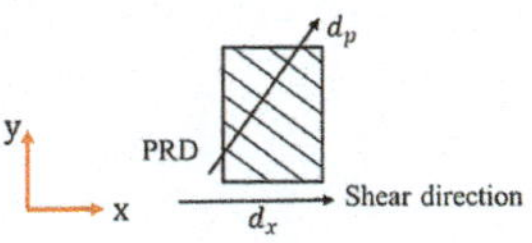
$\theta = 10°$	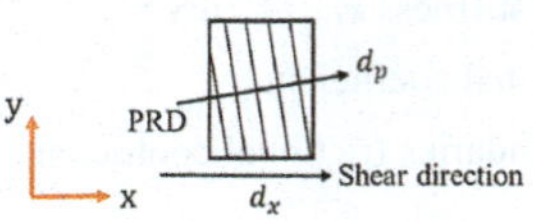	$\theta = 75°$	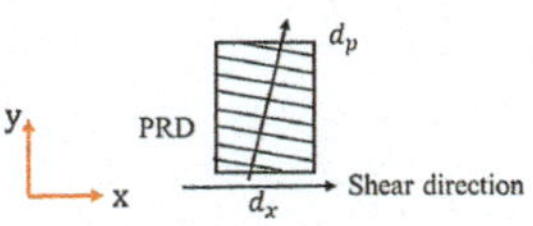
$\theta = 30°$	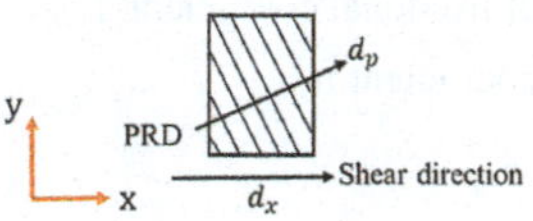	$\theta = 90°$	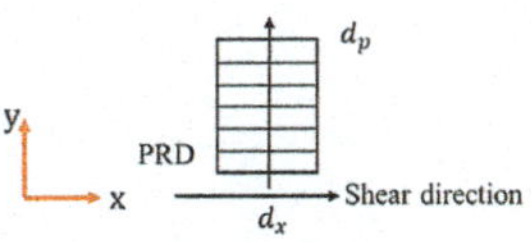
$\theta = 45°$	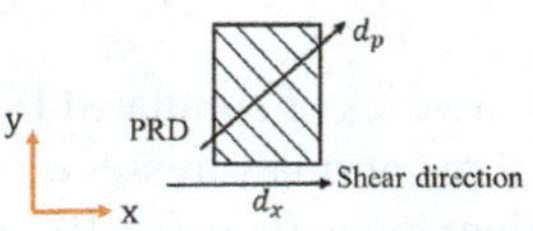		

contact stresses on the two vertical boundaries vertical to the shear direction (i.e. the front and back boundaries). For $\theta = 10$, 30, 45, 60, and 75°, it is observed that the normal contact stresses on the right and left boundaries no longer remains constant. Specifically, the normal contact stress on the left boundary increases at the outset of shearing and stabilizes at the end of shearing, while that on the right boundary suffers an opposite change. Furthermore, this change becomes stronger as θ increases and leads to the occurrence of average shear stress in y-direction.

The evolution of the average shear stress in the x-direction τ_{zx} and that in the y-direction τ_{zy}, with respect to shear displacement d_x, is presented in Fig. 7.4 for the seven specimens ($\theta = 0$, 10, 30, 45, 60, 75, and 90°) under normal stresses of $\sigma_{zz} = 50$ kPa, $\sigma_{zz} = 100$ kPa, and $\sigma_{zz} = 150$ kPa. As can be seen from Fig. 7.4(1), a general observation is that a lower θ gives rise to τ_{zx} and finally produces a higher peak value. When $\theta = 90°$, τ_{zx} increases dramatically at the outset of shearing and peaks at approximate $d_x = 0.15$ mm, following an extremely straight line of τ_{zx} at the end of shearing. In such a case, the sand particles tend to slide more easily along the groove between saw-teeth elements and the failure occurs due to the sliding of the sand particles over the bottom plate. When $\theta = 0°$, namely, the shear direction is completely parallel to the PRD, and the curve of τ_{zx} show extinct fluctuations after climbing to a peak value, with stress-softening behavior. In this case, the translation and rotation of soil particles adjacent to the rough plate are mainly restricted by the saw-teeth element distributing on the bottom plate surface in the process of shearing,

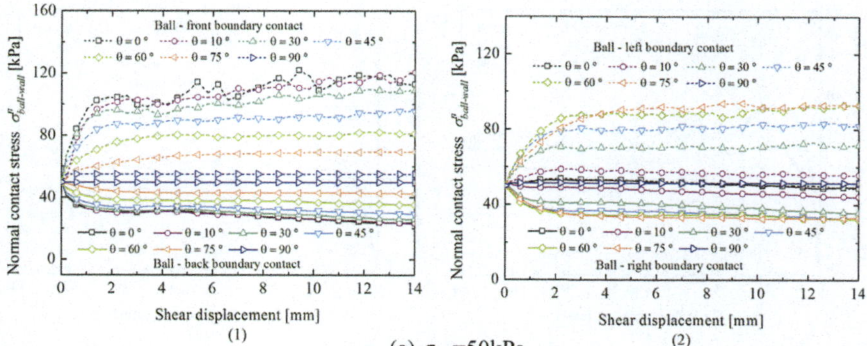

(a) σ_{zz}=50kPa

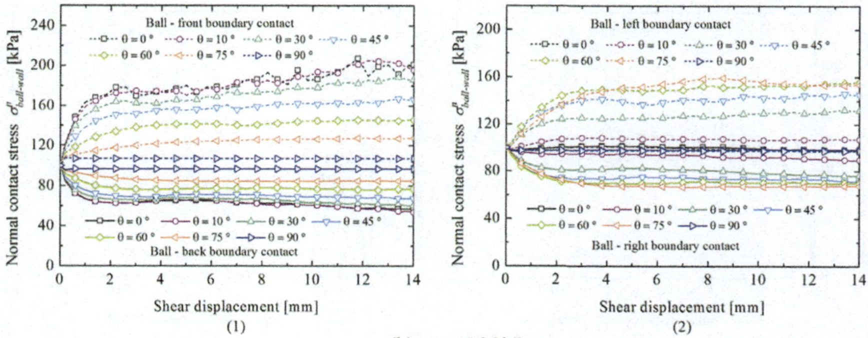

(b) σ_{zz}=100kPa

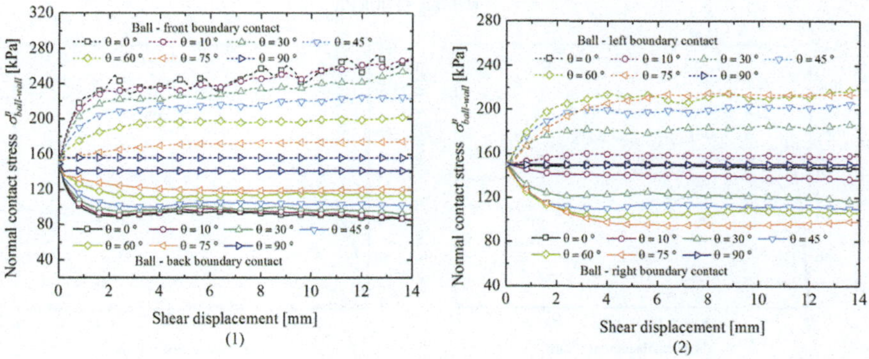

(c) σ_{zz}=150kPa

Fig. 7.3 Relationship between average normal contact stresses on the four lateral boundaries $\sigma^n_{ball-wall}$ and shear displacement d_x for seven specimens ($\theta = 0°$, 10°, 30°, 45°, 60°, 75°, and 90°) under normal stresses: **a** $\sigma_{zz} = 50$ kPa, **b** $\sigma_{zz} = 100$ kPa, and **c** $\sigma_{zz} = 150$ kPa

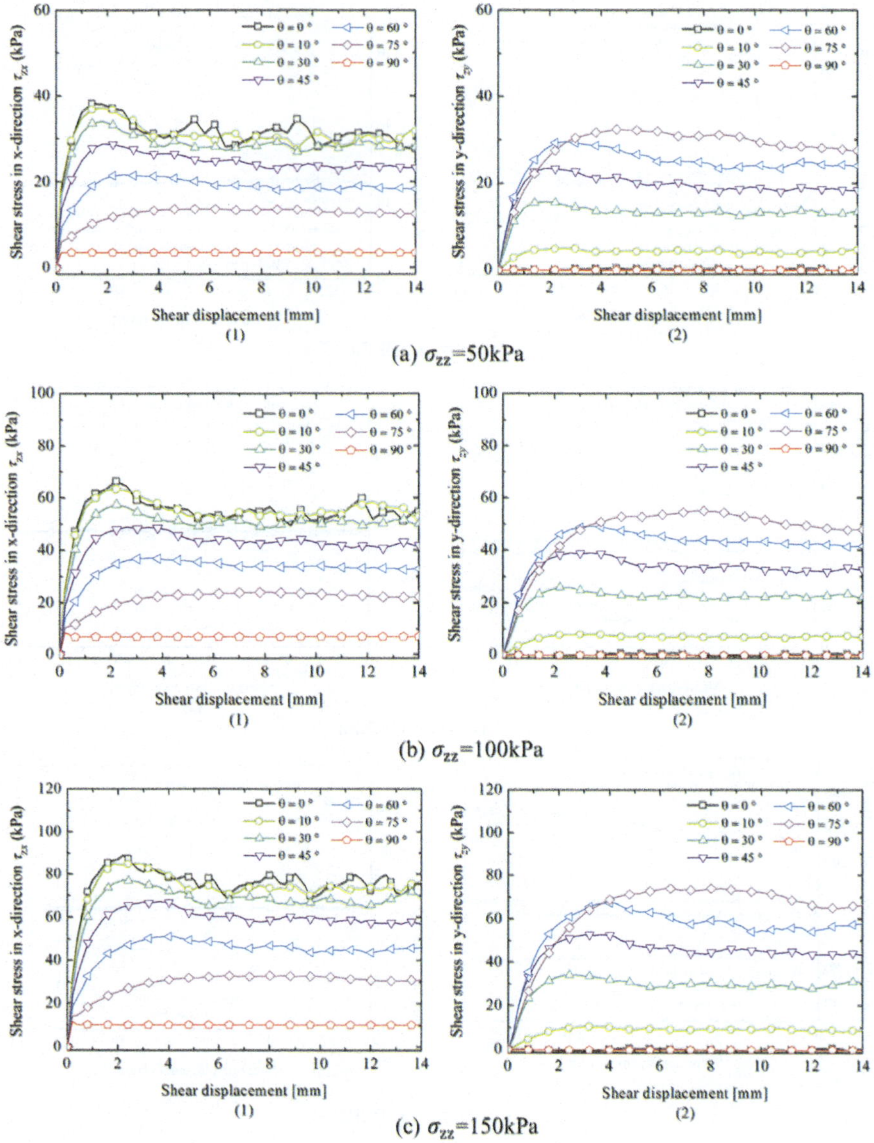

Fig. 7.4 Average shear stress in x-direction and that in y-direction with respect to shear displacement d_x for seven specimens under normal stresses: **a** $\sigma_{zz} = 50$ kPa, **b** $\sigma_{zz} = 100$ kPa, and **c** $\sigma_{zz} = 150$ kPa

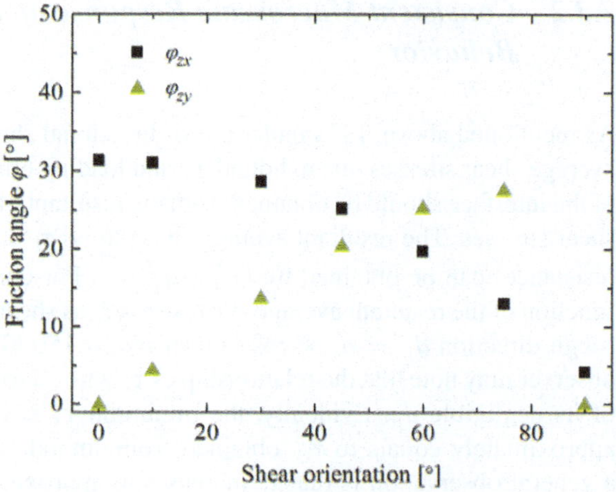

Fig. 7.5 Variation of the friction angles in ***x***-direction and ***y***-direction with the shear orientation

which results in fully mobilizing shear resistance. The reader may see from Fig. 7.4 that for $\theta = 0°$ and $\theta = 90°$, τ_{zy} remains negligible throughout shearing. However, τ_{zy} for the other cases increases at the beginning of shearing and remains stable at a large shear displacement. Moreover, τ_{zy} levels up as θ increases.

The relationship between the shear strength τ_{zx}^{peak} in x-direction and τ_{zy}^{peak} in y-direction and normal stress σ_{zz} for all ISTs were fitted using Mohr–Coulomb line: $\tau = c + \sigma_{zz} \times tan(\varphi)$, where c is the cohesion and φ is the interface friction angle. c was set to zero in the present chapter, and the fitting coefficient R^2 ranges from 0.92 to 0.98. Figure 7.5 shows the variation of the corresponding friction angles φ_{zx} and φ_{zy} with the shear orientation θ for all ISTs. As shown in Fig. 7.5, the maximum φ_{zx} is observed when $\theta = 0°$ and decreases with an increase in θ up to a minimum value ($\theta = 90°$). The increase in θ decreases the restraint against the particle sliding along the direction of the texture line. Consequently, more and more particles adjacent to the asperities may slide along the direction of the texture line and lead to the mobilization of particle-bottom plate friction. Until $\theta = 90°$, that is, the shear direction is parallel to the PRD, most sand particles in the vicinity of the surface slide along the groove between saw-teeth elements and fully mobilize the friction between the particle and bottom plate. The failure occurs due to the sliding of the sand particles over the interface. It may further be inferred that the shear strength for such a case depends, to a large extent, on the friction force between the particles and bottom plate. The fact is verified by the subtle difference between the friction angle ($\varphi_{zx}^{smooth} = 3.89°$) obtained from the smooth interface shear test and $\varphi_{zx} = 4.16°$ for $\theta = 90°$. It can be also observed that for $\theta = 0°$ and $\theta = 90°$, φ_{zy} remains negligible due to no displacement of particles in y-direction. However, φ_{zy} levels up as θ increases.

7.3.2 Consistent Macroscale Response of True Shear Behavior

As mentioned above, ISTs under multi-directional shearing possess two directional average shear stresses due to boundary and loading effects. The true shear resistance of the interface should be obtained from the resultant of these two directional average shear stresses. The resultant average shear stress τ_r on the interface (i.e., true shear resistance) can be obtained by $\left(\tau_{zx}^2 + \tau_{zy}^2\right)^{1/2}$. For convenience of comparison, the function of the resultant average shear stress τ_r as shear displacement in the principal rough direction $d_p = d_x \times \cos\theta$ under $\sigma_{zz} = 150$ kPa is shown in Fig. 7.6a. The observer may note that the relationship of τ_r with d_x for $\theta = 90°$ is presented because of its negligible d_p. Evidently, the minimum τ_r is observed for $\theta = 90°$, which approximately equals to τ_{zx} obtained from smooth shearing. For the other cases, a general observation is that τ_r increases to approximately the same peak value at the same displacement d_p of 2.2 mm before gradually decreasing to a steady state. According to the relationship of d_x and d_p, we can know that d_x, corresponding to the peak resultant average stress ratio, increases as θ increases. Figure 7.6b shows that dilative behavior occurs for all ISTs except $\theta = 90°$ and that the volumetric strain increases at a similar rate as d_p. A subtle difference exists in the magnitude of both the peak resultant average stress ratio and volumetric strain among these ISTs. Notably, a negligible shear dilatancy is evident when $\theta = 90°$. The consistency of results in Fig. 7.6 indicates that the magnitudes of both these stress–strain quantities are independent of the shear orientation θ. In other words, a consistent shear behavior exists in multi-directional shear tests. Moreover, the specimen with lower θ requires larger d_x to achieve the same value compared to the one with higher θ. The macroscale responses for the cases of $\sigma_{zz} = 50$ kPa and $\sigma_{zz} = 100$ kPa are not shown but similar consistent response was also found in general.

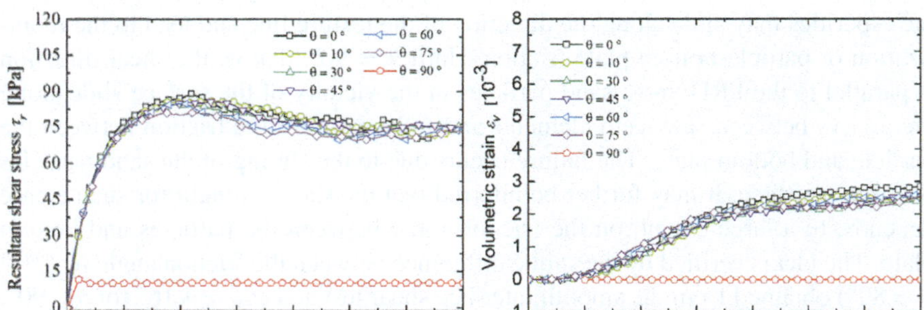

Fig. 7.6 **a** Evolutions of resultant average shear stress τ_r in terms of d_p for $\theta = 0°, 10°, 30°, 45°, 60°$ and $75°$ and versus d_x for $\theta = 90°$ under $\sigma_{zz} = 150$ kPa and **b** evolutions of volumetric strain ε_v in terms of d_p for $\theta = 0, 10, 30, 45, 60$ and $75°$ and versus d_x for $\theta = 90°$ under $\sigma_{zz} = 150$ kPa

The relationship between the resultant shear strength τ_r^{peak} and normal stress σ_{zz} for all ISTs was also fitted using Mohr–Coulomb line. The corresponding friction angles φ_r versus the shear orientation for all ISTs are shown in Fig. 7.7. Two additional series of ISTs under $R_n = 0.5$ and $R_n = 2.0$ were conducted to check the effect of R_n on consistent macroscale response in multi-directional shear tests, while the dependency of particle size distribution on consistent macroscale response in multi-directional shear tests was checked using the results obtained from two additional series of ISTs on samples with C_u of 1.5 and C_u of 1.9. Their corresponding results are all plotted in Fig. 7.7. Clearly, a slight difference of φ_r can be observed among the ISTs under $\theta = 0$, 10, 30, 45, 60, and 75° for the cases of $R_n = 0.5$, $R_n = 1.0$, and $R_n = 2.0$, which confirms that the resultant shear resistance is not affected by θ. For the results of ISTs on specimens with different C_u, it is reconfirmed that the resultant shear resistance is independent on θ. The minimum φ_r appears for the IST with $\theta =$ 90°, that is, $\varphi_r = 4.25°$ for $R_n = 0.5$, $\varphi_r = 4.16°$ for $R_n = 1.0$, and $\varphi_r = 4.2°$ for R_n = 2.0 respectively, which are closed to the friction angle ($\varphi_{zx}^{smooth} = 3.89°$) obtained from the smooth interface shear test. Some results selected from other literatures were also presented in Fig. 7.7 for comparison. As reported by the literatures, the friction angles of $R_n = 0.5$, $R_n = 1.0$, and $R_n = 2.0$ in the present chapter are also closed to that of the corresponding soil–soil direct shear tests. It implies that the shear failure occurs inside the soil and the interaction between particles and interface like the interaction among pure particles.

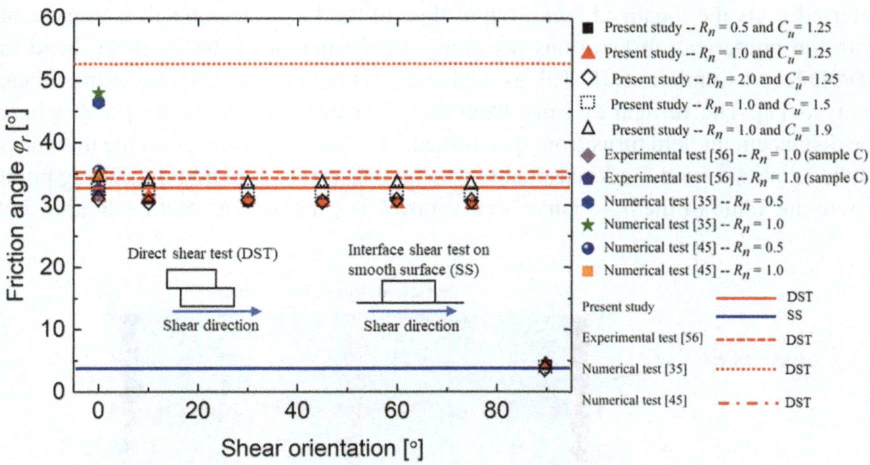

Fig. 7.7 Comparison between the present chapter and literature: relationship between shear orientation and friction angle under different $\boldsymbol{R_n}$ and different $\boldsymbol{C_u}$

7.3.3 Localized Band Thickness and Particle Displacement Vector

Soil failure in SSI is often accompanied by strain localization occurring close to the interface. The localized band can be analyzed by tracing the movements of each particle at a specific displacement state. At a given moment of the shearing process, the displacement field d_r in the granular system is represented by displacements on particle centers. To average the displacement field, we set a series of measuring units at different heights of the specimen at the end of shearing (Fig. 7.8). The average shear displacement $\overline{d_r}$ on layers of various depths can be calculated. As a result of the negligible d_p for $\theta = 90°$, the values of $\overline{d_r}$ as a function of z for $\theta = 90°$ at different shear displacements d_x under normal stress $\sigma_{zz} = 150$ kPa (R_n= 1.0) are plotted in Fig. 7.9a. The values of $\overline{d_r}$ as a function of z for six specimens ($\theta = 0$, 10, 30, 45, 60 and 75°) at different shear displacements in the principal rough direction d_p are also plotted in Fig. 7.9b. Each dot in the Fig. 7.9 represents one measurement unit at a given height of the specimen. Spline interpolation is applied to attain a smooth $\overline{d_r}$–z curve $f(z)$. As can be seen, with the progress of shearing, the $\overline{d_r}$–z curves, except $\theta = 90°$, are structured into two distinct zones (Fig. 7.9): Zone-I, the domain with a highly non-linear displacement above the interface surface, and Zone-II, the remaining domain, which possesses a relatively linear displacement. The phenomenon of localized band becomes most evident at $d_p = 7$ mm. The displacement induced by the interface shearing shows a large concentration in the former, referred to as the localized band, rather than in the latter. This result is consistent with the numerical observations regarding the formation of the localized band in 2D/3D DEM simulations [8, 10], as well as the laboratory experiments using image analysis [7]. The vertical distance from the interface to the boundary point, where the displacement field turns from quasi-linearity to non-linearity, gives the thickness of the localized band T_s. A reasonable way to measure T_s is to find the cutting point where the value of the $\overline{d_r}$–z curve's curvature k is equal to a given threshold a, i.e.

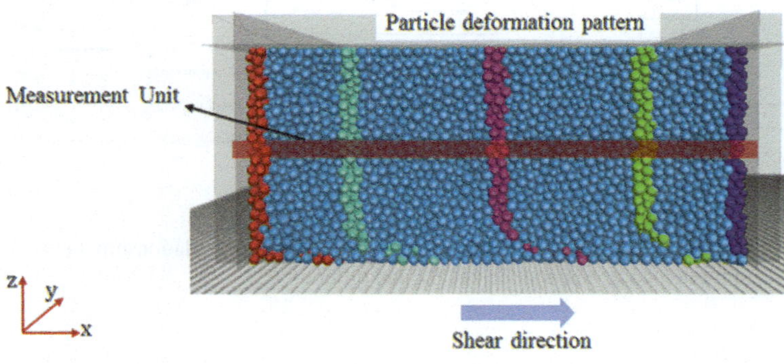

Fig. 7.8 Measurement unit with dimensions of 80 mm × 80 mm × 2 mm in the specimen at the end of shearing

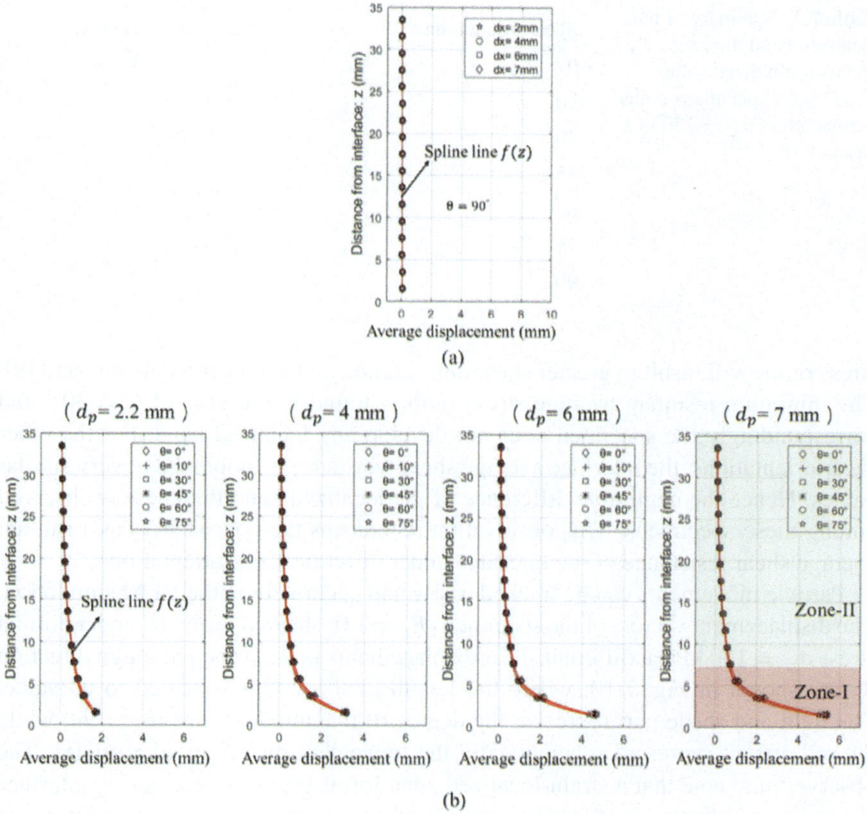

Fig. 7.9 **a** Values of $\overline{d_r}$ as a function of z for $\theta = 90°$ at different d_x and **b** values of $\overline{d_r}$ as a function of z for six specimens (θ = 0, 10, 30, 45, 60 and 75°) under σ_{zz} = 150 kPa (R_n = 1.0) at different d_p

$k = \frac{|f''(z)|}{(1+f'(z)^2)^{3/2}} < a$, where $f'(z)$ and $f''(z)$ are, respectively, the first- and second-order derivatives of $f(z)$. Jing et al. [8] suggested that the inflection point is where k equals 0.02.

According to the previously described definition method, the localized band thicknesses T_s for the seven specimens are determined and listed in Table 7.3. In general, the thicknesses range from 6.8 to 7.2 mm, which are approximately 4.0–4.2 times the D_{50} = 1.7 mm. As seen in Table 7.3, for $\theta = 90°$, no localized band is evident. The observation confirms that the particles for $\theta = 90°$ tend to slide along the groove, with negligible disturbance to the soil sample, which is consistent with the negligible contraction observed in the case of $\theta = 90°$ (see Fig. 7.6b). As θ decreases, the shear failure plane may develop from interface to soil; eventually, the IST fails inside the soil. θ increases from 0° to 75°, and correspondingly the localized band thicknesses range from 6.8 to 7. 2 mm. Generally speaking, the surfaces that can sustain higher

Table 7.3 Summary of the localized band thickness $\boldsymbol{T_s}$ and its normalized value $\boldsymbol{T_s/D_{50}}$ for specimens under normal stress $\sigma_{zz} = 150$ kPa ($\boldsymbol{R_n} = 1.0$)

Shear orientation θ(°)	T_s(mm)	T_s/D_{50}
0	7.2	4.0–4.2
10	7.1	
30	7.1	
45	7.0	
60	6.8	
75	6.8	
90	0	0

stress ratios will result in greater shear deformation and a thicker localized band [9]. The minimum resultant average stress ratio is found in the case of $\theta = 90°$, and correspondingly the specimen does not develop any localized band. For the other shear orientations, the interface average shear resistance is mobilized to very similar values. Hence the negligible difference of the localized band thickness is observed among these specimens. This observation reconfirms the consistent peak resultant average shear resistance of the interface under different shear orientations.

Particle movements can be tracked easily and accurately in the DEM simulation. The displacement vectors of the specimen ($R_n = 1.0$) sheared by $\theta = 0°$ under normal stress $\sigma_{zz} = 150$ kPa at different shear displacements in the principal rough direction d_p are shown in Fig. 7.10, where the length of the vector is scaled to its value. The light and shade part represent the degree of magnitude of displacement, while the red arrows represent schematically the movement directions of particles. The observer may note that a strain-localized zone forms progressively during interface shearing, as indicated by the high gradient of displacement at the bottom interface. Soil movements gradually reduces and become negligible around the middle of the shear box, while a counterclockwise particle flow occurs inside the specimen at large values of d_p. Such a phenomenon results from the lateral boundary effect, wherein soil close to the front boundary experiences stress concentration and becomes dense, while that near the back boundary becomes loosened and is accompanied by stress relaxation, especially at the basal interface. A similar observation is reported by Gu et al. [33]. Figure 7.11 exhibits the vertical views of displacement vectors for six samples ($\theta = 0$, 10, 30, 45, 60, and 75°) under normal stress $\sigma_{zz} = 150$ kPa at the different d_p of 2.2 mm, 4 mm, and 7 mm, respectively, while that for $\theta = 90°$ is not shown herein due to reasons already explained before. As expected, for $\theta = 0°$, most of the soil particles move along the x-direction in the process of shearing, especially in the vicinity of the bottom interface. For $\theta = 10, 30, 45, 60$, and 75°, the movement directions of particles in said area are not consistent with the x-direction but are close to the principal rough direction during shearing. Such directional deviation becomes stronger. It indicates that the movement directions of particles under an interface shearing mainly depends on the interface shearing in the PRD other than shear direction, in other words, the interface shearing in the PRD can mobilize particles movements, forcing particles' movement directions to be closed to its direction. This

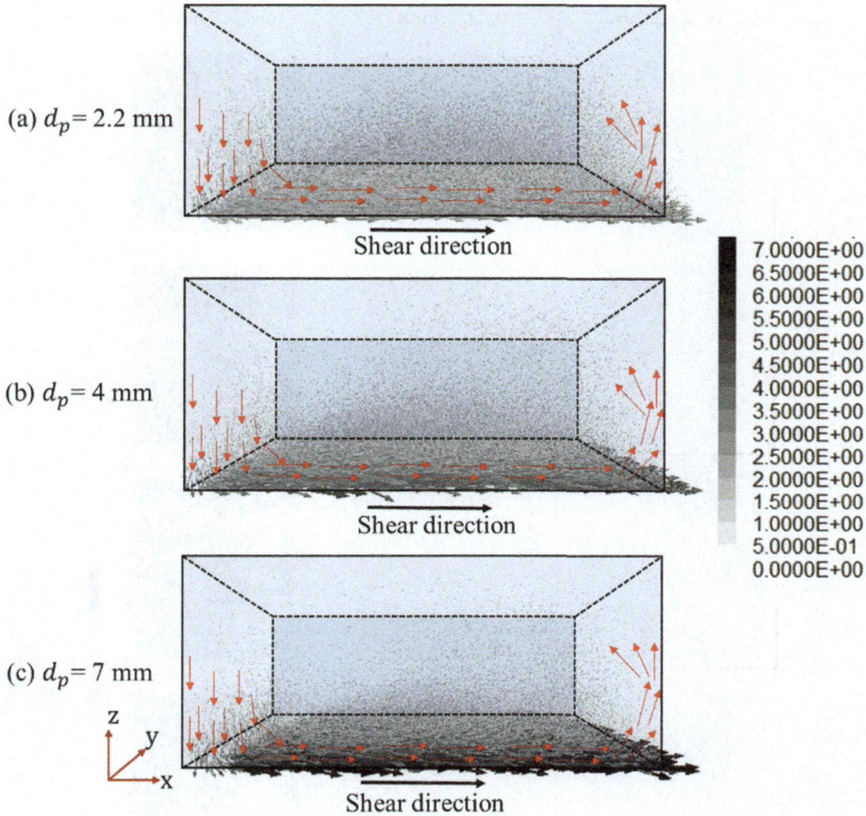

Fig. 7.10 Displacement vectors of the specimen ($R_n = 1.0$) sheared by $\theta = 0°$ under $\sigma_{zz} = 150$ kPa: **a** at $d_p = 2.2$ mm, **b** $d_p = 4$ mm, and **c** $d_p = 7$ mm

movement pattern will lead to the rearrangement of the normal contact force on the right and back boundaries and then the occurrence of the average shear stress in y-direction, as mentioned in Fig. 7.3.

7.3.4 Porosity and Coordination Number

The coordination number C_n is an important index of the stability potential of a granular assembly and is linked to the volumetric response at the macroscale behavior. It is defined as the ratio of $\sum_{N_p} N_c^p$ with N_p, where $\sum_{N_p} N_c^p$ is the number of contacts in particle p and N_p is the numbers of particles in the measured region. As mentioned earlier, after the localized band is fully developed, the granular assembly is divided in two regions, that is, within the localized band and outside its domain. The evolutions of the coordination number for $\theta = 0$, 10, 30, 45, 60, and 75° inside the localized

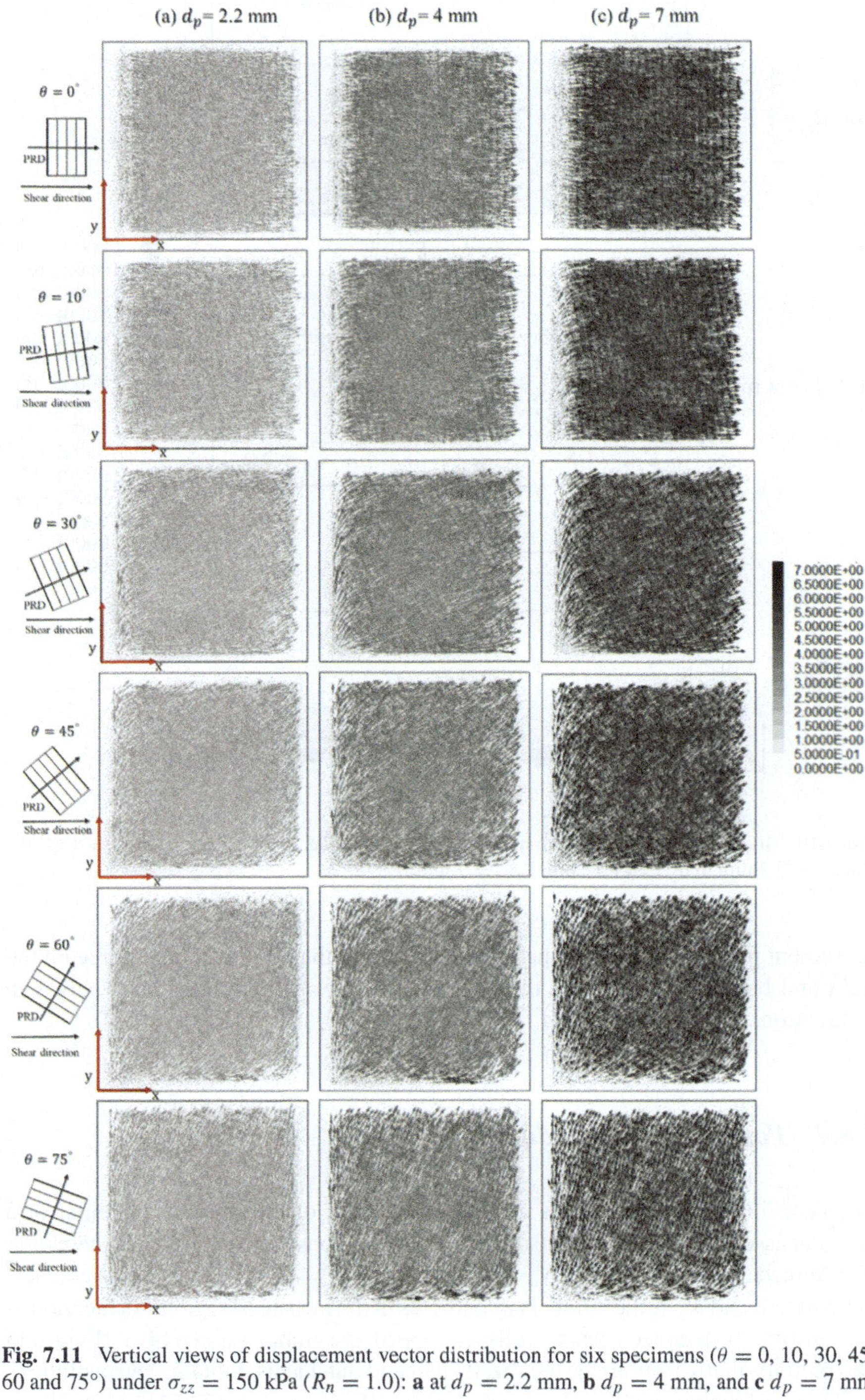

Fig. 7.11 Vertical views of displacement vector distribution for six specimens ($\theta = 0$, 10, 30, 45, 60 and 75°) under $\sigma_{zz} = 150$ kPa ($R_n = 1.0$): **a** at $d_p = 2.2$ mm, **b** $d_p = 4$ mm, and **c** $d_p = 7$ mm

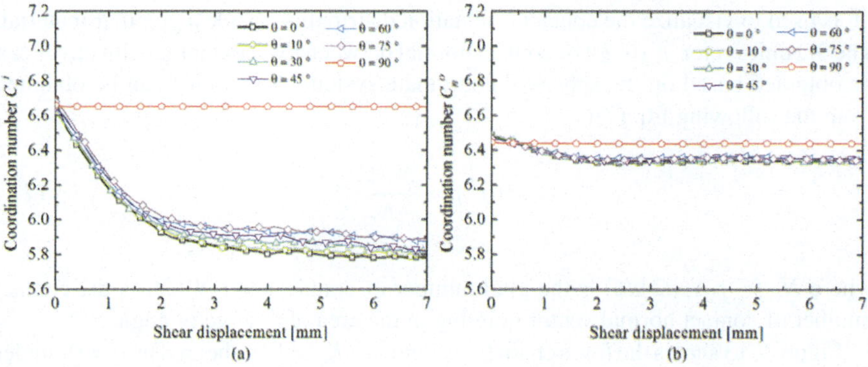

Fig. 7.12 Evolutions of coordination number C_n in terms of d_p for $\theta = 0°, 10°, 30°, 45°, 60°$ and $75°$ and with respect to d_x for $\theta = 90°$ under $\sigma_{zz} = 150$ kPa ($R_n = 1.0$) measured: **a** inside localized band C_n^i and **b** outside localized band C_n^o

band C_n^i and outside the localized band C_n^o, in terms of the shear displacement in the principal rough direction d_p ($\sigma_{zz} = 150$ kPa and $R_n = 1.0$), are illustrated in Fig. 7.12. For convenience of comparison, the relationship of the coordination number for $\theta = 90°$ inside the localized band C_n^i and outside the localized band C_n^o versus d_x is also presented. Noticeably, C_n suffers no change for $\theta = 90°$. However, C_n^i as well as C_n^o for other cases decreases at a similar rate during the process of shearing, respectively. The decreasing in coordination number implies that the granular material at the interface shearing loses the contact. The results explain the dilative response in the six ISTs (Fig. 7.6b) and why a similar degree of dilation exists among these six specimens, while the IST for $\theta = 90°$ suffers negligible volumetric change, as shown in Fig. 7.6b. This observation reconfirms that a consistent volumetric strain behavior exists in an interface shearing that is sheared by different shear orientations. Furthermore, C_n that is produced by a higher θ requires more shear displacement to achieve a similar value to those caused by a lower θ. The reader may note that the values of the coordination number, measured within and outside the localized band, have a slight difference in the initial state of shearing, which is probably caused by the normal pressure applied to the top of the specimen before shearing [8]. As the shearing progresses, the different characteristics at these two domains become increasingly obvious, and the microscale of the granular assembly takes a turn for the heterogeneous. Clearly, the change of C_n^i in the entire process of shearing largely exceeds the variations of C_n^o.

7.3.5 *Material Fabric*

The distribution of the contact orientation is used frequently to describe the material fabric. The probability density distribution $P(\vec{n})$ of a unit vector of contact normal

$\overrightarrow{n}$ is used to visualize the contact orientation distribution inside a granular material. The 3D unit vector $\overrightarrow{n}(\theta, \varphi)$ of contact normal between two contacting objectives can be obtained based on the spherical coordinate system. The $P(\overrightarrow{n})$ can be obtained from the following Eq. (7.1)

$$P(\overrightarrow{n}) = \frac{N_c(\Omega)}{N_c} \tag{7.1}$$

where $N_c = \int N_c(\Omega)d\Omega$ is the total number of contacts and $N_c(\Omega)$ is the contact number of contact normal vector pointing in the area of a range of angle $d\Omega$.

Figure 7.13 shows the force chain for specimen ($R_n = 1.0$) sheared by $\theta = 0°$ under normal stress $\sigma_{zz} = 150$ kPa at different shear displacements and their corresponding

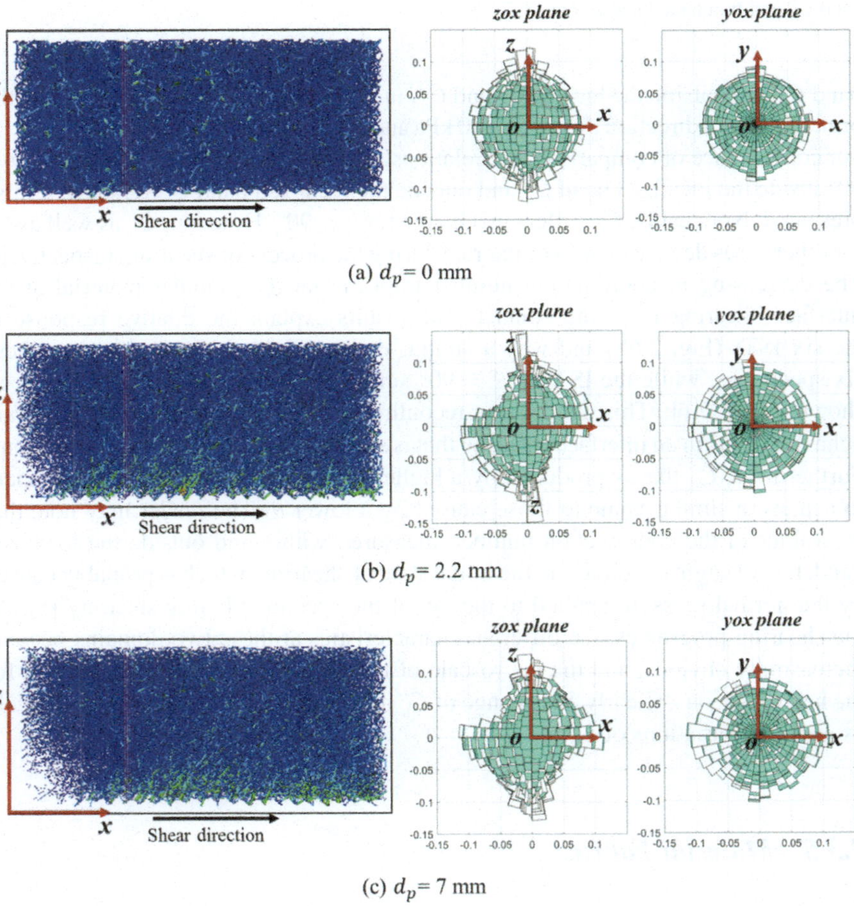

Fig. 7.13 Force chain for specimen ($\boldsymbol{R_n} = 1.0$) sheared by $\boldsymbol{\theta} = 0°$ under $\boldsymbol{\sigma_{zz}} = 150$ kPa at different shear displacements and their corresponding $\boldsymbol{P(\overrightarrow{n})}$ measured in the localized band: (**a**) at $\boldsymbol{d_p} = 0$ mm, (**b**) $\boldsymbol{d_p} = 2.2$ mm, and (**c**) $\boldsymbol{d_p} = 7$ mm

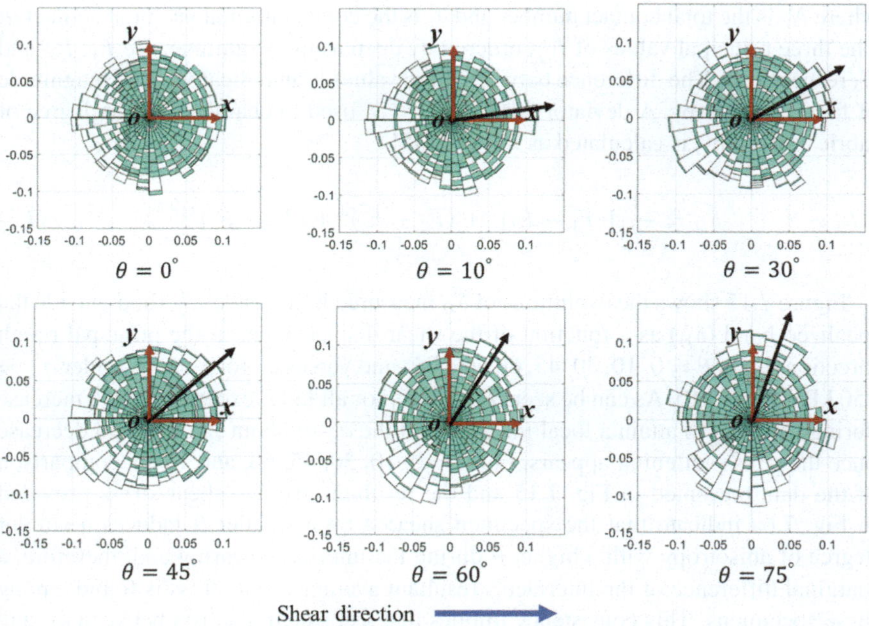

Fig. 7.14 Contact normal distributions (*yox* plane) in the interface zone of six specimens under $\sigma_{zz} = 150$ kPa ($R_n = 1.0$) at $d_p = 7$ mm

$P(\vec{n})$ measured in the localized band. The shape of $P(\vec{n})$ is closed to be spherical at initial state because of its approximately isotropic distribution of force chain. As shearing progresses, the force chains distributions and their corresponding $P(\vec{n})$ shapes gradually accumulate in a certain direction. The concentration of contact orientation is the result of particle contact rearrangement. Figure 7.14 presents $P(\vec{n})$ (*yox* plane) in the localized band of six specimens under normal stress $\sigma_{zz} = 150$ kPa ($R_n = 1.0$) at $d_p = 7$ mm. Unlike $\theta = 0°$, the $P(\vec{n})$ shapes in *yox* plane for $\theta = 10$, 30, 45, 60, and 75° rotate toward not only in *x*-direction but also in *y*-direction and such rotation is stronger for the higher θ, as shown in the black arrow in Fig. 7.14. Accordingly, the changes of interparticle contacts orientations can be attributed to the particle movement pattern mentioned in Fig. 7.11.

The interface that is subjected to shear loading can induce the movements of surrounding soil particles and subsequent changes of interparticle contacts. A second-order tensor F_{ij}, introduced by Satake and Masao [34], is used to quantitatively characterize the distribution in normal contact orientation:

$$F_{ij} = \frac{1}{N_c}\sum_{\alpha}^{N_c} n_i^{\alpha} n_j^{\alpha} (i, j = x, y, z) \tag{7.2}$$

where N_c is the total contact number and n_i is the contact normal vector at contact α. The three principal values of F_{ij}, ordered by decreasing magnitude, are F_1, F_2, and F_3 respectively. The difference between these values is an indicator of the magnitude of fabric anisotropy. A deviator fabric δ_D of F_{ij}, used to characterize the degree of fabric anisotropy, is calculated as follows [35].

$$\delta_D = \frac{1}{\sqrt{2}}[(F_1 - F_2)^2 + (F_2 - F_3)^2 + (F_1 - F_3)^2]^{0.5} \tag{7.3}$$

Figure 7.15 shows the evolutions of δ_D measured both inside (δ_D^i) and outside the localized band (δ_D^o) as a function of the shear displacement in the principal rough direction d_p for $\theta = 0$, 10, 30, 45, 60, and 75° and versus d_x for $\theta = 90°$ under $\sigma_{zz} =$ 150 kPa ($R_n = 1.0$). As can be seen, δ_D^i and δ_D^o for all ISTs, except $\theta = 90°$, increase during shearing to attain a local maximum at the d_p of about 2.2 mm and decrease once the stress softening appears. For θ = 0, 10, 30, 45, 60, and 75°, a comparison of the data presented in Fig. 7.15 and the resultant average shear stress provided in Fig. 7.6a indicate that the specimen sheared by a smaller θ induces a similar degree of anisotropy with a higher θ during the interface shearing and, therefore, a marginal difference of the interface's resultant average shear stress is found among these specimens. This consistency implies that a correlation exists between δ_D and the interface's resultant shear resistance. Such a correlation in direct shear tests was also observed by Cui and O'sullivan [36]. This observation further confirms the consistent shear strength behavior obtained from these ISTs. Moreover, δ_D^i suffers more evident changes compared with the variation of δ_D^o.

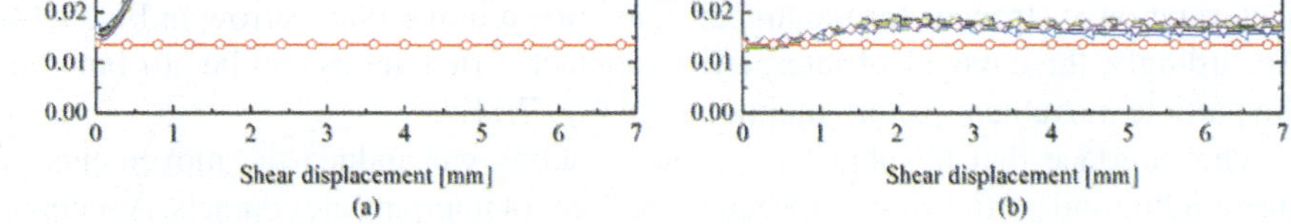

Fig. 7.15 Evolutions of δ_D in terms of d_p for $\theta = 0°$, 10°, 30°, 45°, 60° and 75° and versus d_x for $\theta = 90°$ under $\sigma_{zz} = 150$ kPa ($R_n = 1.0$) measured: (a) inside localized band (δ_D^i) and (b) outside localized band (δ_D^o)

7.4 Conclusions

This chapter analyzes the consistent granular material-structure interface response in multi-directional shear tests using DEM. The main conclusions are summarized as follows:

(1) The interface shearing under different shear orientations should be analyzed considering the average shear stresses in two directions. The friction angle φ_{zx} in x-direction decreases with an increase in the shear orientation θ. In contrast, the friction angle φ_{zy} in y-direction remains negligible for $\theta = 0°$ and $\theta = 90°$ and φ_{zy} levels up as θ increases. The average shear stresses in the x- and y-directions contribute to the resultant average shear stress. The minimum resultant friction angle φ_r and a negligible shear dilatancy appears for $\theta = 90°$. For $\theta = 0$, 10, 30, 45, 60, and 75°, a consistent interface shear behavior is evident.

(2) The movement directions of particles under an interface shearing mainly depend on the interface shearing in the principal rough direction other than shear direction. For $\theta = 90°$, no localized band is evident. As θ decreases, IST's failure plane moves from the interface into inside soil. For $\theta = 0, 10, 30,$ 45, 60, and 75°, the localized band thicknesses are 4.0–4.2 times the D_{50}. This observation confirms the consistent shear resistance observed in the interface under different shear orientations.

(3) Coordination number C_n suffers no change for $\theta = 90°$. However, the coordination number inside the localized band C_n^i and outside the localized band for other cases decreases at a similar rate during the process of shearing, respectively. This confirms the consistent volumetric response found in the interface under different shear orientations.

(4) A comparison of the resultant average shear stress and material fabric implies that a correlation exists between deviator fabric δ_D and the interface's resultant shear resistance. This observation reconfirms the consistent shear behavior obtained from the ISTs.

(5) All the previous statements point out the importance of measuring the horizontal load in x- and y-directions when conducting experimental interface shear tests on rough surface with highly directional properties. The resultant resistance corresponds to the true resistance of the granular assembly.

In this chapter, the granular soil was modelled with spherical ball and the rolling resistance coefficient was imposed on the inter-particle contact to compensate the effect of real particle angularity. Besides, the present chapter only examined a specific type of artificial interface, namely one that has regular pattern. To improve the model's accuracy, both irregular particle shape and random interface should be considered in future investigations. Despite these limitations, the conclusions can improve our understanding of soil-structure interface response in multi-directional shear tests at macro and micro levels.

References

1. Barton N, Choubey V (1977) The shear strength of rock joints in theory and practice. Rock Mech 10:1–54
2. Chen X, Zhang J, Xiao Y, Li J (2015) Effect of roughness on shear behavior of red clay–concrete interface in large-scale direct shear tests. Can Geotech J 52:1122–1135
3. Dove JE, Jarrett JB (2002) Behavior of dilative sand interfaces in a geotribology framework. Journal of geotechnical and geoenvironmental engineering 128:25–37
4. Ebeling RM, Peters JF, Mosher RL (1997) The role of non-linear deformation analyses in the design of a reinforced soil berm at Red River U-frame lock No. 1. Int J Numer Anal Meth Geomech 21:753–787
5. Gómez JE, Filz GM, Ebeling RM, Dove JE (2008) Sand-to-concrete interface response to complex load paths in a large displacement shear box.
6. Hryciw RD, Irsyam M (1993) Behavior of sand particles around rigid ribbed inclusions during shear. Soils Found 33:1–13
7. Hu L, Pu J (2004) Testing and modeling of soil-structure interface. Journal of Geotechnical and Geoenvironmental Engineering 130:851–860
8. Jing XY, Zhou WH, Zhu HX, Yin ZY, Li Y (2018) Analysis of soil-structural interface behavior using three-dimensional DEM simulations. International Journal for Numerical and Analytical Methods in Geomechanics
9. Wang J, Gutierrez MS, Dove JE (2007) Numerical studies of shear banding in interface shear tests using a new strain calculation method. Int J Numer Anal Meth Geomech 31:1349–1366
10. Wang J, Jiang M (2011) Unified soil behavior of interface shear test and direct shear test under the influence of lower moving boundaries. Granular Matter 13:631–641
11. Uesugi M, Kishida H (1986) Influential factors of friction between steel and dry sands. Soils Found 26:33–46
12. Uesugi M, Kishida H (1986) Frictional resistance at yield between dry sand and mild steel. Soils Found 26:139–149
13. Samanta M, Punetha P, Sharma M (2018) Influence of surface texture on sand–steel interface strength response. Géotechnique Letters 8:40–48
14. Majmudar TS, Behringer RP (2005) Contact force measurements and stress-induced anisotropy in granular materials. Nature 435:1079–1082
15. Yang Z, Lit X, Yang J (2008) Quantifying and modelling fabric anisotropy of granular soils. Géotechnique
16. Itasca PD (2008) Particle Flow Code in 3 Dimensions, User's Guide.
17. Claquin C, Emeriault F (2001) Interface behavior of granular materials: Discrete numerical simulation and statistical homogenization. Powders and grains 2001:323–326
18. Frost J, DeJong J, Recalde M (2002) Shear failure behavior of granular–continuum interfaces. Eng Fract Mech 69:2029–2048
19. Frost J, Han J (1999) Behavior of interfaces between fiber-reinforced polymers and sands. Journal of geotechnical and geoenvironmental engineering 125:633–640
20. Jensen RP, Bosscher PJ, Plesha ME, Edil TB (1999) DEM simulation of granular media—structure interface: effects of surface roughness and particle shape. Int J Numer Anal Meth Geomech 23:531–547
21. Zhu H, Nguyen HN, Nicot F, Darve F (2016) On a common critical state in localized and diffuse failure modes. J Mech Phys Solids 95:112–131
22. Zhou W-H, Jing X-Y, Yin Z-Y, Geng XJAG (2019) Effects of particle sphericity and initial fabric on the shearing behavior of soil–rough structural interface. 14:1699–1716
23. Zhu HX, Zhou WH, Jing XY (2019) Yin ZYJIJfN, Geomechanics AMi. Observations on fabric evolution to a common micromechanical state at the soil-structure interface. 43:2449–2470
24. Lai H-J, Zheng J-J, Cui M-J, Chu JJAG (2020) "Soil arching" for piled embankments: insights from stress redistribution behaviour of DEM modelling. 1–20
25. Lai H-J, Zheng J-J, Zhang J, Zhang R-J, Cui LJC, Geotechnics (2014) DEM analysis of "soil"-arching within geogrid-reinforced and unreinforced pile-supported embankments. 61:13–23

26. Wang Z, Jacobs F, Ziegler MJG, Geomembranes (2014) Visualization of load transfer behaviour between geogrid and sand using PFC2D. 42:83–90
27. Wang Z, Jacobs F, Ziegler MJG, Geomembranes (2016) Experimental and DEM investigation of geogrid–soil interaction under pullout loads. 44:230–246
28. Zhu H, Zhou W-H, Yin Z-YJAG (2018) Deformation mechanism of strain localization in 2D numerical interface tests. 13:557–573
29. Lai H-J, Zheng J-J, Zhang R-J (2018) Cui M-JJC, Geotechnics. Classification and characteristics of soil arching structures in pile-supported embankments. 98:153–171
30. Wang Z, Jacobs F, Ziegler M, Yang GJG (2020) Geomembranes. Visualisation and quantification of geogrid reinforcing effects under strip footing loads using discrete element method. 48:62–70
31. Härtl J, Ooi JY (2008) Experiments and simulations of direct shear tests: porosity, contact friction and bulk friction. Granular Matter 10:263–271
32. Su L-J, Zhou W-H, Chen W-B, Jie X (2018) Effects of relative roughness and mean particle size on the shear strength of sand-steel interface. Measurement 122:339–346
33. Gu X, Chen Y, Huang M (2017) Critical state shear behavior of the soil-structure interface determined by discrete element modeling. Particuology 35:68–77
34. Satake M (1992) A discrete-mechanical approach to granular materials. Int J Eng Sci 30:1525–1533
35. Barreto D, O'Sullivan C, Zdravkovic L (2009) Quantifying the evolution of soil fabric under different stress paths. In: AIP Conference Proceedings. AIP, pp 181–184
36. Cui L, O'sullivan C (2006) Exploring the macro-and micro-scale response of an idealised granular material in the direct shear apparatus. Geotechnique 56:455–468

Chapter 8
Interface Shear Strength Considering Structural Surface Morphology

8.1 Introduction

Apart from soil properties and experimental conditions, surface morphology is one of the structural properties, and the effect of surface morphology on the mechanism of interaction between the granular assembly and the solid surface under numerical and experimental interface-shearing tests has been underlined repeatedly in previous studies, as summarized in Table 8.1. Among these studies, the surfaces used to be simplified or considered to be regular surfaces, i.e., saw-teeth or semi-arch pattern, for convenience of analysis. In addition, roughness was evaluated only qualitatively in several previous studies [1, 2]. The surface morphology was approximately classified from very smooth to very rough through the observation of surface morphology. The disadvantage of this approach is the subjective results. Even though the approach of surface morphology quantification was adopted in most previous studies, the surface roughness has been quantified as one parameter. The most commonly used parameter was the relative roughness ($R_n = R_{max}/D_{50}$, where R_{max} is the maximum height of the surface and D_{50} is the average particle diameter) as proposed by Uesugi and Kishida [3, 4]. However, R_n fails to reflect the local distribution and partial variation of the surface profile. Obviously, the same R_n can be presented using many various profiles, as shown in Fig. 8.1. As shown in Table 8.1, some researchers have performed tests on non-random surfaces and have indicated that it is insufficient to adopt R_n to quantify the surface roughness. Accordingly, more comprehensive roughness parameters were introduced to describe the asperity characteristics [5–9]. Some results showed that these non-random surfaces can yield a higher interface shear strength than pure soil [8, 9]. In addition, given that most of structure surfaces are normally irregular or extremely random, the regular surface may fail to reflect the true behavior of an irregular interface. Several studies have focused on random surface shearing [6, 10–12], which provides a valuable understanding of the interface shear behavior. However, the current parameters may fail to satisfy the roughness quantification of the real surface due to its complexity, and the role of morphology parameters

W.-H. Zhou and Z.-Y. Yin, *Practice of Discrete Element Method in Soil-Structure Interface Modelling*, https://doi.org/10.1007/978-981-19-0047-1_8

Table 8.1 Review of recent studies on the effect of morphology parameters on soil-structure interactions

Authors	Surface pattern	Quantified parameter	Selected results
Chen et al. [13]	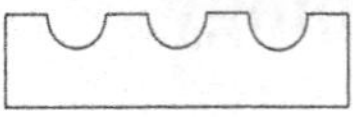	Average depth of the poured sand	The shear strength increases with the surface roughness The shear failure plane is dependent on the confining pressure and surface roughness
Chu and Yin [14]		Roughness angle	The surface roughness has a great impact on the interface shear strength
Dove and Jarrett [5]		Asperity angle Root spacing Asperity spacing Asperity height	The interface behavior is influenced by predictable geometric properties The proposed mechanical equation can be applied to complex manufactured surfaces
Feng et al. [15]		Relative roughness	Stronger interaction between soil and geomembrane was developed under rougher geomembrane shearing
Hu and Pu [16], Jing et al. [17], Su et al. [18], Wang et al. [19], Chen et al. [20], Zhu et al. [21]	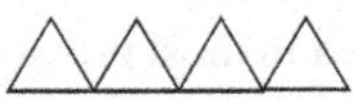	Relative roughness	There exists a critical roughness to affect the interface shear strength
Zhang and Evans [22]	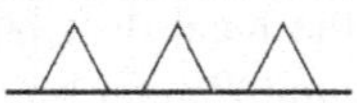	Relative roughness	For smoother interfaces, the contact normal force decays more rapidly compared to rougher interfaces
Wang et al. [7], Guo et al. [8]		Flat width Intersection angle Groove depth	Peak shear stress increases with normal stress and the intersection angle
Han et al. [12], Martinez and Frost [11], Canakci et al. [10], Rui et al. [6]	Random	Normalized average of maximum roughness, Average roughness	The larger interface friction angle was found for the case of rougher interface shearing

Definition of relative roughness: $R_n = \frac{R_{max}}{D_{50}}$

where R_{max} is the max height of structure; D_{50} is the mean particle size.

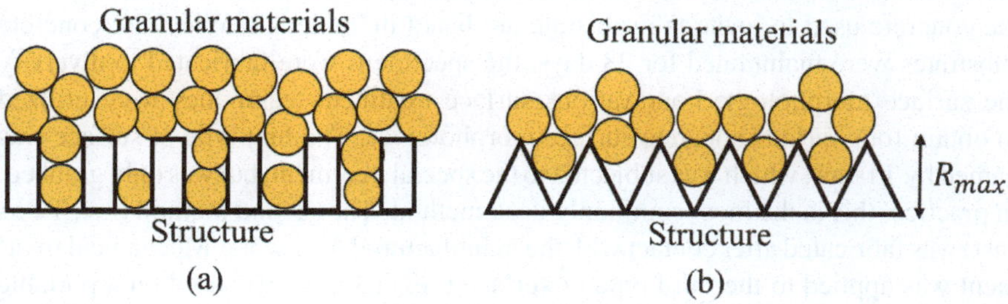

Fig. 8.1 Different rough profiles with the same relative roughness

in the evaluation of interface shear strength under random surface shearing remains ambiguous.

To overcome limitations, the interaction between soil and random surfaces should be evaluated from the perspective of geotribology, and more morphology parameters should be introduced, taking the local information and the spacing between peaks and valleys of the surface into consideration. The tribology theory has generally concentrated on the fields of wear, lubrication, surface characterization, and friction. Likewise, the theory can be applied to the investigation of the strength behavior of interface systems. Previous studies have proposed several techniques to quantify the surface morphology. Among these methods, the optical-based method (i.e., laser scanner device) was most widely used due to its high accuracy and fully non-destructive testing [23–25]. The surface roughness can be characterized based on the scanned random surface and using a full quantitative approach. The 2D and 3D discrete element method (DEM) has been favorably adopted to address a range of geotechnical and geological issues, such as, soil-structure interactions [17, 26–31] and geological hazards [32, 33]. The DEM can also provide an effective technique to simulate and reproduce the interaction between the random surface and soil [26]. Once the scanned random surface is obtained, it is imported into the DEM to conduct a numerical interface shear test. Therefore, the DEM will be used to model the random surface-soil interface shear test in the chapter.

Considering the above, the present chapter focuses on selecting the morphology parameters of random surfaces for evaluating the SSI strength. A complete set of morphology parameters in accordance with standard ISO 4287 [34] was applied to quantifying a series of random profiles of concretes. Based on the 2D DEM simulation, 480 interface shear tests with random profiles were conducted on coarse-grained soils. The relevant morphology parameters were selected using the Spearman's rank correlation coefficient for model selection. In addition, an optimal metaheuristic model was developed using a genetic algorithm and was further expressed by a formula. The proposed formula was compared with the existing predicted formulas.

8.2 Random Surface and Morphology Parameters

The investigated elements manufactured by Sadowski and Stefaniuk [35] and Sadowski et al. [36] consisted of an overlay and a substrate. The components of the concrete used to make the substrate are listed in Table 8.2. When the concrete substrates were maintained for 28 days, the specimens were fabricated to diversify the surface morphology. Four various surface treatment techniques were utilized to obtain four various concrete surface morphologies. The first type of surface was named by T1-raw, which was subjected to the special treatment, but was only grabbed. In practice, this is the most commonly used method. The second method (i.e., T2-as cast) was fabricated after contact with the manufactured formwork. Mechanical treatment was applied to the third type of surface (i.e., T3-ground) based on a portable angle grinder with an abrasive cutter and dust removal. After the dust was removed from the fourth type of surface (i.e., T4-shotblasted), T4 was continuously shotblasted by means of a lightweight shotblasting device. After treatments, a developed device based on the 3D triangulation scanner was used to scan the concrete surfaces and more details were presented by Sadowski et al. [36]. The scanning results showed a 3D isometric view of the investigated concrete surface morphology.

In this chapter, a total of 240 profiles were extracted from the four different concrete surfaces, with each concrete surface having 60 profiles. The isometric views of four different concrete surfaces and one of their corresponding profiles in the software MountainsMap [37] are shown in Fig. 8.2. These profiles were analyzed in MountainsMap to obtain the values of the morphology parameters for surface characterization. The surface morphology parameters used were obtained according to standard ISO 4287 [34]. The standard ISO 4287 contains five types of parameters, namely, amplitude parameter, spacing parameter, hybrid parameter, material ratio curves and related parameter, and peak count parameter. Each parameter is carefully presented in Table 8.3. The physical meaning of each parameter was explained by comparing the characteristics of profiles, as shown in Fig. 8.2. Except for amplitude parameters Psk and Pku, the remaining amplitude parameters are more common and easily calculated from the profiles. Profile P1 extracted from T1 has fairly deep valleys and scratches, which makes Psk negative, while the rest of the profiles have fairly high spikes or peaks, leading to the result that the Psk is positive. Profiles possessing comparatively few low valleys and high peaks lead to a Pku of less than 3, whereas profiles with many low valleys and high peaks are reflected in a Pku of more than 3. Clearly, P1 with a Pku of less than 3 has fewer low valleys and high

Table 8.2 Components of the used concrete

Material	Dosage	Material	Dosage
Portland cement (Type II A-LL 42.5 R)	352 kg	Plasticizer Sika® viscoflow®-6920	2.0 L
Water	165 L	Crushed basalt aggregates	1086.6 kg
Fine aggregate	724.4 kg		

Pp	*Pv*	*Pz*	*Pc*	*Pa*	*Pq*	*Psk*	*Pku*	*PSm*	*Pdq*	*Pmr*	*Pdc*	*PPc*
0.461	0.484	0.945	0.526	0.156	0.189	-0.412	2.616	6.435	18.000	0.100	0.338	1.554

(a) T1

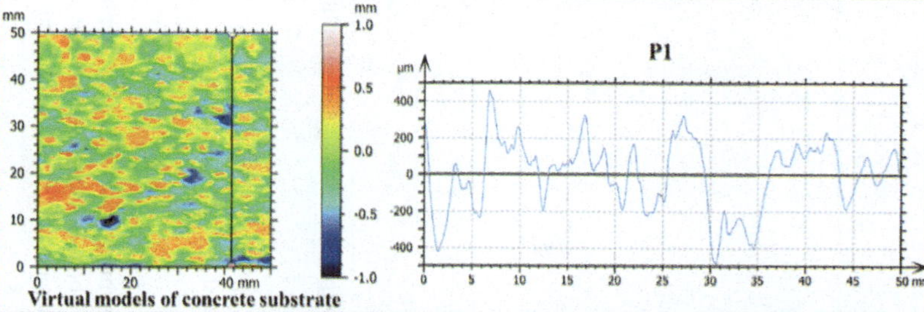

Pp	*Pv*	*Pz*	*Pc*	*Pa*	*Pq*	*Psk*	*Pku*	*PSm*	*Pdq*	*Pmr*	*Pdc*	*PPc*
0.372	0.222	0.594	0.215	0.091	0.115	0.748	3.289	1.840	28 .340	0.200	0.186	5.435

(b) T2

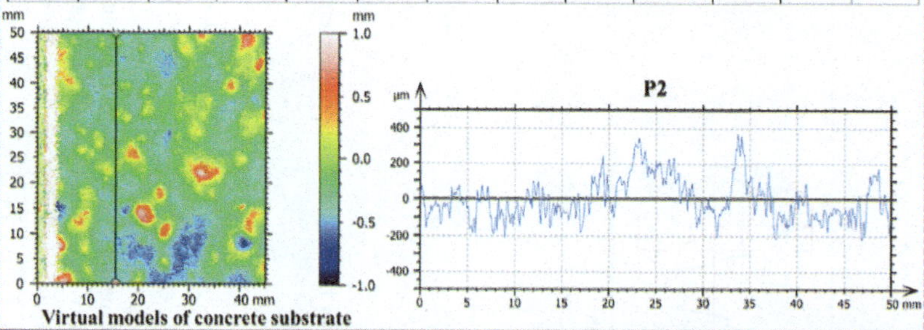

Pp	*Pv*	*Pz*	*Pc*	*Pa*	*Pq*	*Psk*	*Pku*	*PSm*	*Pdq*	*Pmr*	*Pdc*	*PPc*
0.380	0.354	0.734	0.306	0.098	0.126	0.168	3.109	1.383	36.420	0.246	0.205	7.228

(c) T3

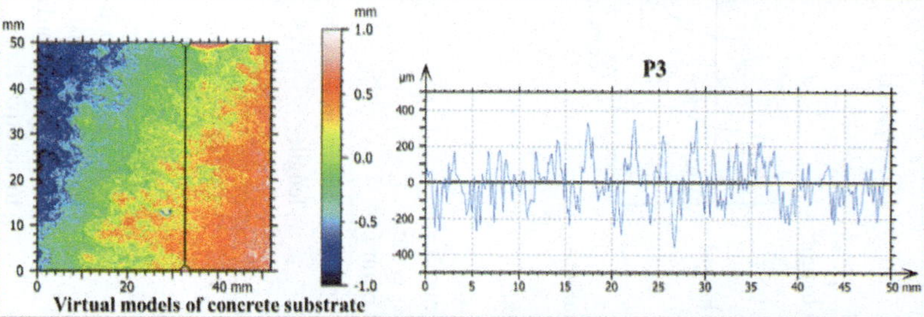

Pp	*Pv*	*Pz*	*Pc*	*Pa*	*Pq*	*Psk*	*Pku*	*PSm*	*Pdq*	*Pmr*	*Pdc*	*PPc*
0.371	0.319	0.690	0.267	0.084	0.112	0.012	3.787	1.052	39.150	0.246	0.165	9.505

(d) T4

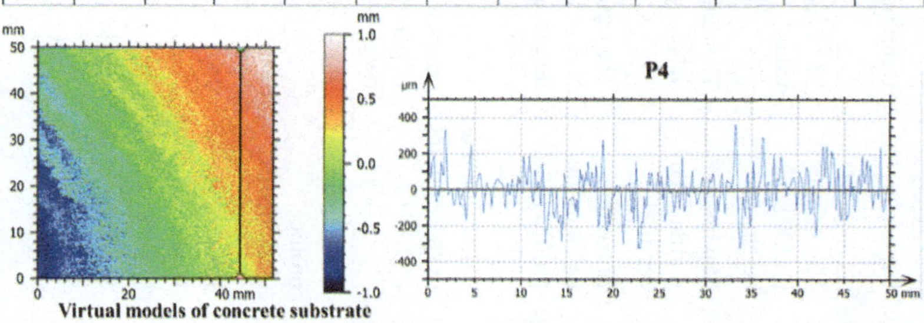

Fig. 8.2 3D isometric views of four types of concrete substrates and their corresponding profiles

Table 8.3 Morphology parameters in accordance with standard ISO 4287 [34]

Morphology parameters	Parameter	Definition	Description
Amplitude parameters	Maximum profile peak height: Pp	The maximum height value	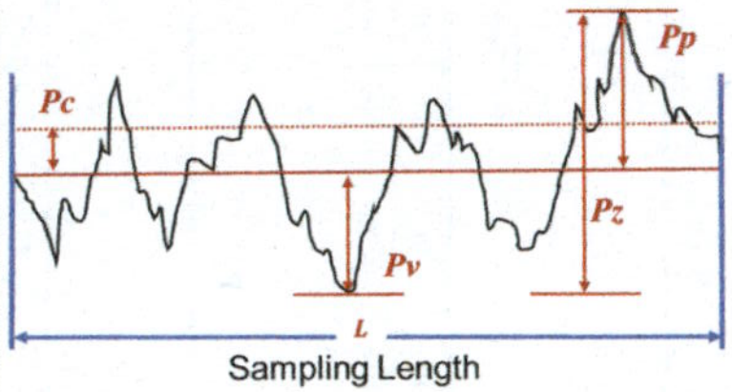
	Maximum profile valley depth: Pv	The minimum height value	
	Maximum height of the profile: Pz	The difference between the maximum height and the minimum height	
	Arithmetic mean deviation of the profile: Pa	$Pa = \frac{1}{n}\sum_{i=1}^{n}\lvert Z_i\rvert$ where n is the number of points and Z_i is the height value at point i	
	Mean height of profile elements: Pc	$Pa = \frac{1}{n}\sum_{i=1}^{n}\lvert Zt_i\rvert$	This represents the mean for the height Zt of profile elements within the investigated sampling length
	Root mean square deviation of the profile: Pq	$Pq = \sqrt{\frac{1}{n}\sum_{i=1}^{n} Z_i^2}$	This is one of the most widely used parameters and is also referred to as the RMS value
	Skewness of the profile height distribution: Psk	$Psk = \frac{1}{Pq^3}\left(\frac{1}{n}\sum_{i=1}^{n} Z_i^3\right)$	If this parameter is zero, it means that the height distribution is symmetric. Positive Psk represents the surfaces possessing fairly high spikes or peaks that protrude above a flatter average. Reversely, Surfaces with fairly deep valleys and scratches in a smoother plateau such as porous surfaces, lead to negative Psk

(continued)

Table 8.3 (continued)

Morphology parameters	Parameter	Definition	Description
	Kurtosis of the profile height distribution: Pku	$Pku = \frac{1}{Pq^4}(\frac{1}{n}\sum_{i=1}^{n} Z_i^4)$	The Pku describes the sharpness of the height distribution. Surfaces normally possess relatively few high peaks and low valleys when kurtosis is smaller than 3. In contrast, surfaces with many high peaks and low valleys lead to a kurtosis value of more than 3
Spacing parameters	Mean width of profile elements: PSm	The mean width of profile elements is neighboring peak-valley pairs	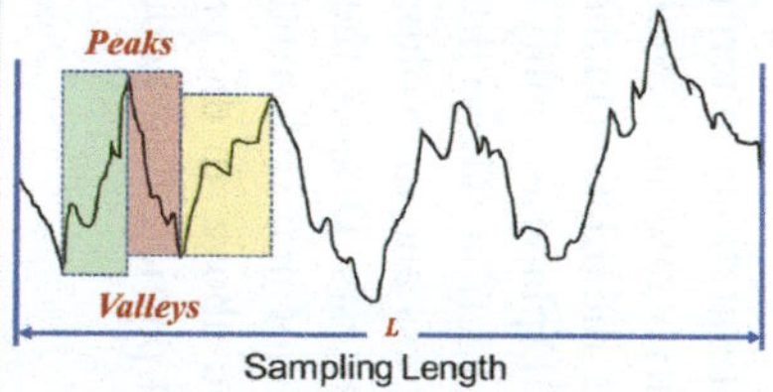
Hybrid parameters	Root mean square slope of the profile: Pdq	$Pdq = \sqrt{\frac{1}{n}\sum_{i=1}^{n}(\frac{dZ_i}{dx})^2}$	This represents the root mean square for the local slope within the sampling length
Material ratio curves and related parameters	Material ratio of the profile: Pmr	The ratio between the material length and the evaluation length for the section height level C (% or μm)	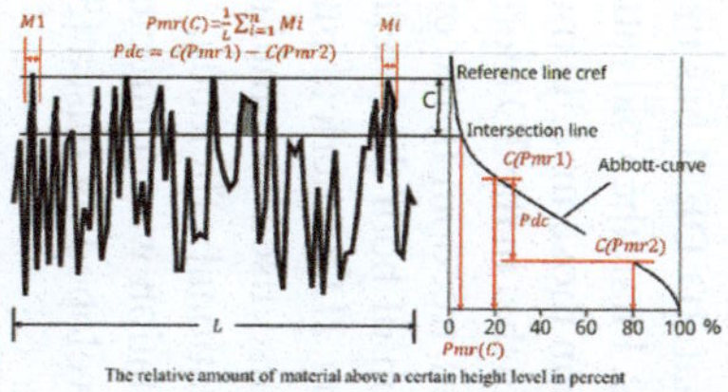
	Relative material ratio of the profile: Pdc	Signifies the height difference in section height level C, matching the two material ratios	
Peak parameter	Number of peaks: PPc	Number of peaks per unit length	

peaks than the other profiles, and the other three with a Pku of more than 3 have a high density of low valleys and high peaks. Pdq describes the root mean square for the local slope of the profile, which is presented as degree in the present chapter. Overall, the local slope of P1 is the most stable, while the local slope of P4 is the steepest. Accordingly, P1 has the smallest Pdq and the largest Pdq is for P4. Evidently, from P1 to P4, the density of the peak gradually increases and the corresponding PPc increases. Conversely, the Psm decreases because the distance between neighboring peak-valley gradually becomes narrow from P1 to P4. The difference in Pmr and Pdc for these four profiles can be easily read from the profiles.

8.3 Numerical Simulation of Interface Shear Test

8.3.1 2D DEM Simulation Limitations

It is accepted that 2D DEM has the following limitations in terms of modeling the behavior of soils. First, the void ratio and porosity values achieved in a 2D model are much smaller than those in a 3D model. Second, the dilation in the 2D model can be much higher than that in the 3D model since the 2D plane assembly can only dilate due to the removal of freedom in the cross-plane direction. These differences may lead to overestimation of both the interface shear stress and the volumetric deformation when conducting 2D numerical interface shear tests. Despite these limitations, the 2D DEM was also used to simulate soil behavior and soil-structure behavior due to high computational efficiency, as mentioned above. In the present chapter, the numerical simulation was conducted with a 2D-particle flow code (PFC^{2D}), based on the DEM developed by Itasca [38].

8.3.2 Numerical Model and Model Process

To compensate for the interlocking influence induced by the various shapes of particles, a rolling resistance linear contact model was applied to contacts [39]. The test apparatus, with a length of 90 mm and a height of 25 mm, is comprised of an upper rigid shear box filled with particles and a rigid lower boundary. The latter one consists of a random surface and two smooth walls of auxiliary zones of 20 mm at each end of the shear box to eliminate the boundary effect. The random surfaces imported into PFC^{2D} were obtained from the profiles of concrete surfaces. There is no particle-to-bottom boundary friction within the auxiliary zones. The concrete surface will suffer some damage and abrasion during shearing, and the corresponding morphology of post shearing surface would change if we investigate this issue using laboratory tests. However, the numerical simulation will avoid damage and abrasion on the imported random surface because the random surface is assumed to be stiff or nondeformable

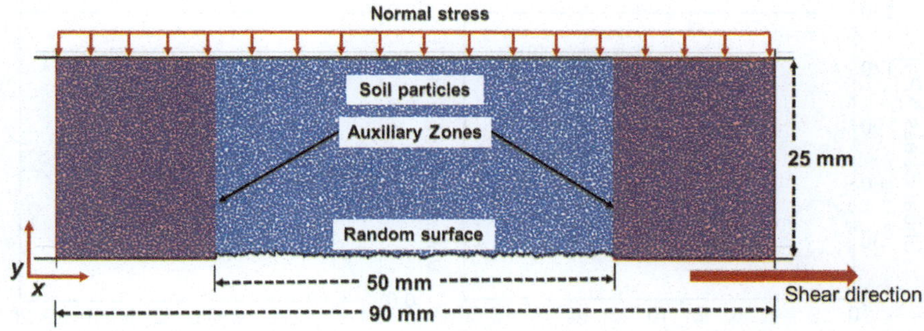

Fig. 8.3 Schematics of the interface shear apparatus with an imported random surface

in PFC^{2D}. The numerical simulation model is shown in Fig. 8.3. The technique of specimen generation called particle size growing proposed by Chareyre and Villard [40] was adopted to obtain a relatively isotropic specimen. Specifically, based on a particle size distribution, a series of particles with a certain diameter range was seeded inside the shear box and their sizes gradually grew. This particle size growing process stopped when the normal pressure applied on the top boundary achieved the targeted value. The radii of particles in the specimen were fixed in the process of shearing. The sample was verified to be uniform at the beginning of the shear phase by checking the spatial distributions of the force chains and void ratio. The bottom random surface started moving horizontally in the x-direction by applying a velocity of 1.0 m $\times$ 10^{-3}/min recommended by ASTM D5321 to fulfill a quasistatic interface shearing. A constant normal pressure applied on the moveable top wall was sustained by a built-in servo control system in PFC^{2D}, while the two lateral counterparts were fixed throughout shearing.

The shearing behavior was obtained by recording the displacements and forces on the walls. The shear stress ratio is the ratio of shear stress to normal stress. The normalized shear displacement is defined as the ratio of shear displacement to D_{50}. The input parameters follow the numerical 2D interface shear test conducted by Zhu et al. [21], as demonstrated in Table 8.4. The effect of normal stress on interface

Table 8.4 The selected input parameters [21]

Parameters	Value
Ball density (kg/m^3)	2650
Inter-particle normal stiffness k_n (N/m)	5.0×10^9
Inter-particle shear stiffness k_t (N/m)	2.5×10^9
Particle–wall normal stiffness k_{nw} (N/m)	9.0×10^9
Particle–wall shear stiffness k_{tw} (N/m)	4.5×10^9
Inter-particle frictional coefficient f_p	0.5
Particle-boundaries frictional coefficient f_{pw}	0.9
Rolling resistance coefficient μ_r	0.1

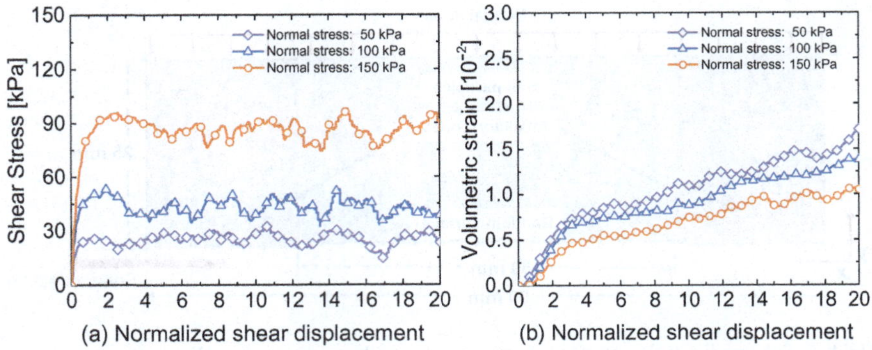

Fig. 8.4 Interface shearing behavior under various normal stresses: **a** shear stress versus normalized shear displacement and **b** volumetric strain versus normalized shear displacement

shearing behavior in preliminary tests is shown in Fig. 8.4. The peak shear stress is shown to increase with the normal stress. The specimens dilate with a decreasing rate and the degree of dilatancy decreases with increasing normal stress. All the preliminary results are consistent with the previous numerical and experimental results [7, 8, 17, 41].

8.3.3 Numerical Test Schemes

To increase the database, three different mean particle sizes, i.e., $D_{50} = 0.35$ mm, $D_{50} = 0.53$ mm, and $D_{50} = 0.80$ mm, but the same uniformity coefficient was adopted for the interface shear test. The detailed three particle size distributions (PSD1, PSD2, and PSD3) are demonstrated in Fig. 8.5. Each specimen for the three different mean particle sizes contains 4,444, 10,000, and 22,500 particles. Accordingly, a total of 480 simulated interface shear tests with random surfaces were sheared under

Fig. 8.5 Three particle size distributions with different mean particle sizes

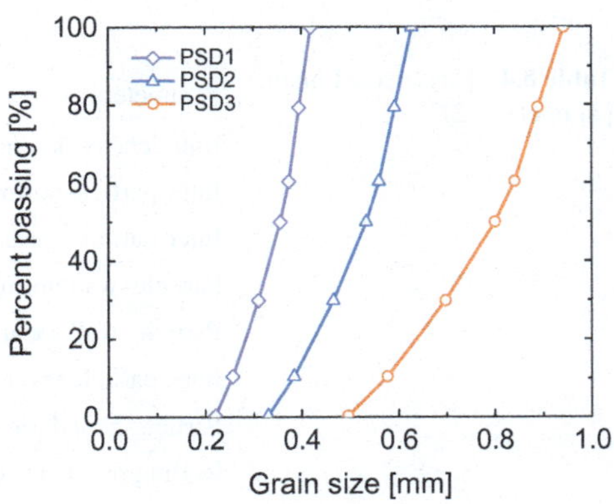

Table 8.5 Experimental program for 480 simulated interface shear tests

Type of concrete profile	Number of profiles	Mean diameter D_{50}	Uniformity coefficient	Total groups
T1	60	D_{50}= 0.35 mm (30 groups) D_{50} = 0.53 mm (60 groups) D_{50} = 0.80 mm (30 groups)	1.46	120
T2	60	D_{50}= 0.35 mm (30 groups) D_{50} = 0.53 mm (60 groups) D_{50} = 0.80 mm (30 groups)	1.46	120
T3	60	D_{50}= 0.35 mm (30 groups) D_{50} = 0.53 mm (60 groups) D_{50} = 0.80 mm (30 groups)	1.46	120
T4	60	D_{50}= 0.35 mm (30 groups) D_{50} = 0.53 mm (60 groups) D_{50} = 0.80 mm (30 groups)	1.46	120

a normal stress of 100 kPa, as shown in Table 8.5. The initial void ratios of the specimens range from 0.182 to 0.184. Direct shear tests on the specimen with varying mean diameters were also conducted to obtain the shear strength of each soil. The normalized interface stress was designated by the efficiency parameter (IE), which was proposed by Koerner [42]. IE is the ratio of $tan\delta$ to $tan\varphi$, where $tan\delta$ presents the interface friction coefficient, whereas $tan\varphi$ is the friction coefficient of pure soil. The efficiency at the peak state IE_P can be calculated using the peak friction coefficients of the interface and pure soil. The value of the efficiency parameter ranges from 0.0 (small interface strength) to 1.0 (fully mobilized soil strength). The database of the 480 numerical test results are available at Github server: https://github.com/umgeotech/Database/tree/master/Soil-structure%20Interface.

8.3.4 Typical Interface Shearing Behavior

Four typical macroscopic interface shearing behaviors versus normalized shear displacement are illustrated in Fig. 8.6. The general curve trend shows that a post-peak strain softening to steady state because of the low initial void ratio. Continuous dilation is also observed in Fig. 8.6b. Similar curves were also observed from the results of other interface shear tests with different surface geometries, but their corresponding curves are not shown. The peak interface efficiency of each simulation IE_P was collected for further analysis using Spearman's rank correlation coefficient and by developing metaheuristic relationships between peak interface efficiency and surface morphology parameters.

8.4 Results and Discussion

The analysis of Spearman's rank correlation coefficient was conducted to eliminate the irrelevant morphology parameters with the peak interface efficiency IE_P. Furthermore, the relevant morphology parameters were combined to form the potential models in the process of model selection. The optimal model was selected among the potential models based on the result of Bayesian nonparametric general regression. The obtained optimal model was formulated and validated with previous studies.

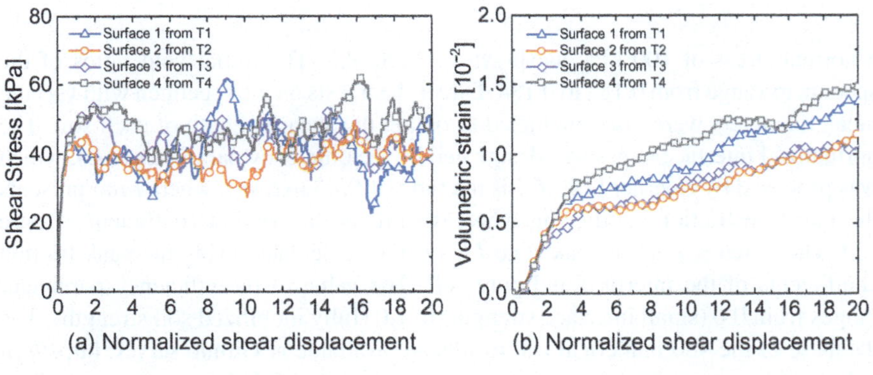

Fig. 8.6 Four typical macroscopic interface shearing behaviors: **a** interface efficiency versus normalized shear displacement and **b** volumetric strain versus normalized shear displacement

8.4.1 Analysis of Correlation by Means of Spearman's Rank Correlation Coefficient

Compared with the Pearson product-moment correlation, Spearman's rank-order correlation belongs to the nonparametric analysis. Spearman's rank correlation coefficient, ρ_s, presents the degree and direction of dependence between two ranked variables. Because ρ_s is only mildly sensitive to divergent results, it is especially effective in analyzing the data where the distribution does not follow the normal distribution. According to Kowalczyk et al. [43], the two random variables x and y in the analysis of Spearman's rank correlation coefficient can be obtained from the following equation:

$$\rho_s = corr \cdot (f(x)_x, f(y)_y) \tag{8.1}$$

where $corr$ is the Pearson correlation coefficient; $f(x)_x$ is the distribution function of variable x at point x, while $f(y)_y$ is the distribution function of variable y at point y. The value of ρ_s ranges from -1 to 1. Kowalczyk et al. [43] also assumed that the data investigated are appropriate to perform an artificial neural network analysis, when the values of ρ_s range from either -1 to -0.4 or from 0.4 to 1.

In this part, the correlation between peak interface efficiency IE_P and 13 morphology parameters were determined using the Spearman's rank correlation coefficient. The parameters with length units were normalized by average diameter D_{50}. The statistics of 13 morphology parameters and peak interface efficiency IE_P based on the results of 480 interface shear tests are presented in Table 8.6. The correlations between the 13 morphology parameters and IE_P with the calculated values of ρ_s are shown in Table 8.7. From Table 8.7, it appears that ρ_s is in a range of 0.4 to 1 or -1 to -0.4 for parameters Pp/D_{50}, Pv/D_{50}, Pz/D_{50}, Pc/D_{50}, Pa/D_{50}, Pq/D_{50}, Psm/D_{50}, Pdq, Pdc/D_{50} and PPc, indicating the highest value (0.777) for parameter Pdq. A positive ρ_s indicates an increase in the values of the parameter with the increase of IE_P, whereas a negative ρ_s indicates a decrease in the values of the parameter with IE_P increased. The rotation and translation of granular material over the random surface are affected by the surface morphology, which indirectly influences the mobilized efficiency of soil. The hybrid parameter Pdq includes not only the amplitude information but also the surface slope information on a random surface; therefore, its effect on IE_P is most productive. PPc signifies the peak number per unit length of a profile and is able to positively affect the soil movement over an interface. More peaks per unit length will lead to decreasing the distance between neighboring peak-valley pairs in a profile, which indicates that PPc is inversely proportional to PSm. Accordingly, the value of ρ_s for PSm/D_{50} is negative and IE_P has a negative relationship with PSm/D_{50}. The ρ_s is less than 0.4 for the remaining parameters, which suggesting that the correlation between the remaining parameters and IE_P is insignificant. As presented in Table 8.3, Psk is used to express the symmetry of peaks and valleys, while Pku is utilized to describe the sharpness of a surface. Both parameters are normally used for the evaluation of

Table 8.6 Statistical characteristics of 13 morphology parameters and measured peak interface efficiency IE_P based on the results of 480 interface shear tests

Morphology parameter		Mean	Standard deviation	Skewness	Kurtosis	Minimum value	Maximum value
Amplitude parameters	Pp/D_{50}	1.5024	1.84427	3.288	12.763	0.23	13.80
	Pv/D_{50}	1.3171	1.59315	3.264	11.681	0.28	10.59
	Pz/D_{50}	2.8193	3.34616	3.135	10.777	0.62	21.08
	Pc/D_{50}	1.1783	1.46368	2.902	9.360	0.23	9.46
	Pa/D_{50}	0.4177	0.50005	3.143	11.451	0.09	3.63
	Pq/D_{50}	0.5269	0.62203	3.014	10.226	0.11	4.29
	Psk	0.1497	0.59633	0.288	2.081	−1.63	2.44
	Pku	3.3258	1.22181	2.705	9.626	2.12	10.29
Spacing parameters	PSm/D_{50}	3.4195	3.06364	3.667	17.643	0.76	28.12
Hybrid parameters	Pdq	36.3505	13.39251	1.186	2.455	9.68	75.98
Material ratio curves and related parameters	Pmr	0.2126	0.06687	0.521	3.312	0.10	0.49
	Pdc/D_{50}	0.8729	1.07105	3.277	12.591	0.18	7.97
Peak count parameter	PPc	6.1984	3.77133	0.464	−0.255	0.63	16.46
Predicted peak interface efficiency	IE_P	0.7051	0.08673	0.283	−0.240	0.52	0.94

Table 8.7 Calculated spearman's rank correlation coefficient for 13 morphology parameters

Morphology parameters		SRCC
Amplitude parameters	Pp/D_{50}	0.554
	Pv/D_{50}	0.541
	Pz/D_{50}	0.562
	Pc/D_{50}	0.551
	Pa/D_{50}	0.546
	Pq/D_{50}	0.548
	Psk	0.102
	Pku	−0.055
Spacing parameters	Psm/D_{50}	−0.485
Hybrid parameters	Pdq	0.777
Material ratio curves and related parameters	Pmr	−0.029
	Pdc/D_{50}	0.543
Peak count number	PPc	0.459

gloss and luster, but not for the evaluation of frictional force [44, 45]. Given all that, the ten parameters p/D_{50}, Pv/D_{50}, Pz/D_{50}, Pc/D_{50}, Pa/D_{50}, Pq/D_{50}, Psm/D_{50}, Pdq, Pdc/D_{50}, and PPc are the relevant input variables for metaheuristic model development.

8.4.2 Data Description and Model Selection

A total of 480 databases were prepared for analysis based on the Bayesian nonparametric general regression (BNGR) carefully presented in Appendix A. The ten selected morphology parameters were used to establish a metaheuristic model to estimate the peak interface efficiency IE_P. The parameters, i.e., Pp/D_{50}, Pv/D_{50}, Pz/D_{50}, Pc/D_{50}, Pa/D_{50}, Pq/D_{50}, Psm/D_{50}, Pdq, Pdc/D_{50}, and PPc, are designated as x_1, x_2, x_3, x_4, x_5, x_6, x_7, x_8, x_9, and x_{10}, respectively.

The traditional regression approaches have an obvious weakness that a large number of the model candidates are generated due to a large number of combinations of potential function structures from the same set of input variables. However, the BNGR method vanquishes this disadvantage from traditional methods. Based on the BNGR algorithm descripted in the study [46], $2^{10} - 1 = 1023$ models were generated as the potential models. Because the prior distribution choice for the vector θ was independent of the model selection [47], both the perdition error scale parameter and the smoothing scale parameter were analyzed based on the uniform prior distributions of [0, 100]. According to the current database, 70% of the database (336 points) was randomly selected as the training database which was used to obtain the optimal model while the remaining 30% of the database (144 points) was used for verification of the selected models. Based on the training database, both the smoothing scale parameter and the perdition error scale parameter were calculated. The results of the optimal model using the BNGR algorithm is carefully summarized in Table 8.8. The first column represents the selected input variables and the sixth column shows the plausibility of the corresponding model, which is ranked in order. According to the value of plausibility, model (x_8) is the optimal model. Figure 8.7 demonstrates that the optimal model shows a high fitting capability in the training phase, which means that it has an efficient learning ability. The most plausible model (x_8) is made of the hybrid parameter, Pdq. Pdq is a combination of amplitude and spacing information of surface and has been proven to be most relevant to IE_P according to the Spearman's rank correlation coefficient analysis.

Table 8.8 Result of the optimal model based on the BNGR algorithm

Model	Smoothing scale parameter υ_1	Perdition error scale parameter υ_2	Maximum likelihood	Evidence $p(y\|X, C^{(k)})$	Plausibility $p(C^{(k)}\|y, X)$
(x_8)	0.0287	33.3831	2.91×10^{-189}	3.76×10^{-196}	0.999

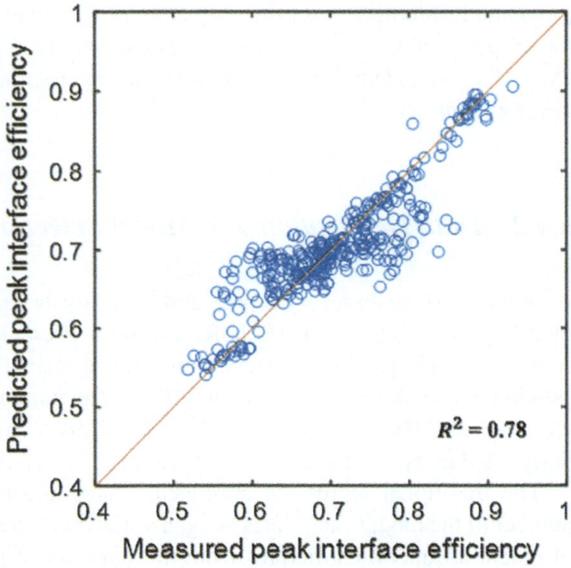

Fig. 8.7 Predicted peak interface efficiency based on the training dataset

8.4.3 Optimal Model Validation

In this section, the remaining 30% of the database (144 points) was used for verification of the optimal model. Figure 8.8 shows the measured peak interface efficiency IE_P from the remaining 30% of the database versus the predicted IE_P with the perfect matched line, i.e., the 45° line. The selected optimal model possesses the highest prediction capability based on the least number of morphology parameters compared with those for the models with the two input variable combinations (x_8 combined with the other six parameters). Figure 8.9 shows that the full model (i.e., model with all input variables (x_1,x_2,x_3, x_4, x_5, x_6 x_7, x_8, x_9 and x_{10}) has a lower fitting capacity than that for an optimal model, indicating that some input parameters are redundant for metaheuristic model development and cannot increase the fitting capacity but instead decrease the accuracy of model prediction.

In previous studies, the most commonly used parameter for the description of surface morphology was the relative roughness $R_n = R_{max}/D_{50}$ (i.e., Pz/D_{50} in this chapter). The model with R_n or Pz/D_{50} was also used to predict IE_P. Figure 8.10 shows that the models with R_n or Pz/D_{50} has a much lower predictive capacity than that for the optimal model. Because the input variable R_n can only represent the local maximum height of the surface, the precited points disperse in a larger domain compared to the distribution of precited points for the optimal model.

For a more objective comparison, two traditional indicators, the mean absolute relative error (MARE) and mean absolute error (MAE), were calculated for the comparison of predictive capability among all selected models, as shown in Table 8.9. Compared with the other models, the optimal model has the smallest values of

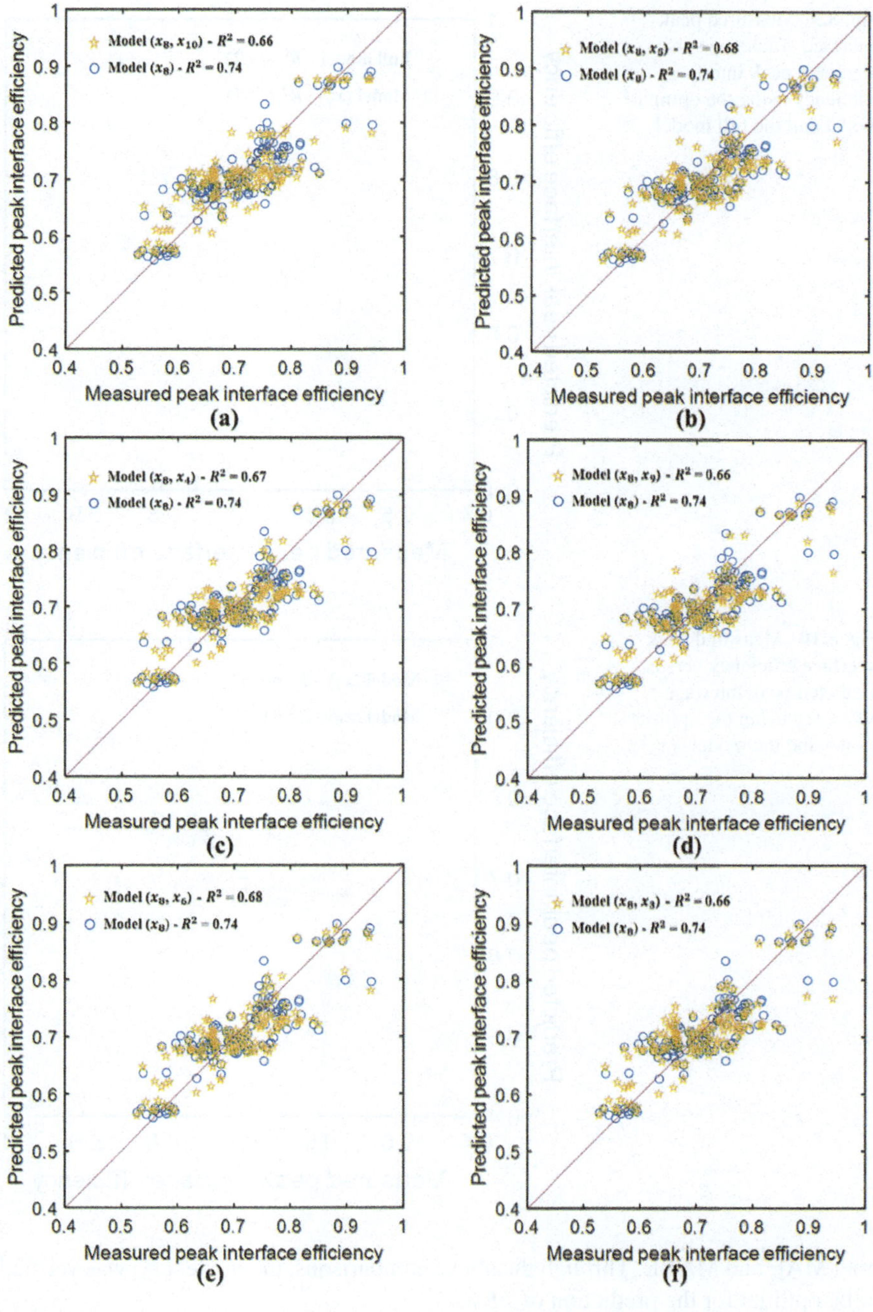

Fig. 8.8 Measured peak interface efficiency versus predicted peak interface efficiency using the optimal model and the models with two inputs

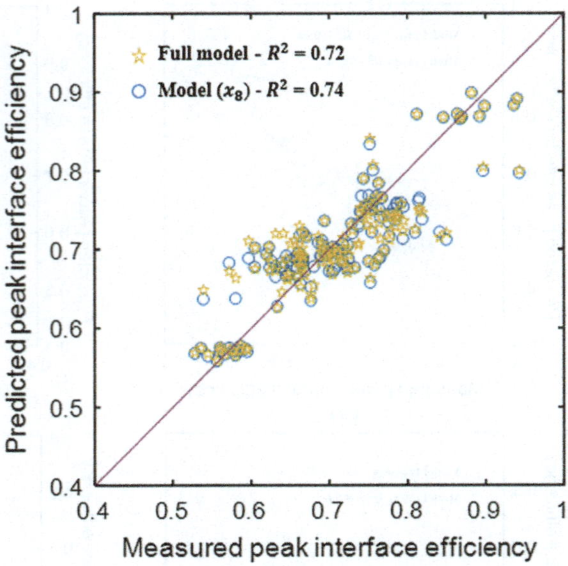

Fig. 8.9 Measured peak interface efficiency versus predicted peak interface efficiency using the optimal model and the full model

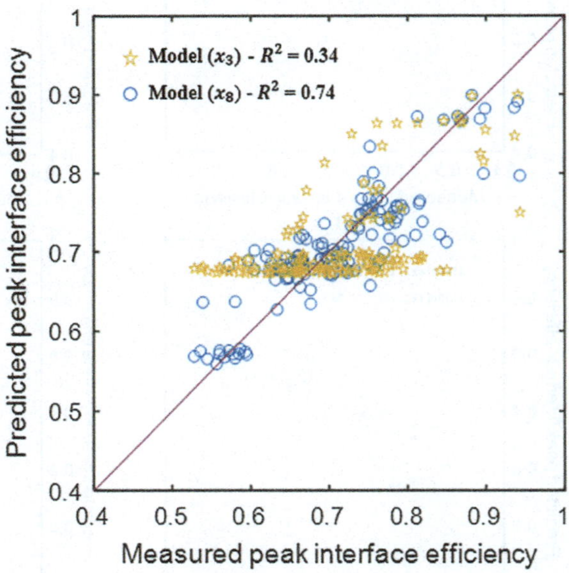

Fig. 8.10 Measured peak interface efficiency versus predicted peak interface efficiency using the optimal model and the models (x_3)

both MAE and MARE. Through the above comparisons, the model (x_8) was verified to be optimal for the prediction of IE_P.

Table 8.9 Accuracy measurement for some selected models

Model	MAE	MARE
Optimal model (x_8)	0.0345	0.0487
(x_8, x_{10})	0.0420	0.0596
(x_5, x_8)	0.0410	0.0583
(x_4, x_8)	0.0417	0.0589
(x_8, x_9)	0.0417	0.0593
(x_6, x_8)	0.0407	0.0580
(x_3, x_8)	0.0428	0.0609
(x_3)	0.0588	0.0853
Full model	0.0367	0.0520

8.4.4 Predicted Formula Development and Comparison with Existing Predicted Formulas

The most plausible model (x_8) is made of the hybrid parameter, Pdq. For the convenience of engineering applicability, the optimal model (x_8) was further expressed by a formula. By incorporating the observations of the relationship between Pdq and peak interface efficiency IE_P, the optimal model class for estimation of IE_P is proposed as follows

$$IE_P = a + Pdq * b \tag{8.2}$$

where a and b are all obtained from curve fitting. Seventy percent of the database was used to develop the predicted formula based on the hybrid parameter, Pdq. Figure 8.11 shows the learning capability of the proposed formula and its expression.

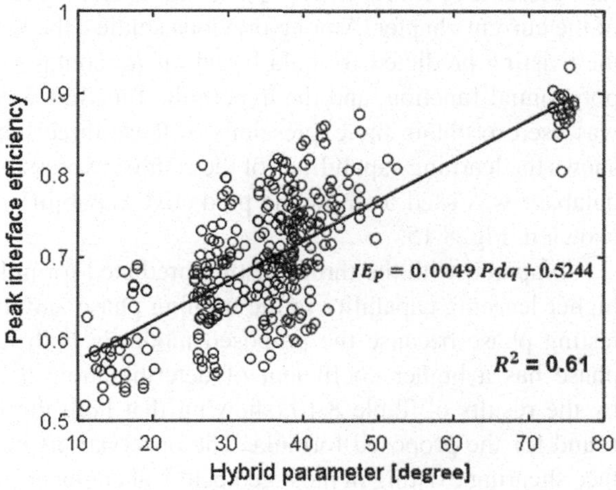

Fig. 8.11 Learning capability of the proposed formula and its expression

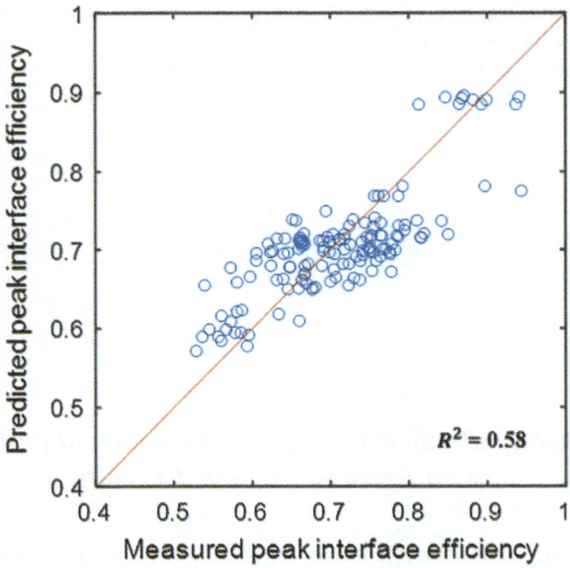

Fig. 8.12 Measured peak interface efficiency IE_P versus predicted peak interface efficiency IE_P using the proposed model

Figure 8.12 shows the measured IE_P from the remaining 30% of the database versus the calculated IE_P from the proposed formula as well as the perfect matched line.

As mentioned in Sect. 8.4.3, the most commonly used parameter for estimation of interface strength is the relative roughness R_n in previous studies. In the present chapter, the relationship between R_n and the peak shear stress ratio is plotted in Fig. 8.13a. To verify the findings of the present chapter, several relevant existing studies [3, 17, 18, 48–51] have been selected for comparison, as shown in Fig. 8.13b. As reported by Su et al. [18], the peak shear stress ratio remains nearly constant as R_n is more than approximately 0.25, while the critical value was reported to be approximately 0.375 by Jing et al. [17]. For the other studies, generally, the peak stress ratio increases with R_n when R_n is less than 1.0, which is consistent with the findings of the current chapter. Among previous studies, the most commonly used pattern of the existing predicted formula based on R_n contains the exponential function, the polynomial function, and the hyperbolic function. Seventy percent of the database was used to obtain the expressions of these three functional patterns. Figure 8.14 shows the learning capabilities of these three expressions. The remaining 30% of the database was used to show the predictive capability of these three expressions, as shown in Fig. 8.15.

Compared with the three existing predicted formulas, the proposed model has a higher learning capability in the training phase and higher predictive ability in the testing phase because the proposed model in both the training phase and testing phase has a higher coefficient of determination R^2. This fact was also verified by the results of Table 8.10, showing that both the lowest MAE and MARE are found for the proposed formula. The observations indicate that for irregular interface shearing existing in most geotechnical engineering, using R_n for characterizing

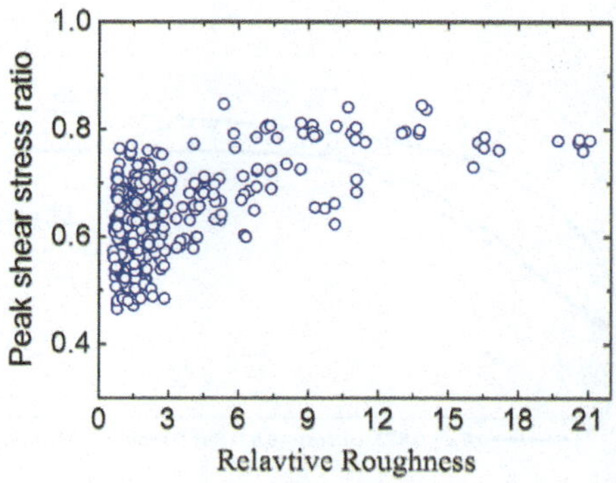

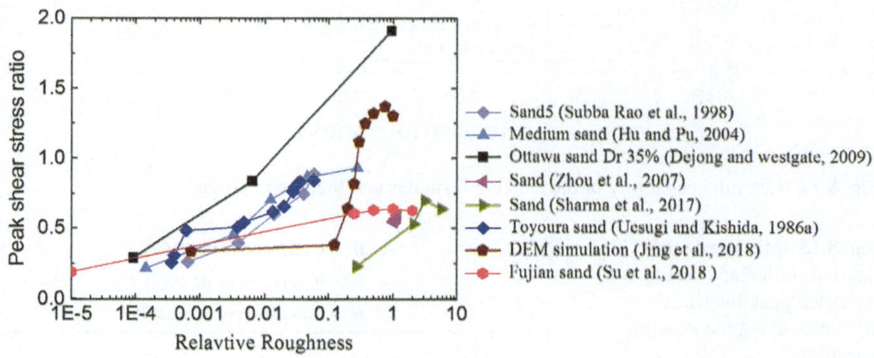

Fig. 8.13 Relationship between the peak interface strength parameter and relative roughness: **a** the present chapter and **b** other studies

surface morphology and estimation of IE_P is inadequate but hybrid parameter Pdq is more efficient and accurate to estimate IE_P, instead, explained as follows. R_n, as the local morphology parameter of an interface, is only evaluated by the maximum height of a profile. It will be efficient for the morphology evaluation of regular surfaces but not for the morphology evaluation of random surfaces. Pdq, as a global morphology parameter of an interface, provides not only height information but also spacing information along the full investigated profile. Accordingly, it is more efficient to evaluate the interface strength. It may be noted that the conclusion was reached based on the 2D DEM simulation, and widespread application needs to conduct more 3D numerical and experimental investigations.

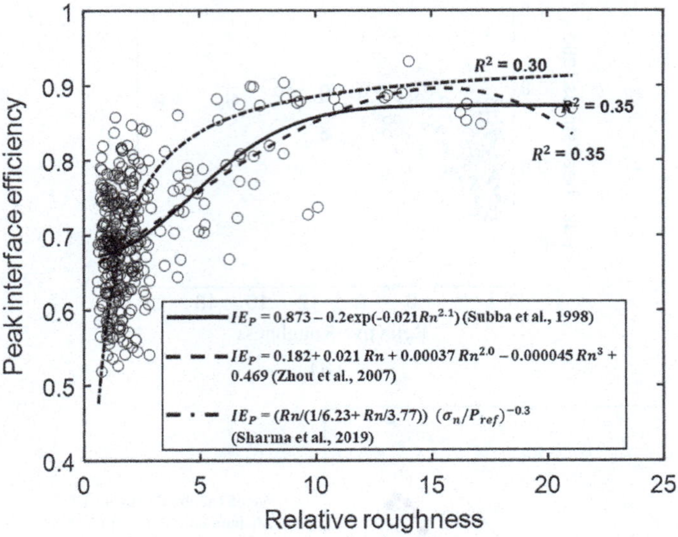

Fig. 8.14 Learning capability of the existing formulas and their expressions

Fig. 8.15 Measured peak interface efficiency versus predicted peak interface efficiency using the existing formulas

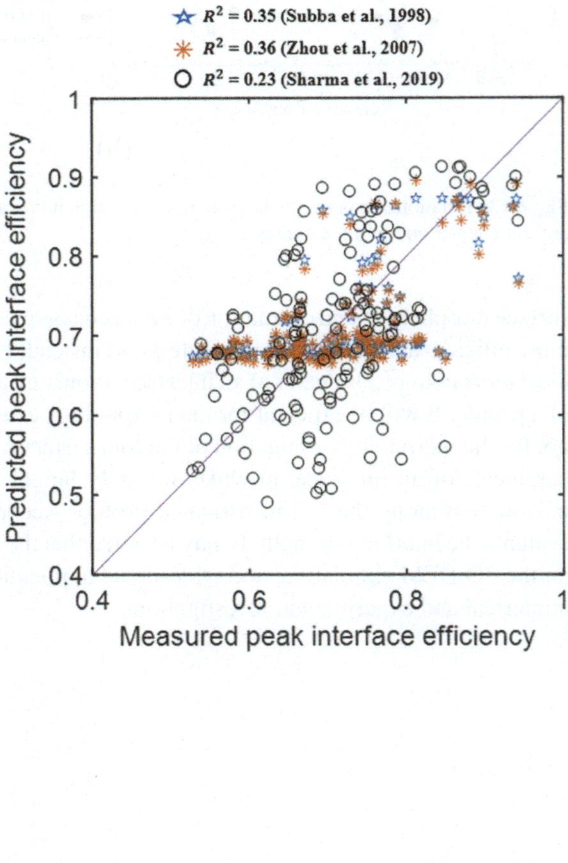

Table 8.10 Accuracy measurements for the proposed formula and existing formulas

Equation	MAE	MARE
The proposed formula	0.0442	0.0631
Formula referred by Subba et al. [51]	0.0594	0.0841
Formula referred by Zhou et al. [50]	0.0589	0.0834
Formula referred by Sharma et al. [49]	0.0760	0.107

8.5 Conclusions

Based on the 2D DEM simulation, 480 interface shear tests with random profiles were conducted on coarse-grained soils. The relevant morphology parameters were selected using Spearman's rank correlation coefficient. BNGR was applied to forming a metaheuristic model for estimation of the SSI shear strength. The key observations are summarized as follows:

(1) ρ_s is in a range of 0.4 to 1 or −1 to −0.4 for parameters Pp/D_{50}, Pv/D_{50}, Pz/D_{50}, Pc/D_{50}, Pa/D_{50}, Pq/D_{50}, Psm/D_{50}, Pdq, Pdc/D_{50} and PPc. The highest Spearman's rank coefficient, amounting to 0.788, has been obtained for hybrid parameter Pdq, which represents not only the amplitude information but also the surface slope information on a random surface.

(2) One significant input variable (Pdq) was effectively selected from 10 potential candidates (Pp/D_{50}, Pv/D_{50}, Pz/D_{50}, Pc/D_{50}, Pa/D_{50}, Pq/D_{50}, Psm/D_{50}, Pdq, Pdc/D_{50} and PPc) by using the BNGR algorithm. The optimal model selected was verified on the testing data and compared with the prediction results of some selected models, the full model, and the model with the most commonly used parameter.

(3) Based on the 2D DEM results, the proposed formula was compared with the existing predicted formulas. For irregular interface shearing, using R_n for characterizing surface morphology and estimating IE_P is inadequate, but the hybrid parameter Pdq is more efficient and accurate for estimating IE_P.

It is noteworthy that the results were achieved based on the 2D DEM simulation, which has inherent limitations in investigating real 3D problems. The granular soil was modeled with circular disks in this chapter, and the rolling resistance was adopted to compensate for the lack of angularity of circular particles. The simplifications in this study surely cause differences between the real and DEM investigations, and 2D numerical investigations still provide helpful results to illustrate the correlation between morphology parameters and interface shear strength. To make the conclusions broad, 3D DEM simulations will be the future work.

References

1. Junaideen S, Tham L, Law K, Lee C, Yue Z (2004) Laboratory study of soil nail interaction in loose, completely decomposed granite. Can Geotech J 41:274–286
2. Sharma M, Samanta M, Sarkar S (2017) Laboratory study on pullout capacity of helical soil nail in cohesionless soil. Can Geotech J 54:1482–1495
3. Uesugi M, Kishida H (1986) Influential factors of friction between steel and dry sands. Soils Found 26:33–46
4. Uesugi M, Kishida H (1986) Frictional resistance at yield between dry sand and mild steel. Soils Found 26:139–149
5. Dove JE, Jarrett JB (2002) Behavior of dilative sand interfaces in a geotribology framework. J Geotech Geoenvironmental Eng 128:25–37
6. Rui S, Wang L, Guo Z, Cheng X, Wu B (2020) Monotonic behavior of interface shear between carbonate sands and steel. Acta Geotech
7. Wang X, Wang X-Z, Zhu C-Q, Meng Q-S (2019) Shear tests of interfaces between calcareous sand and steel. Mar Georesour Geotechnol 37:1095–1104
8. Guo J, Wang X, Lei S, Wang R, Kou H, Wei D (2020) Effects of groove feature on shear behavior of steel-sand interface. Advances in Civil Engineering 2020
9. Wu X-Y, Yang J (2016) Direct shear tests of the interface between filling soil and bedrock of Chongqing airport. Advances in civil, envrionmental, and materials research (ACEM16), Jeju Island
10. Canakci H, Hamed M, Celik F, Sidik W, Eviz F (2016) Friction characteristics of organic soil with construction materials. Soils Found 56:965–972
11. Martinez A, Frost J (2017) The influence of surface roughness form on the strength of sand–structure interfaces. Géotechnique Letters 7:104–111
12. Han F, Ganju E, Salgado R, Prezzi M (2018) Effects of interface roughness, particle geometry, and gradation on the sand–steel interface friction angle. J Geotech Geoenvironmental Eng 144:04018096
13. Chen X, Zhang J, Xiao Y, Li J (2015) Effect of roughness on shear behavior of red clay–concrete interface in large-scale direct shear tests. Can Geotech J 52:1122–1135
14. Chu L-M, Yin J-H (2005) Comparison of interface shear strength of soil nails measured by both direct shear box tests and pullout tests. J Geotech Geoenvironmental Eng 131:1097–1107
15. Feng S-J, Liu X, Chen H-X, Zhao T (2018) Micro-mechanical analysis of geomembrane-sand interactions using DEM. Comput Geotech 94:58–71
16. Hu L, Pu J (2004) Testing and modeling of soil-structure interface. J Geotech Geoenvironmental Eng 130:851–860
17. Jing XY, Zhou WH, Zhu HX, Yin ZY, Li Y (2018) Analysis of soil-structural interface behavior using three-dimensional DEM simulations. Int J Numer Anal Methods Geomech
18. Su L-J, Zhou W-H, Chen W-B, Jie X (2018) Effects of relative roughness and mean particle size on the shear strength of sand-steel interface. Measurement 122:339–346
19. Wang H-L, Zhou W-H, Yin Z-Y, Jie X-X (2019) Effect of grain size distribution of sandy soil on shearing behaviors at soil–structure interface. J Mater Civ Eng 31:04019238
20. Chen W-B, Zhou W-H, dos Santos JA (2020) Analysis of consistent soil–structure interface response in multi–directional shear tests by discrete element modeling. Transp Geotech 24:100379
21. Zhu H, Zhou W-H, Yin Z-Y (2017) Deformation mechanism of strain localization in 2D numerical interface tests. Acta Geotechnica:1–17
22. Zhang N, Evans TM (2018) Three dimensional discrete element method simulations of interface shear. Soils Found 58:941–956
23. Santos PM, Julio EN (2007) Correlation between concrete-to-concrete bond strength and the roughness of the substrate surface. Constr Build Mater 21:1688–1695
24. Hoła J, Sadowski Ł, Reiner J, Stach S (2015) Usefulness of 3D surface roughness parameters for nondestructive evaluation of pull-off adhesion of concrete layers. Constr Build Mater 84:111–120

25. Sadowski Ł, Mathia TG (2016) Multi-scale metrology of concrete surface morphology: Fundamentals and specificity. Constr Build Mater 113:613–621
26. Wang J, Jiang M (2011) Unified soil behavior of interface shear test and direct shear test under the influence of lower moving boundaries. Granular Matter 13:631–641
27. Zhu HX, Zhou WH, Jing XY, Yin ZY (2019) Observations on fabric evolution to a common micromechanical state at the soil-structure interface. Int J Numer Anal Meth Geomech 43:2449–2470
28. Wang J, Gutierrez MS, Dove JE (2007) Numerical studies of shear banding in interface shear tests using a new strain calculation method. Int J Numer Anal Meth Geomech 31:1349–1366
29. Wang Z, Jacobs F, Ziegler M, Yang G (2020) Visualisation and quantification of geogrid reinforcing effects under strip footing loads using discrete element method. Geotext Geomembr 48:62–70
30. Chen W-B, Xu T, Zhou W-H (2021) Microanalysis of smooth Geomembrane-Sand interface using FDM–DEM coupling simulation. Geotext Geomembr 49:276–288
31. Lai H-j, Zheng J-j, Zhang R-j, Cui M-j (2016) Visualization of the formation and features of soil arching within a piled embankment by discrete element method simulation. J Zhejiang Univ-SCIENCE A 17:803–817
32. Shen W, Zhao T, Dai F, Jiang M, Zhou GG (2019) DEM analyses of rock block shape effect on the response of rockfall impact against a soil buffering layer. Eng Geol 249:60–70
33. Utili S, Zhao T, Houlsby G (2015) 3D DEM investigation of granular column collapse: evaluation of debris motion and its destructive power. Eng Geol 186:3–16
34. ISO E (2009) 4287: 2009. Geometrical Product Specifications (GPS)-Surface texture: Profile method–Terms, definitions and surface texture parameters (ISO 4287: 1997+ Cor 1: 1998+ Cor 2: 2005+ Amd 1: 2009)(includes Corrigendum AC: 2008 and Amendment A1: 2009)
35. Sadowski Ł, Stefaniuk D (2017) Microstructural evolution within the interphase between hardening overlay and existing concrete substrates. Appl Sci 7:123
36. Sadowski Ł, Żak A, Hoła J (2018) Multi-sensor evaluation of the concrete within the interlayer bond with regard to pull-off adhesion. Arch Civ Mech Eng 18:573–582
37. Map M (2014) Software (Digital Surf, Besançon, France)
38. Itasca PD (2008) Particle Flow Code in 3 Dimensions, User's Guide
39. Iwashita K, Oda M (1998) Rolling resistance at contacts in simulation of shear band development by DEM. J Eng Mech 124:285–292
40. Chareyre B, Villard P (2002) Discrete element modeling of curved geosynthetic anchorages with known macro-properties. In: Proc, First Int PFC Symposium, pp 197–203
41. Gu X, Chen Y, Huang M (2017) Critical state shear behavior of the soil-structure interface determined by discrete element modeling. Particuology 35:68–77
42. Koerner RM (2012) Designing with geosynthetics. Xlibris Corporation
43. Kowalczyk T, Pleszczynska E, Ruland F (2004) Grade models and methods for data analysis: with applications for the analysis of data populations. Springer Science & Business Media
44. Vik M, Vikova M, HRUBOSOVA Z (2014) Surface topography of textile braille labelling. In: Proceedings of the 7th international textile, clothing & design conference-magic world of Textiles, Dubrovnik. https://www/researchgate/net/publication/270957631_Surface_Topography_of_Textile_Braille_Labelling. Accessed 07 Aug 2018
45. Olympus (2014) Roughness (2D) parameter. https://wwwolympus-imscom/en/knowledge/metrology/roughness/2d_parameter/
46. Zhao L-S, Zhou W-H, Su L-J, Garg A, Yuen K-V (2019) Selection of physical and chemical properties of natural fibers for predicting soil reinforcement. J Mater Civ Eng 31:04019212
47. Yuen K-V, Ortiz GA, Huang K (2016) Novel nonparametric modeling of seismic attenuation and directivity relationship. Comput Methods Appl Mech Eng 311:537–555
48. DeJong JT, Westgate ZJ (2009) Role of initial state, material properties, and confinement condition on local and global soil-structure interface behavior. J Geotech Geoenvironmental Eng 135:1646–1660
49. Sharma M, Samanta M, Punetha P (2019) Experimental investigation and modeling of pullout response of soil nails in cohesionless medium. Int J Geomech 19:04019002

50. Zhou G-Q, Xia H-C, Zhao G-S, Jie Z (2007) Nonlinear elastic constitutive model of soil-structure interfaces under relatively high normal stress. J China Univ Min Technol 17:301–305
51. Subba Rao K, Rao K, Allam M, Robinson R (1998) Interfacial friction between sands and solid surfaces. Proc Inst Civ Eng-Geotech Eng 131:75–82

Chapter 9
Analysis of Smooth Geomembrane–Sand Interface

9.1 Introduction

The interface between smooth geomembrane and soil, as one of the soil–structure interfaces (SSIs), commonly exists at the boundary between a drained layer and an adjacent undrained layer and in landfill camping projects. These interfaces are potential weakness planes if inadequate friction is developed in the soil–geomembrane interface that could resist shear failure. Previous studies have mainly focused on investigating the macro–shear behaviors of rough and rigid interface [1–7] or rough and low–hardness interface [8–14]. In contrast, micro–shear responses under rough interface shearing have also been studied [15–19]. However, soil–geomembrane interface is typically a smooth and low–hardness interface. Previous studies specific to the shear behavior of smooth and deformable interfaces are limited, particularly in the microscopic aspect. The interaction mechanism involved in such a smooth and deformable interface may be different from that involved in a rough and undeformable interface. Understanding the fundamental interaction mechanism of a smooth and deformable interface can aid in the development of accurate soil–smooth geomembrane interaction models.

Experimental studies of the sand–smooth geomembrane interaction have shown that particle morphology, initial density, and normal stress significantly influence the interface shear behavior. Based on some interface direct shear tests on Ottawa sand placed on a smooth geomembrane, Dove and Frost [20] demonstrated that the normal stress highly influences the interface friction coefficient. These experimental results showed that a critical normal stress exists in affecting the interface friction coefficient. Fuggle [21] pointed out that the interface friction angle and the critical normal stress level are strongly influenced by the particle size and mixing proportions of sand samples. Furthermore, quantitative study of surface change induced by the intrusion of angular sands was carried out by Frost et al. [22] and Vangla and Gali [23] using micro–topographical analysis, demonstrating that the angularity of sand particles and the normal stress can directly lead to indentations on smooth geomembrane. However, very limited attention was paid on the micro–response of sand. Sand is one

W.-H. Zhou and Z.-Y. Yin, *Practice of Discrete Element Method in Soil-Structure Interface Modelling*, https://doi.org/10.1007/978-981-19-0047-1_9

of the participants in the sand–geomembrane interaction and its response may play a significant role in the shear mechanism. Besides, not many studies are available for the effect of geomembrane hardness on the interface shear behavior. Although it is possible to track the macro–response of the interface in the experimental interface shear test, the micro–behavior of the sand as it interacts with the geomembrane and the shear–induced changes in the smooth geomembrane surface topography are hard to be evaluated experimentally. Accordingly, numerical simulations are preferable for that purpose.

The finite element method (FEM) or finite difference method (FDM) offers a wide range of capabilities useful in solving issues of soil–geosynthetics interactions [24–30]. The real geogrid or geomembrane used to be modeled by either a truss structure (in 2D analysis) or a continuous sheet (in 3D analysis). Nonetheless, tracking the motion of the surrounding sand particles and directly investigating the micro–mechanical interactions are challenging when using this approach, as discrete soil particles are modeled as continuum elements. As a substituted tool to the continuum approach, the discrete element method (DEM) has been applied in many studies to model soil–geosynthetics interaction issues, such as, soil–geotextile interface [31–33], soil–geogrid interaction [9, 34–38], and soil–geomembrane contact[39]. Since DEM simulations could not incorporate directly into the effect of continuum hardness, Frost et al. [22] modeled the effect of hardness indirectly by varying the particle–continuum surface friction coefficient, instead. Feng et al. [39] assumed that the geomembrane was made of spherical particles bonded together to reproduce the surface of geomembrane. Those studies contribute some valuable findings to understand the micro–mechanical interaction between geomembrane and sand. However, the complex geomembrane deformation caused by the sand puncture or indentation and the surface topography distribution of geomembrane might not be accurately evaluated in an interface direct shear test owing to the inflexibility of walls or the bonded particles in DEM.

Taking advantage of both the FD and DE methods, a coupled FDM–DEM framework was used to model an interface direct shear test for investigation of soil–smooth geomembrane interaction in this chapter. This numerical framework was intended to help researchers gain new insights into the micro–behavior of the sand as it interacts with the geomembrane and the coupling effect between normal stress and surface hardness on interface shear behavior at the microscopic level. Sand particles and smooth geomembrane were modeled using DEM and FDM, respectively, and the introduction of the coupled model was followed by validation of it. Furthermore, the effects of normal stress and surface hardness on interface shear behavior were analyzed from perspectives of the macro–shearing behavior and micro–responses of smooth geomembrane and sand.

9.2 Coupled Finite Difference–Discrete Element Framework

9.2.1 Discrete and Continuum Simulation

The discrete simulation is performed using Particle Flow Code, based on the DEM developed by Itasca [40]. In PFC3D, each particle interacts with its neighboring particles or walls based on Newton's second law of motion, while the force–displacement law is employed to update each contact. A rolling resistance linear contact model was adopted in the DEM simulation to consider the rolling resistance effect. The translational and rotational motions of particle i are calculated by the following equations:

$$m_i \frac{dU_i^p}{dt} = \sum_c F_{ij}^c \tag{9.1}$$

$$I_i \frac{d\omega_i}{dt} = \sum_c M_{ij}^c \tag{9.2}$$

where U_i^p and ω_i are the translational and angular velocities of particle i, respectively. F_{ij}^c and M_{ij}^C are the contact force and torque acting on particle i by particle j or the walls at contact point c, respectively.m_i is the mass of inertia of particle i, while I_i is moment of that. The incremental rolling resistance moment ΔM_{ij}^C is calculated by the rolling stiffness k_r and the incremental relative rotation $\Delta\theta_r$ between the two particles:

$$\Delta M_{ij}^C = -\mathrm{k}_r \Delta\theta_r \tag{9.3}$$

$$k_r = k_t \overline{r} \tag{9.4}$$

where k_t is the tangential stiffness and $\overline{r} = {}^{r_i r_j}/_{(r_i + r_j)}$ is the effective contact radius. A threshold value for M_{ij}^C is denoted by M_{ij}^{C*}:

$$M_{ij}^{C*} = \mu_r \overline{r} F_n \tag{9.5}$$

where μ_r is the rolling resistance coefficient and F_n is the current normal contact force.

The continuum simulation is conducted with the FLAC3D program. Compared with DEM, large–strain is allowed in FDM. By dividing the continuum into quadrilateral elements, the solution domain in FLAC is substituted with a limited number of grid nodes. FLAC subdivides each element into two overlaid sets of constant–strain triangular sub–element to eliminate the issue of hourglass deformations [41].

9.2.2 Contact Between Particles and Elements

The contact forces at the interface are updated according to the rule defined in the DEM. The velocities at the contact points are involved in the nodal velocities by interpolation, while the contact forces contribute to the nodal forces using weighted values. The data transmission between codes using a socket connection between FEM and DEM is shown in Fig. 9.1. Within each timestep, the boundary node velocities in the continuum domain (FLAC) are delivered to the discrete domain using the socket connections. The discrete domain receives the data from the continuum domain, after which the wall coordinates are updated by the coordinates and velocities in the discrete domain, with the resulting wall forces (and moments) sent back to the continuum domain. In this way, the coordinates, velocities, and forces at the walls in both domains are updated during the cycling, so that the coupling framework is achieved.

When considering the individual contact, element–particle contact can be noted by the contact point $x_i^{[C]}$ at the contact surface. $\boldsymbol{n}_i$ is the unit normal vector from the particle's center to the element. The contact force $F_i^{[C]}$ can be decomposed into two mutually perpendicular components: normal contact force $F_{in}^{[C]}$ and tangential contact force $F_{it}^{[C]}$. The relative displacements in the normal $x_{in}^{[C]}$ and tangential $x_{is}^{[C]}$ directions can be expressed by the ratio of the components of contact force over the normal and tangential contact stiffnesses, respectively (Fig. 9.2).

The unit normal vector $\boldsymbol{n}_i$ at the interface is in the interior normal direction. The position of the contact point can be expressed by

$$x_i^{[C]} = x_i^{[B]} + (R_i^{[B]} - (R_i^{[B]} - l)/2)\boldsymbol{n}_i \tag{9.6}$$

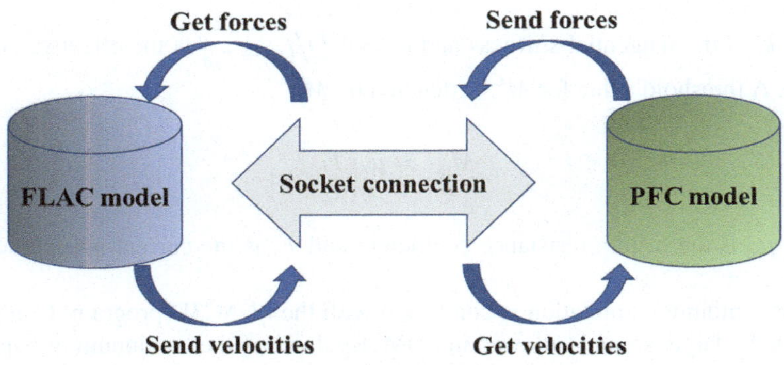

Fig. 9.1 Schematic of data switch between FLAC and PFC

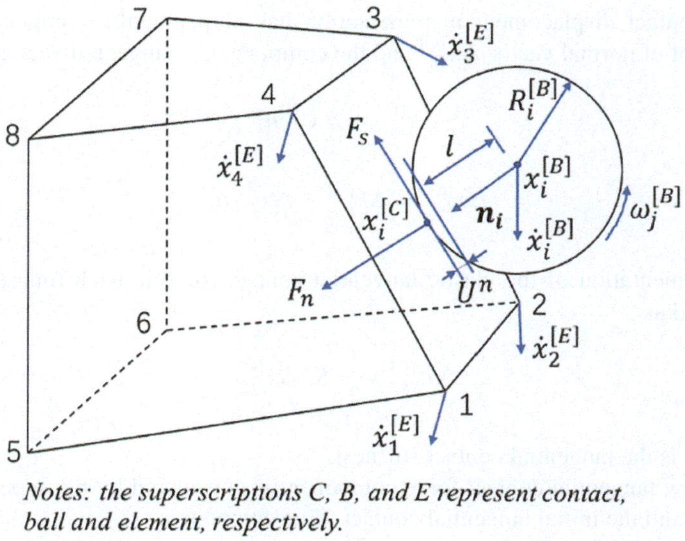

Fig. 9.2 Schematic of the contact geometry between a particle and a continuum element

where l is the distance from the particle's center to the contact point $x_i^{[C]}$, with $R_i^{[B]}$ the diameter of the particle.

The normal contact force $F_{in}^{[C]}$ can be obtained from

$$F_{in}^{[C]} = K^n U^n \boldsymbol{n}_i \tag{9.7}$$

where K^n is the normal contact stiffness, which can be evaluated by the contact stiffness model, and U^n is the contact overlap thickness, which can be formulated by $U^n = R_i^{[B]} - l$. The component in tangential direction can be obtained by incrementing the elastic tangential contact force, with incrementation calculated by the tangential relative displacement. The motion of the contacts can be obtained by calculating the normal vector $\boldsymbol{n}_i$ and the position of contacts $x_i^{[C]}$ in each timestep. The relative contact velocity at interface V_i is given by

$$\mathrm{V}_i = \dot{x}_{i,E}^{[C]} - \dot{x}_{i,B}^{[B]} \tag{9.8}$$

where $\dot{x}_{i,E}^{[C]}$ and $\dot{x}_{i,B}^{[B]}$ are the velocities of element and particle at the interface, respectively.

The contact displacement incrementation for every timestep is given by

$$\Delta x_{in}^{[C]} = \mathrm{V}_i \Delta t \tag{9.9}$$

where Δt is the timestep.

The contact displacement incrementation has perpendicular components: the component of normal vector $\Delta x_{in}^{[C]}$ and the component of tangential vector $\Delta x_{is}^{[C]}$:

$$\Delta x_{in}^{[C]} = \Delta x_i^{[C]} \boldsymbol{n}_i \tag{9.10}$$

$$\Delta x_{is}^{[C]} = \Delta x_i^{[C]} - \Delta x_i^{[C]} \boldsymbol{n}_i \tag{9.11}$$

The incrementation of the elastic tangential contact force in each timestep Δt is formulated as

$$\Delta F_{is}^{[C]} = -K^s \Delta x_{is}^{[C]} \tag{9.12}$$

where K^s is the tangential contact stiffness.

The new tangential contact force component is considered by the superposition of $\Delta F_{is}^{[C]}$ and the initial tangential contact force component

$$F_{is}^{[C]} \leftarrow F_{is}^{[C]} + \Delta F_{is}^{[C]} \le \mu F_{in}^{[C]} \tag{9.13}$$

where μ is the friction coefficient.

Equations (9.7) and (9.13) can be used to obtain the resultant force and resultant moment on contacted particles,

$$F_i^{[B]} \leftarrow F_i^{[B]} - F_i^{[C]} \tag{9.14}$$

$$M_i^{[B]} \leftarrow M_i^{[B]} - e_{ijk}\left(x_j^{[C]} - x_j^{[B]}\right) F_k^{[C]} \tag{9.15}$$

where $F_i^{[B]}$ and $M_i^{[B]}$ are the superposition of contact forces and the moment of contact forces, respectively, on contacted particles; $x_j^{[C]}$ is the coordinates of the contact point; and $x_j^{[B]}$ is the center of the contacted particle.

In the continuum element at the interface, the nodal forces come from the contact forces through a weighted value, and total contact force is updated using

$$F_i^{[E,j]} = F_i^{[E,j]} + F_i^{[C]} K_j \tag{9.16}$$

where $F_i^{[E,j]}$ is the nodal force in the node j of the continuum element at the interface and the weight value K_j is a type function.

Because the overlap between continuum elements and particles is negligible, the effect of the force moment on elements $\mathrm{M} = \left|F_{is}^{[C]} U^n\right|/2$ caused by the tangential contact force is negligible.

Based on the above equations, the contact forces and the positions of particles and nodes can be updated. The contact relationship between particles and elements can be extended to apply to whole coupled interfaces between particles and continuum elements.

9.2.3 Coupling Simulation Mechanism

In 3D numerical modelling, 8–node hexahedral elements are applied in the continuum simulation, whereas particles are created in the discrete simulation. Translational motion equation can be obtained by

$$F_i = \sum_c F_c + R^i = m\ddot{x}_i \tag{9.17}$$

and the rotational motion equation in the discrete element can be expressed as

$$M_3 = \sum_c r_c \times F_c + K^i = I\dot{\omega}_3 \tag{9.18}$$

where F_i represents the total external forces; F_c the contact forces; m the mass; R^i and K^i are the external force and its corresponding external moment; r_c is the vector from the contact point pointing to the particle's center; x_i is the particle's displacement; I the particle's moment of inertia; and ω_3 the particle's angular velocity; M_3 is the total force moment. Since rolling resistance is incorporated into the linear model in the present study, the moment arising from the rolling resistance exists, which means the torque acting on the particle or element takes effect.

The foregoing equations of motion in discrete and continuum elements are solved using the explicit central differencing method:

$$\dot{x}_t^{t+\Delta t/2} = \dot{x}_t^{t-\Delta t/2} + \frac{F_i^t}{m}\Delta t \tag{9.19}$$

$$\omega_3^{t+\Delta t/2} = \omega_3^{t-\Delta t/2} + \frac{M_i^t}{I}\Delta t \tag{9.20}$$

The displacement at the time $t + \Delta t$ can be obtained from

$$x_i^{t+\Delta t/2} = x_i^{t+\Delta t/2}\Delta t + x_i^t \tag{9.21}$$

Within each timestep, the contact force at the interface is updated and then sent back to the two domains. The motion equation is used to solve for velocity and relative displacement. The new position of the elements and particles at the interface will first be obtained through the aforementioned calculation once the next timestep starts, after which the contacts at the interface will be updated and distributed back to the two domains before the next calculation round. More detailed coupling mechanism can be found in the literature [42].

9.3 Coupled Interface Shear Test Model

9.3.1 Interface Shear Test Model

In this study, the experimental interface direct shear tests performed on coarse sand (CS) [43] were modeled using the proposed coupled FDM–DEM framework. Details of the experimental test are summarized as follows:

- Based on the conventional direct shear test apparatus, the bottom half of the shear box was replaced with a steel plate to fix the smooth geomembrane;
- The soil used in the experiment was the high–rounded CS, with mean particle size D_{50}=3.0 mm. The smooth high–density polyethylene (HDPE) geomembrane was adopted in the experiment;
- For the interface shear test, only half the shear box was filled with CS under the required relative density. After sample preparation, normal stresses of 21 kPa, 37 kPa, and 57 kPa were applied above the box. Shear loading was then applied to the bottom plate at a constant rate of 1.0 mm/min.

The smooth geomembrane was modeled using FDM. The smooth geomembrane, which is 200 mm long, 150 mm wide, and 1.5 mm thick, was modeled using 8–node brick elements with integration points. The smooth geomembrane was kept 50 mm longer than the shear box to ensure a length adequate for shearing. Previous numerical studies of reinforcement mechanisms have adopted a linear elastic model for geosynthetics [24, 44, 45]. It has been reported that the maximum shear–induced roughness is far less than the geomembrane thickness and the corresponding strain is too small to achieve the needed strain in the yielding phase of geomembrane [22, 23, 43]. Accordingly, a linear elastic material model was applied for the smooth geomembrane in this study, as summarized in Table 9.1. The full geometry of the smooth geomembrane consisted of 7,500 finite elements, as shown in Fig. 9.3a. To model the deformation behavior of the smooth geomembrane attached to a rigid base in the process of shearing, the smooth geomembrane upper surface could be deformed freely, whereas the opposite side was rigidly fixed. Surface roughness analyses were performed on smooth geomembrane after shearing. Although standards for surface characterization are available in the literature [45, 46], many of the parameters defined in the standards capture certain aspects of the surface that are relevant to specific applications. Surface morphology for the samples in the present study was quantified by the popular amplitude parameter: root mean square roughness value (R_q), as used by [22, 23, 43]. The parameter was defined as shown in Fig. 9.4. Each specimen

Table 9.1 Summary of input parameters used in the coupled FDM–DEM model

Type of elements	Parameter	Value
Discrete particles	Ball density (kg/m^3)	2,600
	Inter–particle normal stiffness k_n (N/m)	0.7×10^7
	Inter–particle shear stiffness k_t (N/m)	0.35×10^7
	Particle–wall normal stiffness k_{nw} (N/m)	1.0×10^8
	Particle–wall shear stiffness k_{tw} (N/m)	0.5×10^8
	Inter–particle frictional coefficient f_p	0.45
	Particle–wall frictional coefficient f_{pw}	0.24
	Rolling resistance coefficient μ_r	0.1
Geomembrane	Young modulus E (pa)	1.5×10^8
	Poisson's ratio v	0.33

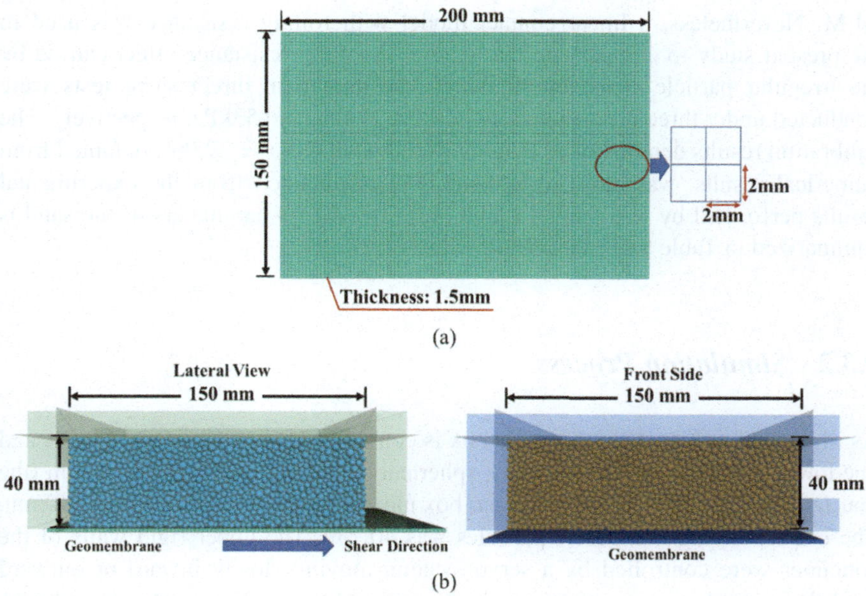

Fig. 9.3 The coupled FDM–DEM interface shear test model: **a** geometry of the smooth geomembrane and **b** schematic of an interface shear test

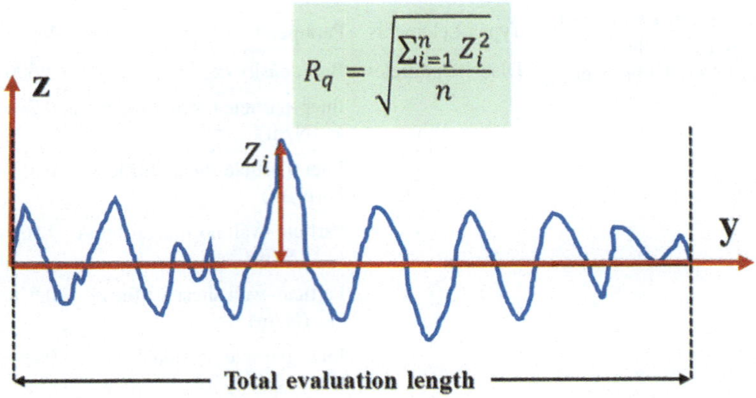

Fig. 9.4 Sketch of rough surface and definition of amplitude parameter

size 150 mm × 150 mm (the global shear area) was selected and the corresponding amplitude parameter was determined in the direction perpendicular to the shear direction, thus, to obtain the closest possible estimates of profile roughness.

Since the CS is high–rounded, spherical particles were used to model it using DEM. Nevertheless, a linear contact model with rolling resistance was used in the present study to compensate the inevitable rolling resistance effect caused by the irregular particle shape interlocking. The numerical direct shear tests were conducted under three normal stresses: 21kPa, 37kPa, and 53kPa, respectively. The calibration results demonstrated that the frictional angle φ_n= 39.96°, obtained from numerical results, was close to φ_n= 41.00°, as obtained from the experimental results performed by Vangla and Latha [43]. The micro–parameters of the sand is summarized in Table 9.1.

9.3.2 *Simulation Process*

As illustrated in Fig. 9.3b, the shear box is contained by five upper rigid walls and one lower flexible wall. In this study, spherical balls for which D_{50} = 3.0 mm (the real D_{50}) were generated inside a shear box measuring 150mm × 150mm × 40mm. The corresponding number of particles was 40,739. The upper rigid walls of the container were controlled by a servo system, moving slowly inward or outward until the normal stress on them reached a targeted value. The bottom plate began to move horizontally in the *x*–direction at a speed of about 0.1mm/min until the maximum shear displacement reached 30mm to guarantee the quasistatic shearing. In the process of shearing, four upper vertical walls were fixed, and the top and bottom walls were able to move only vertically and horizontally, respectively. The friction coefficient between sand and geomembrane was adjusted iteratively until the numerical results under three different normal stresses (21kPa, 37kPa, and 51kPa)

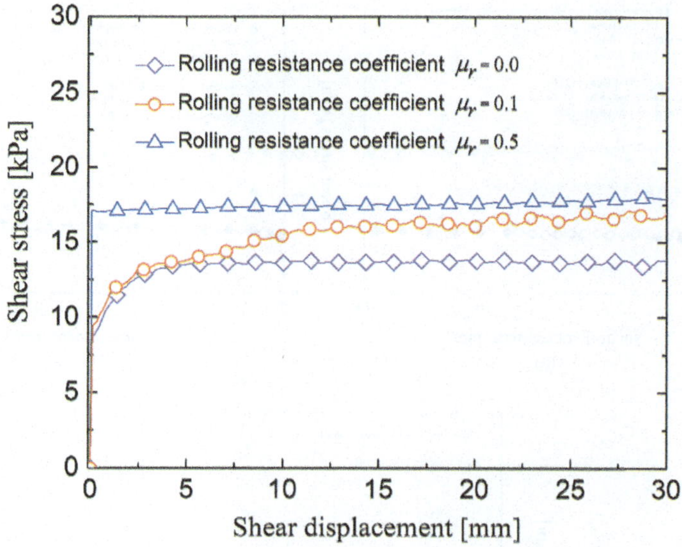

Fig. 9.5 Relationship between shear stress and shear displacement for interface shear tests with different rolling resistance coefficients

showed reasonable agreement with the corresponding experimental data. The input parameters are summarized in Table 9.1. To evaluate the role of rolling friction coefficient μ_r on the numerical results, the shearing results with different μ_r are presented in Fig. 9.5. The results show the maximum shear stress increases with μ_r.

9.3.3 Validation of the Numerical Model

Previous studies have successfully used the DEM to model a granular assembly sheared by a non–deformed rough plate [16, 47, 48]. To validate the possibility of using a coupled FDM—DEM interface shear test, additional numerical interface shear tests using a pure DEM simulation were conducted. In the coupled interface shear test model, one side of the bottom wall—that is, the upper surface of the smooth geomembrane—was able to deform during shearing. For convenience of comparison, the stiffness of the smooth geomembrane was set to an infinite value, ensuring a rigid wall such as that used in pure DEM model. It might be noteworthy that the model generation process and input parameters (see Table 9.1) for the DEM model were identical to those used in the coupled model. The shear stress–shear displacement curve obtained from both the DEM model and the coupled simulations is shown in Fig. 9.6a. The results of DEM models generally agree well with those of the coupled model, indicating the coupled model's suitability for dealing with issues of soil–structure interface.

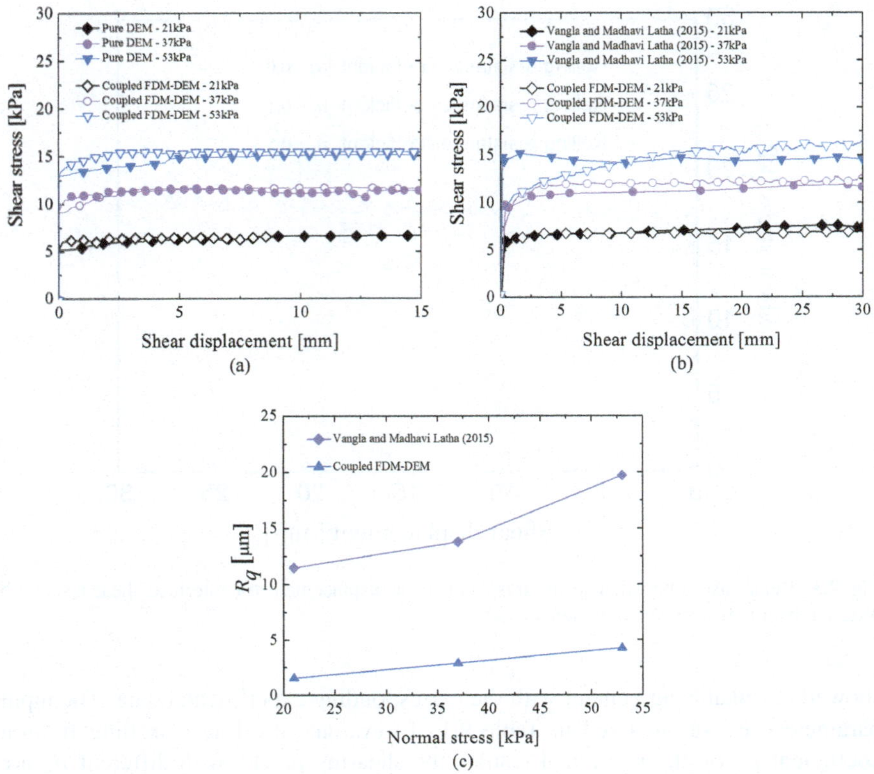

Fig. 9.6 Validation results: **a** relationship between shear stress and shear displacement obtained from both the pure PFC and the coupled model tests, **b** variation of shear stress with shear displacement obtained from both the experimental tests and the coupled model tests, and **c** root mean square roughness value (R_q) versus normal stress obtained from both the experimental tests and the coupled model tests

Figure 9.6b shows variations in shear stress with shear displacement gained from both the experimental test conducted by Vangla and Latha [43] and the coupled model test involving the deformable smooth geomembrane. Vangla and Latha [43] have suggested that coarse sands (compared with our case) are where peak strength is sustained for large deformations. For numerical results, shear stress initially increases with shear displacement, after which the peak value is sustained for large deformations. Even though the initial trends obtained from the experimental tests and the numerical tests show small differences, they share the same peak value and residual value. A plot of root mean square roughness value (R_q) versus normal stress obtained from both the experimental test conducted by Vangla and Latha [43] and the coupled model with deformable smooth geomembrane is shown in Fig. 9.6c. The DEM simulation from the present study shows a consistent trend to that seen in the experiment, even though the values are much lower in the DEM simulation, probably because the rolling resistance used to compensate for the effects of particle asperity in the simulation is still greatly underestimated compared with the rolling resistance of real particles. Particle asperity is an important factor affecting the response of smooth geomembrane [23, 43], but more investigation of the effects of particle shape is out of the scope of this study. In general, good agreement of the coupled model results with the corresponding experimental data demonstrates that the models with the selected micro–input parameters in the numerical simulation can satisfactorily represent laboratory CS sand–smooth geomembrane interface properties.

9.4 Microanalysis of Smooth Geomembrane–Sand Interface

The numerical interface shear test with the calibrated parameters (as summarized in Table 9.1) was adopted to conduct parametric study. The role of normal stress on the sand–smooth geomembrane interface was separately investigated under eight normal stresses (σ_n=10, 21, 30, 37, 53, 75, 100, and 180kPa). Analogous to the surface hardness that is considered as a governing factor in the mechanical interface performance in tribology [49], geomembrane hardness was selected as a quantified parameter. It can be estimated by $log_{10}E = 0.0235S - 0.6403$ (where E is the Young's modulus in MPa and S is the ASTM D2240 type D hardness), with reference to ASTM [50]. To investigate the influence of surface hardness D and its coupled effect with normal stress on the interaction between geomembrane and sand, the granular assembly under six normal stresses (σ_n=10, 30, 53, 75, 100, and 180 kPa) were sheared by the smooth geomembranes with five surface hardnesses ($D = 142$, $D = 240$, $D = 338$, $D = 632$, and $D = 828$), respectively. The macro–interface shear behavior and micro–response of geomembrane and soil were presented to evaluate the micro–interaction between smooth geomembrane and sand.

9.4.1 Effect of Normal Stress

9.4.1.1 Shear Behavior at the Geomembrane–Sand Interface

Figure 9.7a shows the shear stress–shear displacement relationship under eight various normal stresses. As expected, the shear stress increases with the shear displacement. The shear displacement corresponding to the peak shear stress is larger for specimens under the larger normal stress, and the peak shear stress increases with application of normal stress.

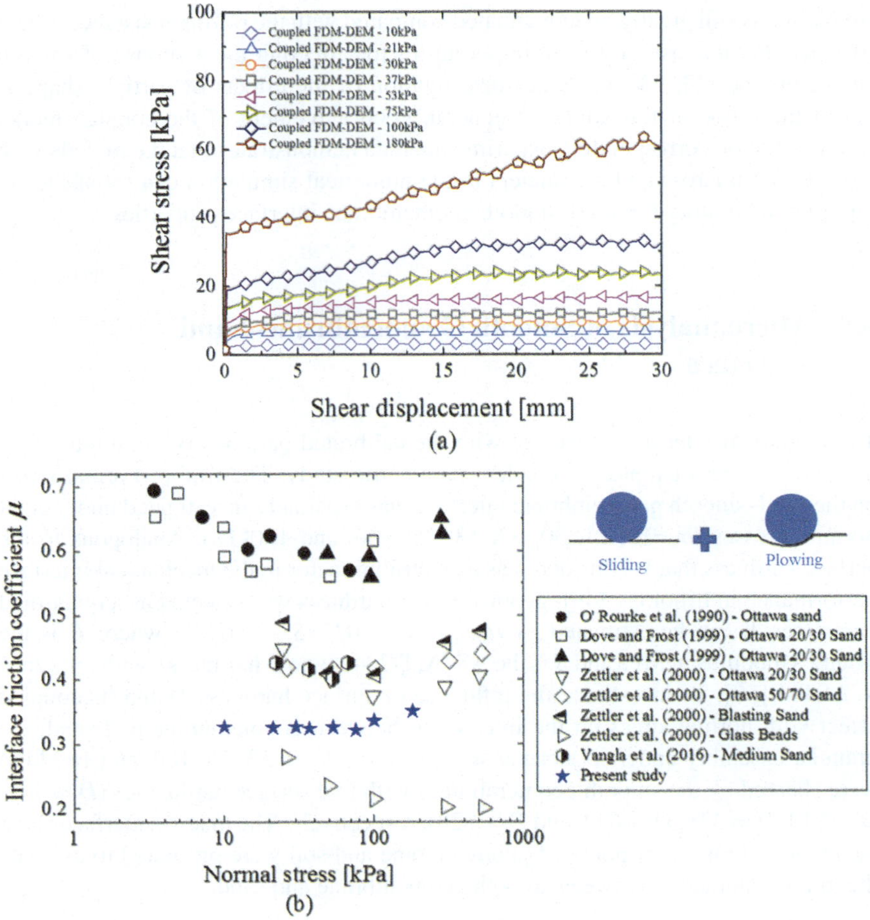

Fig. 9.7 Shear mechanism of interface between sand and smooth geomembrane under eight different normal stresses: **a** shear stress—shear displacement response and **b** variation of peak interface friction coefficient with respect to normal stress

Interface shear strength was expressed by the coefficient of interface friction μ, defined as the ratio of peak shear stress τ_{peak} over normal stress σ_n. Figure 9.7b shows variations in the interface friction coefficient with normal stress. From Fig. 9.7b, with the increase of normal stress, the interface friction coefficient μ initially fluctuates within a very small range (0.31–0.325). Up to a normal stress of 75kPa, after which it shows an increasing trend. It may be that the shear strength of granular assembly–smooth geomembrane interface originates from the relative contributions of the particles sliding and/or plowing along or into the smooth geomembrane surface [51]. At the low normal stress, the shear strength is determined by the particles sliding along the surface of the smooth geomembrane, resulting in that the interface friction coefficient mainly depends on soil particle–smooth geomembrane friction. As the normal stress increases to a critical normal stress at which the contact stress per particle may reach the yield pressure of the softer materials with which it is in contact, the particles eventually possess enough energy to plow into the smooth geomembrane surface and provide a higher interface shear resistance, increasing the interface friction coefficient. The critical normal stress is the phase marked by a change from a predominant sliding mechanism to a predominant plowing mechanism, resulting in higher shear resistance after the critical normal stress. In his study, it is observed that the critical normal stress is about 75 kPa, as shown in Fig. 9.7b.

Some previous studies [23, 43, 51, 52] have focused on granular soil–smooth geomembrane interfaces, which can lead to fundamental understandings of the soil–smooth geomembrane interface shear mechanism. To verify the findings of the present study, several relevant existing studies were selected for comparison of macroscopic shear mechanism. Variations in the interface friction coefficient with normal stress for these studies are also plotted in Fig. 9.7b. Overall, no consistent behavior is observed among these studies. In the present study, the interface shear resistance initially remained nearly constant as normal force increased, consistent with the case of Ottawa 50/70 sand reported by Zettler et al. [51]. As the normal stress increases to the critical value, the contribution of plowing increases, so that the interface friction coefficient increases with normal stress. The critical normal stress was reported to be around 50 kPa by O'rourke et al. [52], around 100–190kPa by Zettler et al. [51], and about 53 kPa by Vangla and Gali [23]. The differences between these studies might indicate that critical stress is affected by soil type, soil particle shape, geomembrane type, and the like. Comparison between the present study and previous studies in Fig. 9.7b offers an overall analysis of the existing data for a particulate material–smooth geomembrane interface. In general, the published studies still provide strong support for the variations of interface friction coefficient observed in the present study.

9.4.1.2 Micro-Response of Smooth Geomembrane

To more clearly clarify the effect of normal stress on interface shear mechanisms, Fig. 9.8a, b present the 3D topographical images of the smooth geomembranes after shearing under five different normal stresses and their corresponding post shear surface profiles, respectively. The virgin geomembrane was assumed to be smooth. It is apparent that the applied normal stress significantly produces surficial damage. Very small indentations and deformation in the smooth geomembrane caused by the sand particles are seen under the lowest normal stress of 10 kPa. With the increase in the applied normal stress, more indentations appear in the smooth geomembrane surface (Fig. 9.8a) and the maximum vertical deformation increases (Fig. 9.8b) correspondingly. As shown in Fig. 9.8b, the maximum vertical deformation under the normal stress of 10 kPa, 30 kPa, 53 kPa, and 100 kPa are around 33.1μm, 62.4μm, 141.2μm, and 185.4μm, respectively, whereas the maximum deformation for the highest loading condition, normal stress of 180 kPa, reaches about 304μm. The

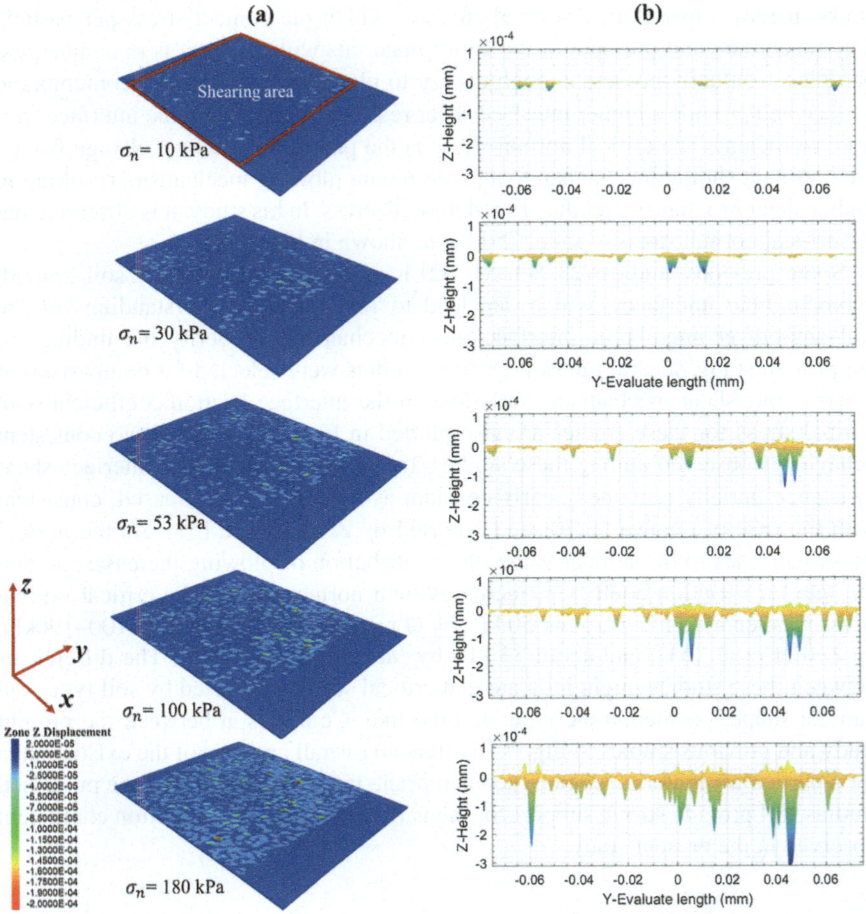

Fig. 9.8 Micro–topographical analysis of sheared smooth geomembrane sample under five different normal stresses at the last state: **a** 3D topographical images and **b** 2D shear surface profiles

amplitude parameter described in Fig. 9.4 was used to quantify the smooth geomembrane surface damage. Figure 9.9 shows variations in root mean square roughness (R_q) within the smooth geomembrane under normal stress. On the whole, as can be seen in Fig. 9.9, surface roughness increases with normal stress. The least surface roughness values are seen in the case of σ_n=10 kPa, as a result of particles' sliding without forming much deeper grooves. As the normal stress increases to 180 kPa, maximum roughness is achieved. The variation of R_q with normal stress evidently shows the effect of critical stress on surface changes. The slope in the plot of R_q with normal stress demonstrates a sudden jump exceeding the critical normal stress of 75 kPa, manifesting deeper indentations beyond the critical normal stress. These results clearly highlight the role of critical normal stress on the shear mechanism of

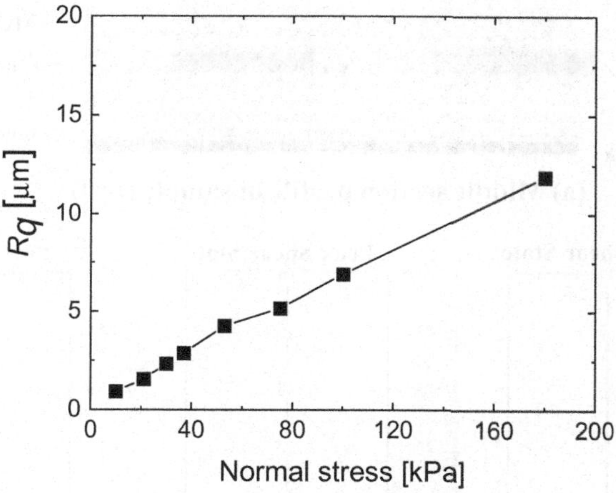

Fig. 9.9 Variation of amplitude parameter R_q at the end of shearing (shear displacement = 30mm) with normal stress

sand–smooth geomembrane interfaces and the phase between predominant sliding and predominant plowing.

9.4.1.3 Micro–Response of Granular Soil

The localized band thickness can be obtained by setting a series of strain measurements inside specimen, as shown Fig. 9.10a. The average shear strain on different layers of sand specimen can be calculated. The vertical average shear strain distributions of each layer along the depth of specimen at initial, peak, and post shear states under normal stress σ_n=53kPa are shown in Fig. 9.10b. It is shown that with the increase of shear displacement, the vertical distribution remains as found that in the initial state, so that no formation of shear band is localized throughout shearing. The findings for other normal stresses are not shown but similar observations were obtained.

Contact force chain (Fig. 9.11a) can reflect the state of contact force distribution. To better understand the evolution of contact force inside a soil sample, polar histogram was used to visualize contact force distribution. The contact force has two components, i.e., tangential contact force and normal contact force. The x–z plane polar histograms of the normal and the tangential contact force distributions were normalized by the average normal contact force over the total number, as shown in Fig. 9.11b. Meanwhile, Fourier series expression was used to fit the normalized histograms and conduct calculations of their anisotropies and principal directions in Fig. 9.11b. The expressions proposed by Bathurst and Rothenburg [53] are

$$\overline{f}_n(\theta) = \overline{f}_0[1 + a_n\cos2(\theta - \theta_n)] \tag{9.22}$$

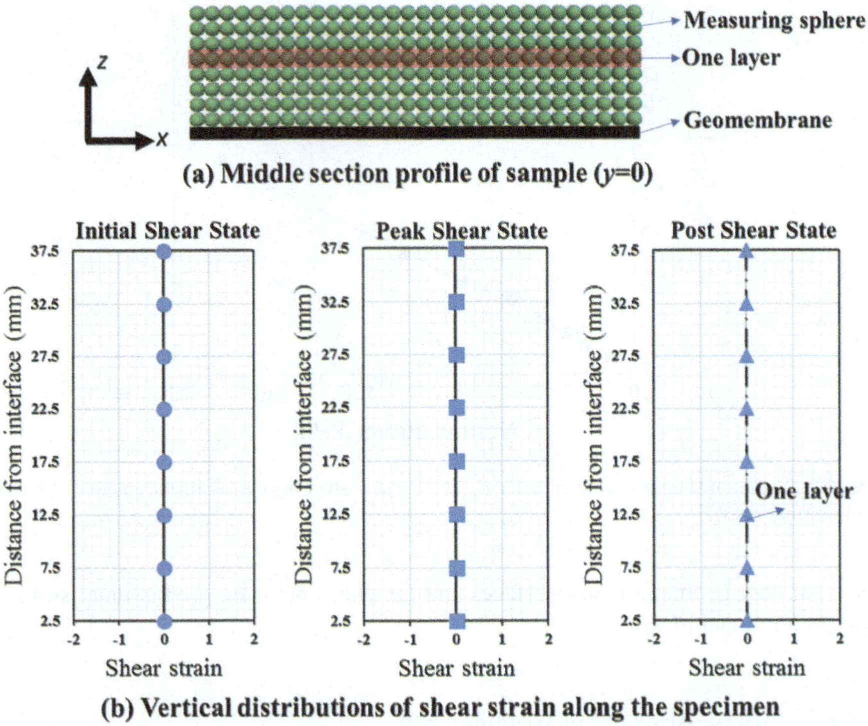

Fig. 9.10 Vertical distributions of shear strain along the specimen at different shear states under $\sigma_n = 53$ kPa

$$\overline{f}_s(\theta) = \overline{f}_0[-a_s \sin 2(\theta - \theta_s)] \tag{9.23}$$

$$\overline{f}_0 = \frac{1}{2\pi}\int_0^{2\pi} \overline{f}_n(\theta)d\theta = \frac{1}{N}\sum_{k=1}^{N} f_n^k \tag{9.24}$$

where $\overline{f}_n(\theta)$ and $\overline{f}_s(\theta)$average normal- and shear-force density distributions, respectively; a_n and$a_s$coefficients of contact normal- and shear-force anisotropies, respectively; θ_n andθ_s = principal directions of contact normal- and shear-force, respectively; $\overline{f}_0$average contact normal force over all contacts N; andf_n^k = contact normal force. The anisotropy parameters for the selected states were summarized in Table 9.2. It is seen from Fig. 9.11b that the principal directions of the normal and shear contact forces are all close to 0.0. As the shearing progress into the peak phase and even last phase, the contact force is distributed in the same manner as those at the initial stage. The forces distributions for other normal stresses are not shown but similar observations were obtained.

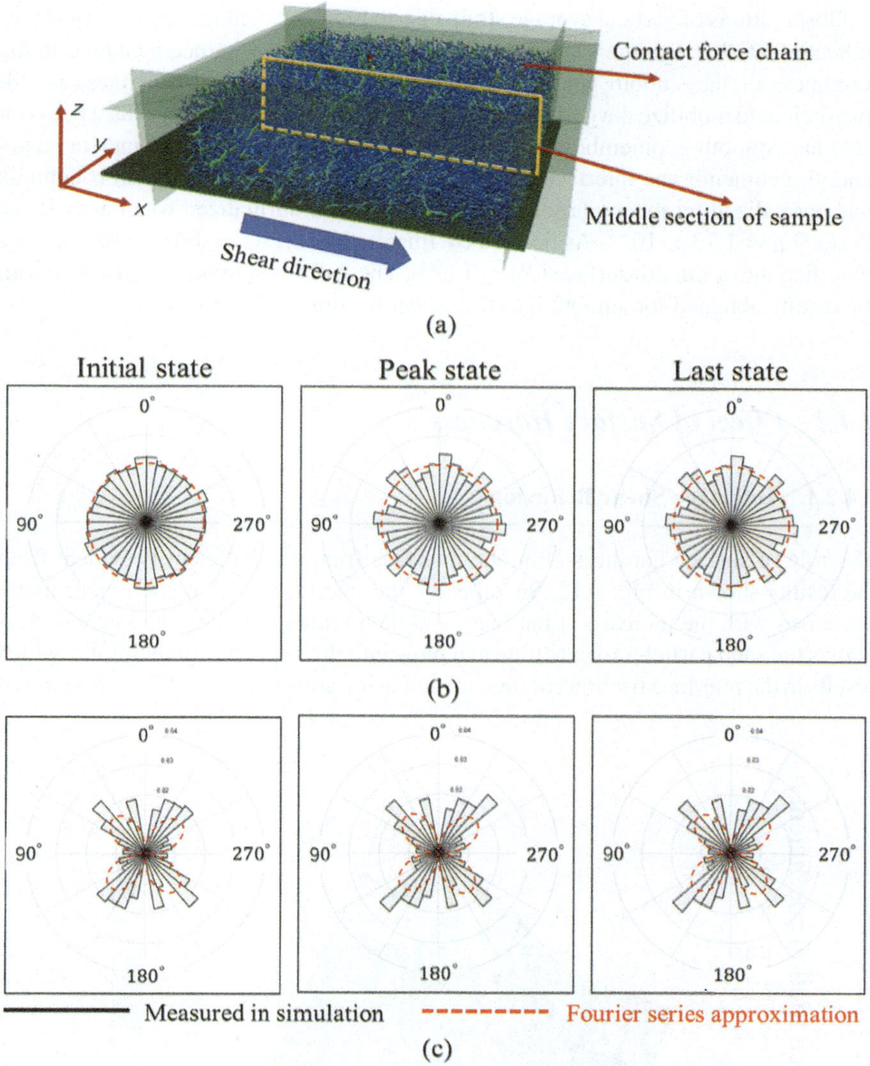

Fig. 9.11 **a** Schematic diagram of contact force chain and location of the middle section profile of sample and *x*–*z* plane distributions of **b** contact normal force and **c** contact shear force for the sample under $\boldsymbol{\sigma}_n = 53$ kPa at different shear states

Table 9.2 Anisotropy and principal direction parameters for the selected states

Anisotropy and principal directions parameters	Initial state	Peak state	Last state
a_n	0.03	0.02	0.02
θ_n	1.33°	1.63°	1.63°
a_s	0.016	0.017	0.017
θ_s	0.06°	0.06°	0.06°

Observations of vertical average strain distribution and contact force distributions indicate that although the sand–smooth interface shear can induce a change in the roughness of the smooth geomembrane surface, shear–induced roughness is still insufficient to mobilize the granular soil. Shear failure occurs in the interface between sand and smooth geomembrane, preventing the formation of shear band for sand–smooth geomembrane interface. In this study, shear–induced maximum roughness was normalized by mean particle size and maximum normalized roughness $R_n = R_{\max}/D_{50} = 1.13 \times 10^{-3}$. As defined by Jing et al. [16], $R_n = 1.13 \times 10^{-3}$ can be classified into a smooth surface ($R_n \leq 0.125$). These observations are consistent with the results obtained for smooth interface shear by Jing et al. [16].

9.4.2 Effect of Surface Hardness

9.4.2.1 Interface Shear Behavior

The role of surface hardness D on the interface friction coefficient is evident from the results shown in Fig. 9.12. Specifically, the interface friction coefficient firstly decreases with the increase of hardness and then remains stable. The low hardness allows the sand particles to penetrate and plow into the smooth geomembrane, which results in the interface friction coefficient increasing as the additional force is required

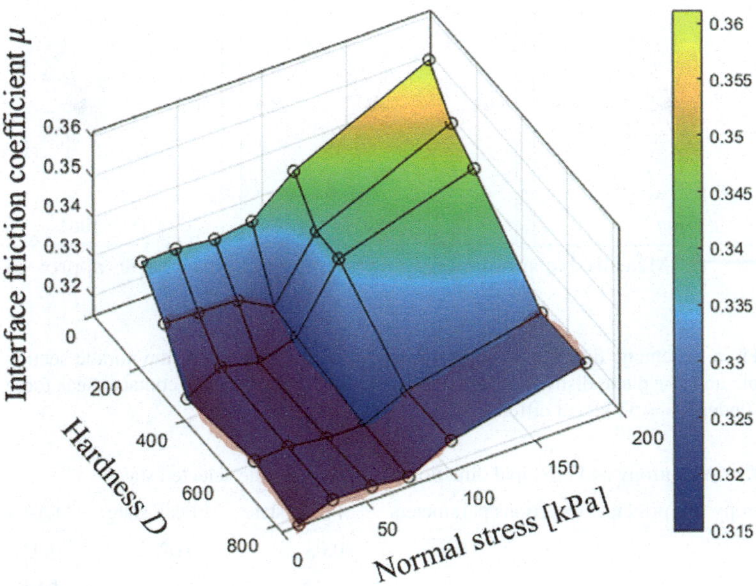

Fig. 9.12 Variation of peak interface friction coefficient with hardness under different normal stresses at the last state (shear displacement = 30 mm)

to damage and wear the smooth geomembrane surface. When the hardness increases to a critical value, such value is large enough to resist the damage of sand particles, leading to the sand particles sliding along the smooth geomembrane surface. Hence, the interface friction coefficient would be kept constant when the hardness is beyond the critical value. The coupled influence of hardness and normal stress on the interface friction coefficient is also evident from Fig. 9.12. It is observed that a simultaneous increase in normal stress and decrease in hardness will generally cause a more significant increase in the interface friction coefficient than if only one of the two factors is to be changed. The maximum interface friction coefficient is achieved for the softest geomembrane under the highest normal stress. When the hardness increases to a certain value large enough to resist the damage of sand particles or the applied normal stress is too low to allow sand particles to penetrate and plow into the smooth geomembrane, sand particles predominantly slide along the smooth geomembrane surface and the interface friction coefficient remains constant at low applied normal stress and high hardness. The constant interface friction coefficient area marked in Fig. 9.12 means a predominant sliding mechanism, while a plowing mechanism is predominant beyond the red area.

9.4.2.2 Micro–Response of Smooth Geomembrane

Figure 9.13a, b present the 3D topographical images of the smooth geomembranes with five different surface hardnesses D after shearing under normal stress $\sigma_n = 53$ kPa and their corresponding post shear surface profiles, respectively. It is apparent that with the increase of hardness D, less indentation appears in the smooth geomembrane surface and the intensity of vertical deformation decreases correspondingly. Almost no deformation and indentation were found in the smooth geomembrane surface when hardness D is more than 338.

Figure 9.14 shows variations in root mean square roughness (R_q) with normal stress and hardness. The damage of geomembrane can directly and quantitatively be reflected by R_q. In general, the geomembrane with lower hardness under higher applied normal stress will generate higher R_q than that with higher hardness under lower applied normal stress. The maximum R_q is obtained for the case of geomembrane with $D = 142$ under $\sigma_n = 180$ kPa, with hardness increased and applied normal stress decreased at the same time, R_q decreases to almost a constant. Compared with the maximum R_q, the constant R_q area marked by red background is negligible. The micro–damage performance of sand particles on the smooth geomembrane, which is dependent on the geomembrane itself property and the external applied normal stress, will be directly relevant with the shear mechanism. It indicates that no deformation and indentation were found in the smooth geomembrane surface and sliding mechanism is predominant in the constant area, which is consistent with the explanations in Fig. 9.12. The constant area is the result of the coupled effect between applied normal stress and hardness.

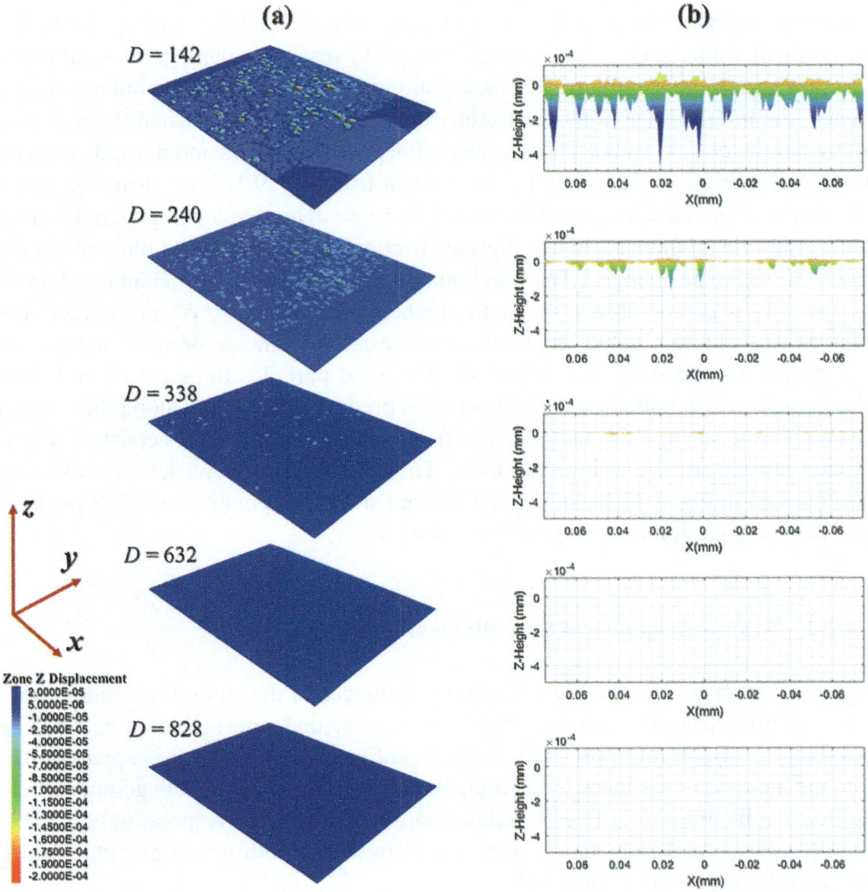

Fig. 9.13 Micro–topographical analysis of sheared smooth geomembrane sample with different hardnesses under σ_n = 75 kPa at the last state (shear displacement = 30 mm): **a** 3D topographical images and **b** 2D shear surface profiles

9.4.2.3 Micro–Response of Granular Soil

Vertical average shear strain distributions at the post shear state under different hardnesses *D* are shown in Fig. 9.15. The average shear strain for any cases linearly distributes along the depth of specimen and non–strain localization occurs in soil particles at the post shear state, which reconfirms no localized band is formed throughout shearing. Even for the geomembrane with the lowest hardness of 142 sheared by the maximum normal stress of 180 kPa, the corresponding shear–induced roughness is still insufficient to mobilize the granular soil. Accordingly, the interface shear test on smooth geomembranes with extremely low vertical deformations can be regraded as non–dilative, as explained in an earlier study of interface shear tests

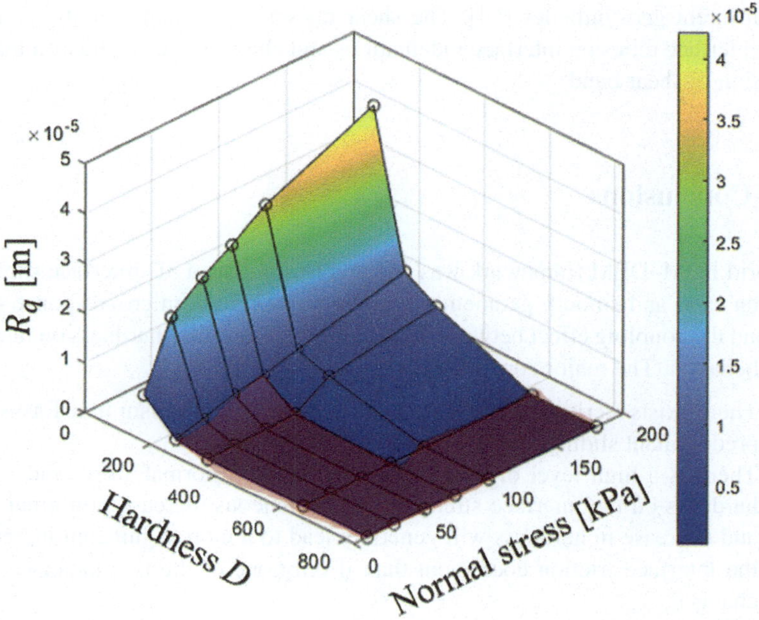

Fig. 9.14 Variation of R_q at the end of shearing (shear displacement = 30 mm) with hardness under different normal stresses

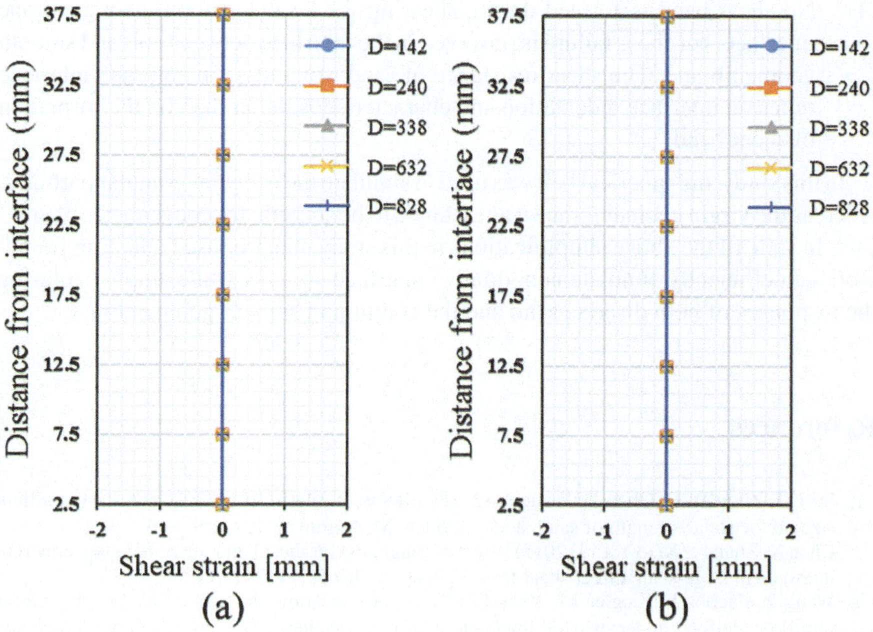

Fig. 9.15 Vertical distributions of shear strain at the post shear state (shear displacement = 30 mm) under different surface hardnesses at: **a** $\sigma_n = 53$ kPa and **b** $\sigma_n = 180$ kPa

with different geosynthetics [54]. The shear resistance of sand–smooth geomembrane interface relies on interface indentations and characteristics rather on than the formation of shear band.

9.5 Conclusions

A hybrid FDM–DEM framework was adopted to simulate a 3D interface shear test between sand and smooth geomembrane. We analyzed the micro–response of the sand and the coupling effect between normal stress and surface hardness on interface shear behavior. The major conclusions were drawn as follows:

(1) There exists a critical normal stress at which the mechanism transforms from predominant sliding to predominant plowing.
(2) There is a high level of coupling effect between normal stress and surface hardness on the interface strength. A simultaneous increase in normal stress and decrease in hardness will generally lead to a more significant increase in the interface friction coefficient than if only one of the two factors is to be changed.
(3) The micro–damage performance of sand particles on the smooth geomembrane, which is dependent on the geomembrane itself property and the external applied normal stress, will be directly relevant to the shear mechanism.
(4) No shear band is formed during shearing for a sand–smooth geomembrane interface, but shear failure is observed in the interface between soil and smooth geomembrane. The shear resistance of sand–smooth geomembrane interface relies on interface indentation and characteristics rather than on the formation of shear band.

In this study, the spherical ball was used to simulate the irregular granular particles, whereas the virgin geomembrane was assumed to be smooth, for convenience of analysis. In spite of the above simplifications in this study, the proposed coupling framework can efficiently simulate non–dilative interface systems while amply capturing the responses of both discrete sand and the continuum smooth geomembrane.

References

1. Su L-J, Zhou W-H, Chen W-B, Jie X (2018) Effects of relative roughness and mean particle size on the shear strength of sand-steel interface. Measurement 122:339–346
2. Chen X, Zhang J, Xiao Y, Li J (2015) Effect of roughness on shear behavior of red clay–concrete interface in large-scale direct shear tests. Can Geotech J 52:1122–1135
3. Wang Z, Jacobs F, Ziegler M, Yang G (2020) Visualisation and quantification of geogrid reinforcing effects under strip footing loads using discrete element method. Geotext Geomembr 48:62–70
4. Yin J-H, Hong C-Y, Zhou W-H (2011) Simplified analytical method for calculating the maximum shear stress of nail-soil interface. Int J Geomech 12:309–317

5. Zhou W-H, Yuen K-V, Tan F (2012) Estimation of maximum pullout shear stress of grouted soil nails using Bayesian probabilistic approach. Int J Geomech 13:659–664
6. Zhou W-H, Yin J-H (2008) A simple mathematical model for soil nail and soil interaction analysis. Comput Geotech 35:479–488
7. Song H, Pei H, Zhu H (2021) Monitoring of tunnel excavation based on the fiber Bragg grating sensing technology. Measurement 169:108334
8. Alkhorshid NR, Araujo GL, Palmeira EM, Zornberg JG (2019) Large-scale load capacity tests on a geosynthetic encased column. Geotext Geomembr 47:632–641
9. Han B, Ling J, Shu X, Song W, Boudreau RL, Hu W, Huang B (2019) Quantifying the effects of geogrid reinforcement in unbound granular base. Geotext Geomembr 47:369–376
10. Liu C-N, Yang K-H, Nguyen MD (2014) Behavior of geogrid–reinforced sand and effect of reinforcement anchorage in large-scale plane strain compression. Geotext Geomembr 42:479–493
11. Sayeed M, Ramaiah BJ, Rawal A (2014) Interface shear characteristics of jute/polypropylene hybrid nonwoven geotextiles and sand using large size direct shear test. Geotext Geomembr 42:63–68
12. Zhao L-S, Zhou W-H, Fatahi B, Li X-B, Yuen K-V (2016) A dual beam model for geosynthetic-reinforced granular fill on an elastic foundation. Appl Math Model 40:9254–9268
13. Zhao L-S, Zhou W-H, Su L-J, Garg A, Yuen K-V (2019) Selection of physical and chemical properties of natural fibers for predicting soil reinforcement. J Mater Civ Eng 31:04019212
14. Zhou J, Chen J-F, Xue J-F, Wang J-Q (2012) Micro-mechanism of the interaction between sand and geogrid transverse ribs. Geosynth Int 19:426–437
15. Chen W-B, Zhou W-H, dos Santos JA (2020) Analysis of consistent soil–structure interface response in multi–directional shear tests by discrete element modeling. Transp Geotech 24:100379
16. Jing XY, Zhou WH, Zhu HX, Yin ZY, Li Y (2018) Analysis of soil-structural interface behavior using three-dimensional DEM simulations. Int J Numer Anal Methods Geomech
17. Palmeira EM (2009) Soil–geosynthetic interaction: modelling and analysis. Geotext Geomembr 27:368–390
18. Wang Z, Jacobs F, Ziegler M (2014) Visualization of load transfer behaviour between geogrid and sand using PFC 2D. Geotext Geomembr 42:83–90
19. Wang Z, Jacobs F, Ziegler M (2016) Experimental and DEM investigation of geogrid–soil interaction under pullout loads. Geotext Geomembr 44:230–246
20. Dove JE, Frost JD (1999) Peak friction behavior of smooth geomembrane-particle interfaces. J Geotech Geoenvironmental Eng 125:544–555
21. Fuggle AR (2011) Geomaterial gradation influences on interface shear behavior. Georgia Institute of Technology
22. Frost J, DeJong J, Recalde M (2002) Shear failure behavior of granular–continuum interfaces. Eng Fract Mech 69:2029–2048
23. Vangla P, Gali ML (2016) Shear behavior of sand-smooth geomembrane interfaces through micro-topographical analysis. Geotext Geomembr 44:592–603
24. Meidani M, Meguid MA, Chouinard LE (2016) Finite-discrete element analysis of interface shear damage to HDPE geomembrane in contact with gravel drainage layer. In: International conference on discrete element methods. Springer, pp 351–359
25. Hussein M, Meguid M (2016) A three-dimensional finite element approach for modeling biaxial geogrid with application to geogrid-reinforced soils. Geotext Geomembr 44:295–307
26. Hussein MG, Meguid M (2019) Improved understanding of geogrid response to pullout loading: insights from three-dimensional finite element analysis. Can Geotech J
27. Jiang Y, Han J, Parsons RL (2020) Numerical evaluation of secondary reinforcement effect on geosynthetic-reinforced retaining walls. Geotext Geomembr 48:98–109
28. Rowe RK, Liu K-W (2015) Three-dimensional finite element modelling of a full-scale geosynthetic-reinforced, pile-supported embankment. Can Geotech J 52:2041–2054
29. Almeida M, Hosseinpour I, Riccio M (2013) Performance of a geosynthetic-encased column (GEC) in soft ground: numerical and analytical studies. Geosynth Int 20:252–262

30. Nagula SS, Nguyen DM, Grabe J (2018) Numerical modelling and validation of geosynthetic encased columns in soft soils with installation effect. Geotext Geomembr 46:790–800
31. Chalak C, Briançon L, Villard P (2019) Coupled numerical and experimental analyses of load transfer mechanisms in granular-reinforced platform overlying cavities. Geotext Geomembr 47:587–597
32. Cheng C, Chen X, Zhang S (2016) Multi-peak deformation behavior of jointed rock mass under uniaxial compression: Insight from particle flow modeling. Eng Geol 213:25–45
33. Chen H-X, Liu X, Feng S-J, Chen J-N, Zhang D-M, Zhou A (2019) Microscale investigation into mechanical behaviors of heat-bonded nonwoven geotextile using DEM. Geotext Geomembr 47:429–438
34. Chen W-B, Zhou W-H, Jing X-Y (2019) Modeling Geogrid Pullout Behavior in Sand Using Discrete-Element Method and Effect of Tensile Stiffness. Int J Geomech 19:04019039
35. Lai H-J, Zheng J-J, Zhang J, Zhang R-J, Cui L (2014) DEM analysis of "soil"-arching within geogrid-reinforced and unreinforced pile-supported embankments. Comput Geotech 61:13–23
36. Lai H-j, Zheng J-j, Zhang R-j, Cui M-j (2016) Visualization of the formation and features of soil arching within a piled embankment by discrete element method simulation. J Zhejiang Univ-SCIENCE A 17:803–817
37. Wang H, Yang G, Wang Z, Liu W (2020) Static structural behavior of geogrid reinforced soil retaining walls with a deformation buffer zone. Geotext Geomembr 48:374–379
38. Gao G, Meguid M (2018) Effect of particle shape on the response of geogrid-reinforced systems: Insights from 3D discrete element analysis. Geotext Geomembr 46:685–698
39. Feng S-J, Liu X, Chen H-X, Zhao T (2018) Micro-mechanical analysis of geomembrane-sand interactions using DEM. Comput Geotech 94:58–71
40. Itasca (2017) PFC User Manual, Version 5.0. Itasca Consulting Group Inc, USA
41. Itasca (2017) FLAC User Manual, Version 6.0. Itasca Consulting Group Inc, USA
42. Zhou J, Jian Q-w, Zhang J, Guo J-j (2012) Coupled 3D discrete-continuum numerical modeling of pile penetration in sand. J Zhejiang Univ, Sci, A 13:44–55
43. Vangla P, Latha GM (2015) Influence of particle size on the friction and interfacial shear strength of sands of similar morphology. Int J Geosynth Ground Eng 1:6
44. Attache S, Mellas M (2017) Numerical study of large-scale pull-out test of horizontal corrugated strips. Int J Geotech Eng 1–9
45. Ward H (1999). In: Thomas TR (ed) Rough surfaces. Longman, London, p 278
46. ASME (1995) Surface texture surface roughness, waviness, and lay, B46.1, New York.:112
47. Zhu H, Nguyen HN, Nicot F, Darve F (2016) On a common critical state in localized and diffuse failure modes. J Mech Phys Solids 95:112–131
48. Zhu HX, Zhou WH, Jing XY, Yin ZY (2019) Observations on fabric evolution to a common micromechanical state at the soil-structure interface. Int J Numer Anal Meth Geomech 43:2449–2470
49. Williams J (2005) Engineering tribology. Cambridge University Press
50. ASTM (2005) Standard test method for rubber property—durometer hardness. ASTM D2240
51. Zettler T, Frost J, DeJong J (2000) Shear-induced changes in smooth HDPE geomembrane surface topography. Geosynth Int 7:243–267
52. O'rourke T, Druschel S, Netravali A (1990) Shear strength characteristics of sand-polymer interfaces. J Geotech Eng 116:451–469
53. Bathurst RJ, Rothenburg L (1990) Observations on stress-force-fabric relationships in idealized granular materials. Mech Mater 9:65–80
54. Dove J, Bents D, Wang J, Gao B (2006) Particle-scale surface interactions of non-dilative interface systems. Geotext Geomembr 24:156–168

Appendix A
PFC Code of Creating and Cutting Wall Under the Periodic Boundary Condition

Please refer to the articles "Zhu, H., Zhou, W. H., & Yin, Z. Y. (2018). Deformation mechanism of strain localization in 2D numerical interface tests. Acta Geotechnica, 13(3), 557–573." and "Zhu, H. X., Zhou, W. H., Jing, X. Y., & Yin, Z. Y. (2019). Observations on fabric evolution to a common micromechanical state at the soil-structure interface. International Journal for Numerical and Analytical Methods in Geomechanics, 43(15), 2449–2470."

```
def truncate(isFirstTime)
    if isFirstTime='true' then        ; if it is the first time, truncate those outside or intersecting with
        the inbox
        local tmp=0.0
        local nBall=0
        loop foreach local ball ball.list
            tmp+=ball.radius(ball)
            nBall+=1
        endloop

        calRoughness
        global WLx = wall.pos.x(wall.find(1))
        global WRx = wall.pos.x(wall.find(3))
        global WLTx = WLx+D50*0.000005
        global WRTx = WRx-10.0*mech.timestep*velShear
```


W.-H. Zhou and Z.-Y. Yin, *Practice of Discrete Element Method in Soil-Structure Interface Modelling*, https://doi.org/10.1007/978-981-19-0047-1

```
loop foreach local wall wall.list
    if wall.isgroup(wall,'plate') then
        ;tmp=wall.pos(wall)
        ;io.out(string(comp.x(tmp))+','+string(comp.y(tmp))+'      '+wall.group(wall))
        ;io.out(string(wallOutbox(wall,WLTx,WRTx,-landa,landa)))
        if wallOutbox(wall,WLTx,WRTx,-landa,landa) then
            wall.delete(wall)
        endif
    endif
endloop

        global WTy=wall.pos.y(wall.find(2))
        wall.delete(wall.find(2))
    else        ; if it is not the first time, truncate with the inbox landa/2 in both side narrower then
        the first time inbox
        global WRTx = WRx-10.0*mech.timestep*velShear

        loop foreach wall wall.list
            if wall.isgroup(wall,'plate') then
                if wallOutbox(wall,WLTx+landa/2.0,WRTx-landa/2.0,-landa,landa) then
                    wall.delete(wall)
                endif
            endif
        endloop
    endif
end

def wallOutbox(wall,left,right,low,up)    ;at least one vertex outside box
    local tmpWallOutbox=false
    loop foreach local vertex wall.vertexlist(wall)
        tmp=wall.vertex.pos(vertex)
        if comp.x(tmp)<left then
            ;io.out(string(comp.x(tmp))+','+string(comp.y(tmp))+'      '+wall.group(wall))
            tmpWallOutbox=true
            exit loop
        endif
```

```
            if comp.x(tmp)>right then
                tmpWallOutbox=true
                exit loop
            endif
            if comp.y(tmp)<low then
                tmpWallOutbox=true
                exit loop
            endif
            if comp.y(tmp)>up then
                tmpWallOutbox=true
                exit loop
            endif
        endloop
          wallOutbox=tmpWallOutbox
    end

    def rebuildWall(isFirstTime)

        if isFirstTime='true' then

            local tmpArray=array.create(2)
            tmpArray[1] = WLTx
            tmpArray[2] = 0.0
            local vp = vector(tmpArray)
            leftEnd=wall.vertex.pos(wall.vertex.near(vp))
            if math.abs(comp.x(leftEnd)-WLTx)>0.25*minRad then
                if math.abs(comp.y(leftEnd))<=1e-8 then        ;upper end
                    local tmpy=WLTx-comp.x(leftEnd)+comp.y(leftEnd)
                else     ;lower end
                    tmpy=comp.x(leftEnd)-WLTx+comp.y(leftEnd)
                endif
                command
                    wall create group 'plate' vertices ...
                        [comp.x(leftEnd)] [comp.y(leftEnd)] ...
                        [WLTx] [tmpy]
                endcommand
            endif
```

```
tmpArray[1] = WRTx
vp = vector(tmpArray)
RightEnd=wall.vertex.pos(wall.vertex.near(vp))
if math.abs(comp.x(RightEnd)-WRTx)>0.25*minRad then
    if math.abs(comp.y(RightEnd))<=1e-8 then        ;upper end
        tmpy=comp.x(RightEnd)-WRTx+comp.y(RightEnd)
    else    ;lower end
        tmpy=WRTx-comp.x(RightEnd)+comp.y(RightEnd)
    endif
    command
        wall create group 'plate' vertices ...
            [comp.x(RightEnd)] [comp.y(RightEnd)] ...
            [WRTx] [tmpy]
                endcommand
            endif

            command
                wall create group 'box' id 2 vertices ...       ;top
                                    [WLx+1e-5] [WTy] ...
                                    [WRx-1e-5] [WTy]
            endcommand
        else

            tmpArray=array.create(2)
            tmpArray[1] = WLTx
            tmpArray[2] = 0.0
            vp = vector(tmpArray)
            leftEnd=wall.vertex.pos(wall.vertex.near(vp))
            if math.abs(comp.y(leftEnd))<=1e-8 then        ;upper end
                local tmpx1=comp.x(leftEnd)-landa/2.0
                local tmpy1=comp.y(leftEnd)-landa/2.0
                tmpy=tmpx1-WLTx+tmpy1
            else    ;lower end
                tmpx1=comp.x(leftEnd)-landa/2.0
                tmpy1=comp.y(leftEnd)+landa/2.0
                tmpy=WLTx-tmpx1+tmpy1
            endif
```

```
        command
            wall create group 'plate' vertices ...
                [comp.x(leftEnd)] [comp.y(leftEnd)] ...
                [tmpx1] [tmpy1]
        endcommand
        if math.abs(tmpx1-WLTx)>0.25*minRad then
            command
                wall create group 'plate' vertices ...
                    [tmpx1] [tmpy1] ...
                    [WLTx] [tmpy]
            endcommand
        endif

        tmpArray[1] = WRTx

            vp = vector(tmpArray)
            RightEnd=wall.vertex.pos(wall.vertex.near(vp))
            if math.abs(comp.y(RightEnd))<=1e-8 then        ;upper end
                tmpx1=comp.x(RightEnd)+landa/2.0
                tmpy1=comp.y(RightEnd)-landa/2.0
                tmpy=WRTx-tmpx1+tmpy1
            else     ;lower end
                tmpx1=comp.x(RightEnd)+landa/2.0
                tmpy1=comp.y(RightEnd)+landa/2.0
                tmpy=tmpx1-WRTx+tmpy1
            endif
            command
                wall create group 'plate' vertices ...
                    [comp.x(RightEnd)] [comp.y(RightEnd)] ...
                    [tmpx1] [tmpy1]
            endcommand
            if math.abs(tmpx1-WRTx)>0.25*minRad then
                command
                    wall create group 'plate' vertices ...
                        [tmpx1] [tmpy1] ...
                        [WRTx] [tmpy]
                endcommand
            endif
        endif
    end
```

Appendix B
PFC Code of Creating Specimens with Different Porosities

Please refer to the articles "Zhu, H., Zhou, W. H., & Yin, Z. Y. (2018). Deformation mechanism of strain localization in 2D numerical interface tests. Acta Geotechnica, 13(3), 557–573." and "Zhu, H. X., Zhou, W. H., Jing, X. Y., & Yin, Z. Y. (2019). Observations on fabric evolution to a common micromechanical state at the soil-structure interface. International Journal for Numerical and Analytical Methods in Geomechanics, 43(15), 2449–2470."

```
def consolidate
    local n=0
    local cumulation=0.0
    local p_last=porosity
    loop while 1
        tmpfric= ball.prop(ball.near(vector(0.0,0.0)),'fric')
        p0=porosity
        command
          ball property fric [0.9*tmpfric]
          cycle 2
          solve arat 3e-5
        endcommand

        cumulation+=math.abs(p0-porosity)    ;decide to exit
        n+=1
        if math.abs(cumulation)>=0.0001 then
            cumulation=0.0
            n=0
        endif
```


W.-H. Zhou and Z.-Y. Yin, *Practice of Discrete Element Method in Soil-Structure Interface Modelling*, https://doi.org/10.1007/978-981-19-0047-1

```
        if n=8 then
            if math.abs(cumulation)<0.0001 then
                exit loop
            else
                cumulation=0.0
                n=0
            endif
        endif                                   ;decide to exit

        if p_last-porosity>0.001 then
            p_last=porosity
            saveState('true')
        endif
    endloop
end
```

Appendix C
PFC Code of Constant Normal Stiffness Control

Please refer to the article "Wang P, Yin Z-Y*, Zhou WH, Chen WB (2021). Micromechanical analysis of soil-structure interface behavior under constant normal stiffness condition with DEM. Acta Geotech., https://doi.org/10.1007/s11440-021-01374-8."

```
[volume_strain_PB = 0.0]
[weyy_PB = 0.0]
def weyy_PB
  weyy_PB = (wall.pos.y(wall.find(1)) - b)/b
end
fish def top_ballwall_stiffness
    local stiffness  = 0.0
    local wp = wall.find(1)
    loop foreach contact wall.contactmap(wp)
        stiffness = stiffness + contact.prop(contact,"kn")
    endloop
    if stiffness = 0.0
        stiffness = 50*b_kn
    endif
    gy = stiffness/a/1 ;pa/m
end
[a_no_converge = 0]
def servo_constant_stiffness
     local count = 50
    converge = 0
    local wp = wall.find(1)
    local start_cycle = mech.cycle
     if a_wsyy_0 = 0.0
         a_wsyy_0 = wsyy_0
     endif
```


W.-H. Zhou and Z.-Y. Yin, *Practice of Discrete Element Method in Soil-Structure Interface Modelling*, https://doi.org/10.1007/978-981-19-0047-1

```
        if constant_stiffness = 0.0
            a_wsyy_0 = confining_pressure
        endif
        a_position_y = top_y0
        a_gy = ab_gy_calculate
        if a_gy = 0
            top_ballwall_stiffness
            a_gy = gy/0.1/1 + constant_stiffness
        endif
        a_count = 0
    loop   local i (1,count)
                a_count = a_count + 1
                compute_wallstress_PB
                a_wsyy_current = wsyy
                a_delta_sigma = a_wsyy_current - a_wsyy_0
                a_stiffness_current   = math.abs(a_gy+constant_stiffness)
                a_delta_d = -a_delta_sigma/a_stiffness_current
                a_delta_d_next = a_delta_d
                solve_time = 1.0e-4
                a_vel_top = a_delta_d_next/solve_time
          a_stress =  a_wsyy_0-constant_stiffness*a_delta_d
          if math.abs(a_vel_top) > 5.0e-3
              a_vel_top = 5.0e-3*a_vel_top/math.abs(a_vel_top)
              a_delta_d_next = solve_time*a_vel_top
              a_delta_d = a_delta_d_next
              a_stress =  a_wsyy_0-constant_stiffness*a_delta_d
          endif
          wall.vel.y(wp_top) = a_vel_top
command
    model solve time [solve_time]
endcommand

  compute_wallstress_PB
  ab_wsyy_current = wsyy

  top_ballwall_stiffness
  ab_gy = gy/0.1/1
  ab_gy_calculate1= ab_wsyy_current - a_wsyy_current
  ab_gy_calculate2 = ab_gy_calculate1/a_delta_d
  if ab_gy_calculate2 <= 0.0
       ab_gy_calculate = a_stiffness_current*0.01
  else
       ab_gy_calculate = ab_gy_calculate2
  endif

  ab_delta_sigma = ab_wsyy_current - wsyy_0 ;total stress
  ab_delta_d_current = wall.pos.y(wp) - top_y0 ;total displacement
  if math.abs(ab_delta_d_current)>0.0
       ab_stiffness = -ab_delta_sigma/ab_delta_d_current
  endif
  if constant_stiffness # 0.0
       if math.abs((ab_stiffness-constant_stiffness)/constant_stiffness) < 1.0e-2
            converge = 1
       endif
```

```
            else
                if math.abs(ab_delta_sigma) <=500
                    converge = 1
                endif
            endif
            if a_count >= count
                if converge # 1
                    a_no_converge = a_no_converge + 1
                endif
            endif
            a_wsyy_0 = a_stress
            a_gy = ab_gy_calculate

  endloop
end
def update_stress_change
  compute_wallstress_PB
  delta_sigma_n = wsyy - wsyy_0
  delta_d = -delta_sigma_n/constant_stiffness
end
wall hist position-y id 1
wall hist force-contact-y id 1
wall hist velocity-y id 1

history interval 1
wall att disp 0
[shear_vel_PB = 0.1]
[hist_i = 0]
[mech_cycle_initial = mech.cycle]
def shear_stage_PB
  shear_strain_PB = 0.0
  loop while shear_strain_PB < shear_strain_target
    hist_i = hist_i + 1
    compute_wallstress_PB
    wsyy_0 = wsyy
    top_y0 = wall.pos.y(wall.find(1))
    cut_right_SawtoothPlate
    command
      wall attribute   velocity-x [shear_vel_PB] range group 'SawtoothPlate'
      model solve time 8.0e-5
      wall attribute   vel 0.0
      model solve time 8.0e-4
    endcommand
    add_left_SawtoothPlate
    servo_constant_stiffness
    shear_strain_PB = shear_strain_PB + shear_vel_PB*8.0e-5/b

    clump_num = clump.num
    contact_num = 0
    mech_cycle = mech.cycle
    loop foreach clump clump.list
      contact_num = contact_num + clump.contactnum(clump)
    endloop
    average_contact_num = contact_num*1.0/clump_num
    porosity_now_PB
```

```
    compute_InterfaceStress_PB
    compute_wallstress_PB
    q_stress_PB = ShearStress_PB
    p_stress_PB = wsyy - ShearStress_PB
    wsyy_1 = wsyy
    top_y1 = wall.pos.y(wall.find(1))
    save_hist(hist_i) = string(mech_cycle) + ' ' + string(porosity_now_PB) + ' ' +
string(q_stress_PB) ...
                        + ' ' + string(p_stress_PB) + ' ' + string(shear_strain_PB) + ' ' +
string(average_contact_num) ...
                        + ' ' + string(volume_strain_PB)+ ' ' + string(wsyy_0,9,9,9,e) + ' ' +
string(wsyy_1,9,9,9,e) ...
                        + ' ' + string(top_y0*100,9,9,9,e) + ' ' + string(top_y1*100,9,9,9,e) + ' ' +
string(gy)

    file.open('Data_SoilStructureInterface_Test_PB_e_q_p_axialstrain_acn.txt',1,1)
    file.write(save_hist,hist_i)
    file.close()

    loop local i(1,6)
      if shear_strain_PB >= save_strain(i)
        filename = 'Data_SoilStructureInterface_Test_PB_Stress_Strain_' + string(save_strain(i)) +
'.txt'
        create_measure_circles(filename)

        filename = 'Data_SoilStructureInterface_Test_PB_Clump_Pos_Disp_' +
string(save_strain(i)) + '.txt'
        get_pebbles_data_initial(filename)

        filename = 'Data_SoilStructureInterface_Test_PB_Pebble_Vel_Rotation_' +
string(save_strain(i)) + '.txt'
        get_vel_rotation(filename)

        num_contact = contact.num('pebble-pebble')
        filename = 'Data_SoilStructureInterface_Test_PB_Force_Chain_' + string(save_strain(i)) +
'.txt'
        get_force_chain(filename)

        filename = 'Data_SoilStructureInterface_Test_PB_Contact_Pos_Force_Direc_' +
string(save_strain(i)) + '.txt'
        contact_force_info(filename)

        filename = 'Data_SoilStructureInterface_Test_PB_' + string(save_strain(i)) + '.sav'
        command
            model save [filename]
        endcommand
        save_strain(i) = 100    ; insure the recorded state won't be recorded in following loop.

        exit loop

      endif
    endloop
  endloop
end
@shear_stage_PB
```

Appendix D
PFC Code of Creating Specimens with the Minimum and the Maximum Densities

Please refer to the article "Zhou, W. H., Jing, X. Y., Yin, Z. Y., & Geng, X. (2019). Effects of particle sphericity and initial fabric on the shearing behavior of soil–rough structural interface. Acta Geotechnica, 14(6), 1699–1716."

```
Define ContactN ; generate the specimen with the maximum density
    Nc = contact.num('pebble-pebble')
    Cn = 2.*Nc/bnum
end

define porosity_specimen
    wall_size
    volume_t = wlx*wly*wlz
    volume_s = 0.0
loop foreach bp clump.list ;number of the balls
    volume_s += clump.vol(bp)
endloop
    Ns = (volume_t - volume_s)/volume_t
end


Define QuickInc(Ifactor)
    ContactN
    fac = Ifactor
    loop while 1
        loop foreach bp clump.list()
            cid = clump.id(bp)
            vol = clump.vol(bp)
            pre_rad = (3. * vol / 4. / 3.14)^(1/3.)
            new_dia = 2*pre_rad +   fac*0.5*(dia_l+dia_h)
            command
                clump scale diameter @new_dia range id @cid
            endcommand
        endloop
        command
            cycle 200
```


W.-H. Zhou and Z.-Y. Yin, *Practice of Discrete Element Method in Soil-Structure Interface Modelling*, https://doi.org/10.1007/978-981-19-0047-1

```
                    endcommand
                    if Cn >= 1.0 then
                        exit loop
                    endif
                endloop

                fac = 0.3*Ifactor
                loop while 1
                    loop foreach bp clump.list()
                        cid = clump.id(bp)
                        vol = clump.vol(bp)
            pre_rad = (3. * vol / 4. / 3.14)^(1/3.)
            new_dia = 2*pre_rad +    fac*0.5*(dia_l+dia_h)
            command
                clump scale diameter @new_dia range id @cid
            endcommand
        endloop
        command
            cycle 400
        endcommand
        if Cn >= 2.0 then
            exit loop
        endif
    endloop
end

set fish callback 1.0 @ContactN
set fish callback 2.0 @porosity_specimen
set fish callback 3.0 @wall_size
set fish callback 4.0 @wall_stress

set timestep scale
cycle 1000 calm 10
calm

[tss = 1.e7]    ;target stress 100kPa
wall servo activate on zforce [tss * wlx *wly] vmax 0.000001 range id 1
wall servo activate on xforce [tss * wlz *wly] vmax 0.000001 range id 3
wall servo activate on yforce [tss * wlx *wlz] vmax 0.000001 range id 5
wall servo activate on zforce [-tss * wlx *wly] vmax 0.000001 range id 2
wall servo activate on xforce [-tss * wlz *wly] vmax 0.000001 range id 4
wall servo activate on yforce [-tss * wlx *wlz] vmax 0.000001 range id 6

define servo_walls_all
  wall_size
  wall.servo.force.z(wall.find(1)) = [tss * wlx *wly]
  wall.servo.force.x(wall.find(3)) = [tss * wlz *wly]
  wall.servo.force.y(wall.find(5)) = [tss * wlx *wlz]
  wall.servo.force.z(wall.find(2)) = [-tss * wlx *wly]
  wall.servo.force.x(wall.find(4)) = [-tss * wlz *wly]
  wall.servo.force.y(wall.find(6)) = [-tss * wlx *wlz]
end
```

```
    define stop_me
      a = mech.solve("aratio")
      if math.abs((sigma_0 - tss)/tss) > 0.005
        exit
      endif
      if mech.solve("aratio") > 5.e-5
        exit
      endif
      stop_me = 1
    end

    set fish callback 5.0 @servo_walls_all
    solve fishhalt @stop_me
Define QuickInc(Ifactor) ; generate the specimen with the minimum density
    ContactN
    fac = Ifactor
    loop while 1
        loop foreach bp clump.list()
            cid = clump.id(bp)
            vol = clump.vol(bp)
            pre_rad = (3. * vol / 4. / 3.14)^(1/3.)
            new_dia = 2*pre_rad +   fac*0.5*(dia_l+dia_h)
            command
                clump scale diameter @new_dia range id @cid
            endcommand
        endloop
        command
            cycle 200
        endcommand
        if Cn >= 1.0 then
            exit loop
        endif
    endloop

    fac = 0.3*Ifactor
    loop while 1
        loop foreach bp clump.list()
            cid = clump.id(bp)
            vol = clump.vol(bp)
            pre_rad = (3. * vol / 4. / 3.14)^(1/3.)
            new_dia = 2*pre_rad +   fac*0.5*(dia_l+dia_h)
            command
                clump scale diameter @new_dia range id @cid
            endcommand
        endloop
        command
            cycle 400
        endcommand
        if Cn >= 2.0 then
            exit loop
        endif
    endloop
end

set fish callback 1.0 @ContactN
set fish callback 2.0 @porosity_specimen
set fish callback 3.0 @wall_size
set fish callback 4.0 @wall_stress
```

```
set timestep scale
cycle 1000 calm 10
solve arat 1e-5
calm

[tss = 1.e7]    ;target stress 100kPa
wall servo activate on zforce [tss * wlx *wly] vmax 0.000001 range id 1
wall servo activate on xforce [tss * wlz *wly] vmax 0.000001 range id 3
wall servo activate on yforce [tss * wlx *wlz] vmax 0.000001 range id 5
wall servo activate on zforce [-tss * wlx *wly] vmax 0.000001 range id 2
wall servo activate on xforce [-tss * wlz *wly] vmax 0.000001 range id 4
wall servo activate on yforce [-tss * wlx *wlz] vmax 0.000001 range id 6

define servo_walls_all
  wall_size
  wall.servo.force.z(wall.find(1)) = [tss * wlx *wly]
  wall.servo.force.x(wall.find(3)) = [tss * wlz *wly]
  wall.servo.force.y(wall.find(5)) = [tss * wlx *wlz]
  wall.servo.force.z(wall.find(2)) = [-tss * wlx *wly]
  wall.servo.force.x(wall.find(4)) = [-tss * wlz *wly]
  wall.servo.force.y(wall.find(6)) = [-tss * wlx *wlz]
end

define stop_me
  a = mech.solve("aratio")
  if math.abs((sigma_0 - tss)/tss) > 0.005
    exit
  endif
  if mech.solve("aratio") > 5.e-5
    exit
  endif
  stop_me = 1
end

set fish callback 5.0 @servo_walls_all
solve fishhalt @stop_me
```

Appendix E
MATLAB Code of 3D Contact Normal Distribution Demonstration

Please refer to the articles "Zhou, W. H., Jing, X. Y., Yin, Z. Y., & Geng, X. (2019). Effects of particle sphericity and initial fabric on the shearing behavior of soil–rough structural interface. Acta Geotechnica, 14(6), 1699-1716." and "Chen, W. B., Zhou, W. H., & dos Santos, J. A. (2020). Analysis of consistent soil–structure interface response in multi–directional shear tests by discrete element modeling. Transportation Geotechnics, 24, 100379."

```
%-------the format of file to be read is (x,y,z)--------------------------------------------------------------
clear;
clc
close all

temp=xlsread('xxx');
stepangle = 10.;
Hbeta =10.;%get the contact in a certain area

counter = zeros(360/stepangle,180/stepangle);

% % read the info at different stages (tota; stage 28)
% stage = 28;
% % get the orientaion
% x = temp(:,1+(stage-1)*7);
% y = temp(:,2+(stage-1)*7);
% z = temp(:,3+(stage-1)*7);


x = temp(:,1);
y = temp(:,2);
z = temp(:,3);
x(isnan(x))=[];
y(isnan(y))=[];
z(isnan(z))=[];
```


W.-H. Zhou and Z.-Y. Yin, *Practice of Discrete Element Method in Soil-Structure Interface Modelling*, https://doi.org/10.1007/978-981-19-0047-1

```
% get the opposite direction (double it)
fx(:,1) = [x(:,1);(x(:,1).*-1)];
fy(:,1) = [y(:,1);(y(:,1).*-1)];
fz(:,1) = [z(:,1);(z(:,1).*-1)];

%---------------------------------------------------------------------
% get the "phi" angle in x-y plane 0-360. from +x
phi = zeros(length(fx),1);

for j = 1:length(fx)
    if fx(j) >= 0   && fy(j) > 0
        phi(j) = atand(fy(j)/fx(j));

        elseif fy(j) >= 0 && fx(j) < 0
           phi(j) = 180. + atand(fy(j)/fx(j));
        elseif fy(j) <= 0 && fx(j) < 0
           phi(j) = 180. + atand(fy(j)/fx(j));
        else
           phi(j) = 360. + atand(fy(j)/fx(j));
        end
    end

    % get the "theta" angle of each contact vector from -z   0~180
    theta = zeros(length(fx),1);
    for j = 1:length(fx)
        if fz(j,1) >= 0
           theta(j) = 90. - asind(fz(j,1));
        else
           theta(j) = 90. +   abs(asind(fz(j,1)));
        end
    end

    %--------------------------------------------------------------------
    for i = 1:1:36
        k = 1;
        for j = 1:length(fx)
            if (phi(j) <= Hbeta*i)   && (phi(j) > (Hbeta*(i-1.))) %get the contact in a certain area
        %              if phi(j) > Lbeta && phi(j) < Hbeta   %for the case not side
                contactNo(k,1) = j;   %record the position of contact
                k = k + 1;
            end
        end

        % get the probability
        for m = 1:1:180/stepangle
            for mm = 1:1:(k-1)
                if  theta(contactNo(mm,1))  >  stepangle*(m-1)  &&  theta(contactNo(mm,1))  <=
            stepangle*m
                    counter(i,m) = counter(i,m) + 1;
                end
            end
        end

        clear contactNo
    end
```

```
TotalNumber = sum(counter(:));
%devide the area sin(theta)d(theta)   xdata is the theta in spherical coordi
xdata = (180-stepangle/2):-stepangle:(stepangle/2.);

for ii = 1:1:180/stepangle
    counter(:,ii) = counter(:,ii) /sind(xdata(ii)) / (stepangle*pi/180.) / (stepangle*pi/180.) / length(x)
        /2;
end

% xlswrite('PdataS10.xlsx',exp_data,'exp')
% xlswrite('PdataS10.xlsx',sumP,'factor','C1:C28')

%------------------3D Contact normal distributions-----------------------------------------------------------
clc
clear all

  [data]=xlsread('xxx')
  theta=0
  phi = data(:,1)
  j=2
  while (j<=19)
      i=1
      R=data(:,j)
    while (i<=36)
      dot0=[0.0,0.0,0.0]
      dot1=[R(i)*sind(theta)*cosd(phi(i)),R(i)*sind(theta)*sind(phi(i)),R(i)*cosd(theta)]
      dot2=[R(i)*sind(theta+10)*cosd(phi(i)),R(i)*sind(theta+10)*sind(phi(i)),R(i)*cosd(theta+10)]
      dot3=[R(i)*sind(theta)*cosd(phi(i+1)),R(i)*sind(theta)*sind(phi(i+1)),R(i)*cosd(theta)]

          dot4=[R(i)*sind(theta+10)*cosd(phi(i+1)),R(i)*sind(theta+10)*sind(phi(i+1)),R(i)*cosd(t
          heta+10)]

      x1=[0.0,R(i)*sind(theta)*cosd(phi(i)),R(i)*sind(theta+10)*cosd(phi(i))]
      y1=[0.0,R(i)*sind(theta)*sind(phi(i)),R(i)*sind(theta+10)*sind(phi(i))]
      z1=[0.0,R(i)*cosd(theta),R(i)*cosd(theta+10)]
      h=patch(x1,y1,z1,'w');grid on;view(3)
      hold on

      x2=[0.0,R(i)*sind(theta)*cosd(phi(i)),R(i)*sind(theta)*cosd(phi(i+1))]
      y2=[0.0,R(i)*sind(theta)*sind(phi(i)),R(i)*sind(theta)*sind(phi(i+1))]
      z2=[0.0,R(i)*cosd(theta),R(i)*cosd(theta)]
      h=patch(x2,y2,z2,'w');grid on;view(3)
      hold on

      x3=[0.0,R(i)*sind(theta+10)*cosd(phi(i+1)),R(i)*sind(theta)*cosd(phi(i+1))]
      y3=[0.0,R(i)*sind(theta+10)*sind(phi(i+1)),R(i)*sind(theta)*sind(phi(i+1))]
      z3=[0.0,R(i)*cosd(theta+10),R(i)*cosd(theta)]
      h=patch(x3,y3,z3,'w');grid on;view(3)
      hold on

      x4=[0.0,R(i)*sind(theta+10)*cosd(phi(i+1)),R(i)*sind(theta+10)*cosd(phi(i))]
      y4=[0.0,R(i)*sind(theta+10)*sind(phi(i+1)),R(i)*sind(theta+10)*sind(phi(i))]
      z4=[0.0,R(i)*cosd(theta+10),R(i)*cosd(theta+10)]
      h=patch(x4,y4,z4,'w');grid on;view(3)
      hold on
```

```
      x6=[R(i)*sind(theta)*cosd(phi(i)),R(i)*sind(theta+10)*cosd(phi(i)),R(i)*sind(theta+10)*c
      osd(phi(i+1)),R(i)*sind(theta)*cosd(phi(i+1))]

      y6=[R(i)*sind(theta)*sind(phi(i)),R(i)*sind(theta+10)*sind(phi(i)),R(i)*sind(theta+10)*sin
      d(phi(i+1)),R(i)*sind(theta)*sind(phi(i+1))]
    z6=[R(i)*cosd(theta),R(i)*cosd(theta+10),R(i)*cosd(theta+10),R(i)*cosd(theta)]
    h=patch(x6,y6,z6,'c','linewidth',0.01);grid on;view(3)
    hold on
    i=i+1
  end
    j = j+1
    theta = 10 + theta
end
axis equal
set(gca,'fontsize',14)
alpha(.95)
xlim([-0.15,0.15])
ylim([-0.15,0.15])
zlim([-0.15,0.15])
axis square
box on
grid on
view(0,90)
saveas(gcf,'90_3xz','png')
view(0,0)
saveas(gcf,'90_3xy','png')
```

Appendix F
PFC Code of Importing a Random Surface into the Model

Please refer to the articles "Chen, W. B., Zhou, W. H., Sadowski, Ł., & Yin, Z. Y. (2021). Metaheuristic model for the interface shear strength between granular soil and structure considering surface morphology. Computers and Geotechnics, 135, 104141."

```
geom import Drawing1.dxf ; Drawing1.dxf from CAD
wall import geom Drawing1 clean id 17 group box
;generate some balls in the box
def inBlender(pos,b)
  inBlender = 0
  if type.pointer.id(b) = ball.typeid
    wp = wall.find(17);
    inBlender = wall.inside(wp,ball.pos(b))
  endif
end

def inBody(pos,b)
  inBody = 0
  if type.pointer.id(b) = ball.typeid
    local tmp = 0
    loop foreach local wp wall.list
      if wall.id(wp) # 17
        tmp = wall.inside(wp,ball.pos(b))
        if tmp = 1
          inBody = 1
          exit
        endif
      endif
    endloop
  endif
end
```


W.-H. Zhou and Z.-Y. Yin, *Practice of Discrete Element Method in Soil-Structure Interface Modelling*, https://doi.org/10.1007/978-981-19-0047-1

```
set random 10001
[iniRadius=math.sqrt(w*d/float(np)/math.pi*0.1)]
ball generate radius [iniRadius*(1.0-Rvar)*0.9] [iniRadius*(1.0+Rvar)*0.9] ...
            number [np]                                   ...
            box [-w/2.0]  [w/2.0] -0.007 [d] ...
            range fish @inBlender ...
                  fish @inBody not

wall delete walls range group box
wall create group 'box' id 1 vertices ...     ;left
                        [-w/2.0] [gap]...
                        [-w/2.0] [d*1.5]
wall create group 'box' id 2 vertices ...     ;top
                        [-w/2.0*1.5] [d] ...
                        [w/2.0*1.5] [d]
wall create group 'box' id 3 vertices ...     ;right
                        [w/2.0] [d*1.5] ...
                        [w/2.0] [gap]
wall create group 'box2' id 4 vertices ...     ;right
                           0.0250 [0] ...
                           0.08 [0]
wall create group 'box2' id 5 vertices ...     ;right
                           -0.025 [0] ...
                           -0.08 [0]

wall import filename Drawing2.dxf nothrow group 'plate' id 500 ; Drawing2.dxf from CAD
```

The manufacturer's authorised representative in the EU is Springer Nature Customer Service Centre GmbH, Europaplatz 3, 69115 Heidelberg, Germany. If you have any concerns regarding our products, please contact ProductSafety@springernature.com

Printed and bound by CPI Group (UK) Ltd, Croydon, CR0 4YY
30/09/2026
02225250-0006